JON H. DAVIS

DIFFERENTIAL EQUATIONS WITH MAPLE

AN INTERACTIVE APPROACH

BIRKHÄUSER
BOSTON • BASEL • BERLIN

Jon H. Davis
Department of Mathematics
and Statistics
Queen's University
Kingston, Ontario
K7L 3N6 Canada

Library of Congress Cataloging-in-Publication Data
Davis, Jon H., 1943-
Differential equations with maple : an interactive approach / Jon H. Davis.
p. cm.
Includes bibliographical references and index.
ISBN 0-8176-4181-5 (alk. paper) – ISBN 3-7643-4181-5 (alk. paper)
1. Differential equations—Data processing. 2. Maple (Computer file) I. Title.

QA371.5.D37D37 2000
515'.35'0285—dc21 00-049842
CIP

AMS Subject Classifications: 34-01, 34D20, 44-01, 65-01, 65L06, 68-04, 93B51

Printed on acid-free paper.

Birkhäuser 

ISBN 0-8176-4181-5 SPIN 10764698
ISBN 3-7643-4181-5

Typeset by the author in LaTeX.
Printed and bound by Hamilton Printing, Rensselaer, NY.
Printed in the United States of America.

9 8 7 6 5 4 3 2 1

Preface

Differential equations is a subject of wide applicability, and knowledge of differential equations topics permeates all areas of study in engineering and applied mathematics. Some differential equations are susceptible to analytic means of solution, while others require the generation of numerical solution trajectories to see the behavior of the system under study. For both situations, the software package *Maple* can be used to advantage.

To the student

Making effective use of differential equations requires facility in recognizing and solving standard "tractable" problems, as well as having the background in the subject to make use of tools for dealing with situations that are not amenable to simple analytic approaches.

The software package *Maple (tm)* is one of the tools that is of use in the study (and subsequent use) of differential equations. Maple is a program for symbolic manipulation of mathematical expressions, and it is shipped with a large library of "canned" routines for solving standard problems. In particular, there are routines for recognizing and "solving" a variety of differential equation problems. One might at this point be tempted think that this would reduce differential equations to a question of "what to type in Maple to solve standard problems".

In fact, this is not the case. What is obtained from solving problems is more than a collection of solutions to set-piece problems. More valuable is the subtext, which consists in the experience of what sorts of problems have simple solutions, and what the forms of those solutions should look like.

It may be the case that Maple will generate closed form solutions to simple problems, but it does what "you said", not what "you meant". Given the ease of generating (even reams) of output, the ability to recognize garbage output is essential. The ability of Maple to handle complicated symbolic calculations actually makes it possible to deal with much more interesting and substantial problems than are possible if only hand calculations are allowed, as long as one is knowledgeable enough about the subject to know what computations should be done.

If the utility of Maple is not to put simple problem solution entirely out of business, where Maple is truly useful is really in solving "hard" problems. These are problems that are "solvable in principle" but in fact are so imposing in scale that hand calculations are just not on. Maple generally does not lose minus signs, and it does not forget factors of three. It tirelessly collects terms.

In order to make such use of Maple (that is, bend it to the solution of *your* problems) what you need is an acquaintance with Maple as a programming vehicle, and not just as a giant symbolic calculator.

This book integrates Maple instruction with differential equations by using it to investigate topics in differential equations that are inaccessible without computational aid. The approach is based on a sequence of integrated Maple labs, with cross links between the " diff-eq theory" and the "Maple hacks" that illustrate and support it. These links are indicated by shaded tags embedded in the text. The format of these links is as follows.

⋮

The differential equation (with the "gain parameter" k)

$$\frac{d^n x}{dt^n} + p_{n-1}\frac{d^{n-1}x}{dt^{n-1}} + \ldots + p_0\, x + k\left(q_m \frac{d^m x}{dt^m} + q_{m-1}\frac{d^{m-1}x}{dt^{m-1}} + \ldots + q_0\, x\right) = 0,$$

occurs in the study of control systems. Maple can easily generate the root locus plots used to study stability of the system as k varies through a range of values.

Maple connection **Page: 358**

⋮

Maple exercises must be regarded as part of the learning process. In addition to describing Maple as a programming language, the example exercises provide techniques that support computations for the differential equation topics of the text.

To the instructor

The differential equations course for which this is the text has a weekly Maple lab in addition to the lectures on differential equation topics. The labs are based on the text exercises, and aim initially to introduce Maple programming and data structures to the point where generating and plotting data become feasible.

This allows for the discussion of numerical methods in the context of first order differential equations before the variety of analytical methods for linear equations and systems. The author feels that you do not understand an algorithm unless you can code it, and that solution curve plots are more informative than columns of numbers when it comes to numerical methods for differential equations.

We stick to Runge–Kutta methods in the discussion of numerical methods. They are generally useful, and also have clear motivation and accuracy properties.

It is important that students undertake learning Maple as a programming tool. This enables them to profitably use Maple to solve their own problems, and also allows them to become comfortable with the tool by seeing how it is actually put together. Even though it is "just software", Maple can be very intimidating to beginning differential equations students, since it seems to "know" enough to solve the standard problems. Seeing that its "knowledge" consists of a huge amount of pattern matching code and standard algorithms is a liberating experience. It also reduces "flaky edge" frustration, since any code that is causing a problem can usually be inspected, and one can always either fix it or write one's own replacement.

Overview

The first part of the text introduces Maple. Maple is much more than an interactive calculator (although that aspect is useful), and so the first chapter introduces Maple as a programming language. On the syntactic level, Maple is a block structured language, but its data types and structures have facilities adapted to symbolic computation. Both of these aspects are discussed, and some of the things that make Maple different from conventional procedural languages are covered. The discussion is self contained, and no "previous programming experience" is required. Maple is conventional enough in design that it can serve as an introduction to procedural programming. Previous programming experience lessens the shock of assignments and loops, but also requires one to deal with the differences between Maple and some previous programming medium.

The second part of the text first covers conventional differential equation topics. That is, first order equations, n-th order equations and systems, periodic solutions, stability, and an introduction to boundary value problems are included. Vector differential equation models are covered throughout the discussion. This approach is needed in any event for numerical methods, and the Maple linear algebra routines are readily applicable to problems formulated in terms of vector variables.

The last part of the text consists of both Maple differential equations applications, and some "hard" Maple programming projects.

The differential equation applications consist of Maple demonstrations and exercises corresponding to topics in the differential equations part. These make use of the Maple linear algebra and integral transform library routines, which can be adapted to solve problems involving constant coefficient systems of differential equations, and to construct plots of solution curves.

The first "hard case" began as an attempt to use Maple to verify accuracy (truncation order) conditions for Runge–Kutta type numerical differential equation solvers. It mushroomed into the present form when the original attempt would barely run due to space problems, and then produced wrong answers to boot. I must thank Jim Verner for assistance with the codes. There is nothing like a trial user who knows what the answers should be for expediting the debugging process.

The other programming project of the last part is included to illustrate the building of a Maple package. It also serves to show that Maple is adaptable to what are commonly thought of as engineering calculations in the area of control systems. The package provides an environment for both classical (frequency domain) calculations, and state space routines for realizations and linear regulator problems. Both continuous and discrete time models are supported, and the fact that Maple is a symbolic computation platform makes it possible to code routines that handle either case by adapting to the user input format.

And . . .

The example Maple code of the text is included on the accompanying CD-ROM disk. This includes sources for the large projects of the final text part, as well as sample worksheets for smaller examples.

Many of the Maple exercises of the text are variants of problems used in a web based computer lab that is run to support the course based on this text. The "web site" associated with that enterprise is included on the CD, since it is as convenient a way to present the sample labs as any. The sample site may be of interest to instructors without experience in running a web based computer lab.

The text contains a number of problems, including a large review Section. Instructors may obtain a solutions manual through the publisher.

The text manuscript was produced using a variety of open source and freely available software, including LaTeX2e, pstricks, ghostview, and the GNU gpic and m4 programs, all running under GNU/Linux. The circuit diagrams were created using the m4 circuit macros package by D. Aplevitch.

Thanks are finally due to my editor Ann Kostant and her staff, who guided the process of turning class notes into a textbook manuscript, and to my wife Susan who put up with all the side effects of the process.

Jon H. Davis

Contents

Preface v

I Maple Use and Programming 1

1 Introduction to Maple 3

1.1 Calculator or What? . . . 4
1.1.1 First Contact . . . 5
1.2 Programming in Maple . . . 7
1.3 Maple Variables . . . 8
1.3.1 Variable Types . . . 8
1.3.2 Maple Expressions . . . 9
1.3.3 Aggregate Types . . . 12
1.3.4 Maple Tables . . . 14
1.4 Maple Syntax . . . 15
1.4.1 Function Declarations . . . 16
1.4.2 Assignments . . . 18
1.4.3 Conditionals . . . 18
1.4.4 Loops . . . 18
1.4.5 Maple Evaluation . . . 19
1.4.6 Details about Maple Procedures . . . 22
1.4.7 Procedure Arguments . . . 23
1.4.8 Programming Gotchas . . . 25
1.4.9 Maple Program Development . . . 26

II Differential Equations 27

2 Introduction to Differential Equations 29

2.1 Example Problems 29
2.1.1 Simple Rate Equations 30
2.1.2 Mechanical Problems 31
2.1.3 Electrical Models 32
2.1.4 Kinetics 34
2.1.5 Beam Models 35
2.2 Order, Dimension and Type 43
2.2.1 First and Higher Order 45
2.2.2 Partial Derivative Problems 46
2.3 Problems 46

3 First Order Equations 51

3.1 Separable Problems 52
3.1.1 Separable Examples 55
3.2 Level Sets and Exact Equations 59
3.2.1 Solving Exact Problems 63
3.2.2 Exact Equation Examples 64
3.3 Linear Equations 65
3.3.1 Linear Examples 70
3.4 Tricky Substitutions 71
3.4.1 Lagrange 71
3.4.2 Riccati 71
3.4.3 Mechanics 72
3.5 Do Solutions Exist? 73
3.5.1 The Issue 73
3.5.2 Existence Theorems 75
3.6 Problems 77

4 Introduction to Numerical Methods 81

4.1 The Game and Euler's Method 82
4.2 Solution Taylor Series 85
4.3 Runge–Kutta Methods 88
4.4 General Runge–Kutta Methods 91
4.5 Maple Numeric Routines 95
4.6 Calling all RK4's (and relatives) 97
4.7 Variable Step Size Methods 100
4.7.1 Running RK4 With RK3 102
4.7.2 Adjusting h 103
4.8 Serious Methods 108
4.9 Problems 109

5 Higher Order Differential Equations **111**
5.1 Equations of Order N 111
5.1.1 Vector Equations 112
5.1.2 Equivalent Formulations 113
5.2 Linear Independence and Wronskians 115
5.2.1 Linear Independence 116
5.2.2 Wronskians 116
5.3 Fundamental Solutions 117
5.3.1 Existence of Fundamental Solutions 118
5.3.2 Constructing Fundamental Solutions 123
5.4 General and Particular Solutions 124
5.5 Constant Coefficient Problems 125
5.5.1 Homogeneous Case 125
5.5.2 Undetermined Coefficients 130
5.6 Problems 136

6 Laplace Transform Methods **141**
6.1 Basic Definition 141
6.1.1 Examples 143
6.2 New Wine From Old 145
6.2.1 Algebra and Tables 148
6.3 Maple Facilities 153
6.4 Derivatives and Laplace 153
6.5 High Order Problems by Laplace 156
6.5.1 Fundamental Solutions 158
6.5.2 Undetermined Coefficients Redux 160
6.6 Convolutions 163
6.6.1 Inhomogeneous Problems 168
6.7 Problems 169

7 Systems of Equations **173**
7.1 Linear Systems 173
7.2 Bases and Eigenvectors 174
7.3 Diagonalization 176
7.3.1 Using Maple 180
7.4 Jordan Forms 181
7.4.1 Using Maple 187
7.5 Matrix Exponentials 188
7.5.1 Laplace Transforms 190
7.5.2 Using Maple 191
7.6 Problems 193

8 Stability **197**
8.1 Second Order Problems 197
8.2 Harmonic Oscillator Again 198

8.3 Nodes and Spirals 199
8.4 Higher Order Problems 203
8.4.1 The Easy Cases 204
8.4.2 Jordan Form 205
8.5 Linearization 206
8.6 Introduction to Lyapunov Theory 210
8.6.1 Linear Problems 212
8.6.2 Applications 214
8.7 Problems 216

9 Periodic Problems **221**
9.1 Periodic Inputs 221
9.2 Phasors 222
9.3 Fourier Series 225
9.4 Time Domain Methods 227
9.5 Periodic Coefficients 234
9.6 Fundamental Matrices 236
9.7 Stability 239
9.8 Floquet Representation 240
9.9 Problems 243

10 Impedances and Differential Equations **245**
10.1 Introduction to Impedances 245
10.2 AC Circuits 246
10.3 Transient Circuits 250
10.4 Circuit Impedance Examples 251
10.5 Loops and Nodes 254
10.6 Problems 255

11 Partial Differential Equations **259**
11.1 Basic Problems 259
11.1.1 Heat Equation 259
11.1.2 Wave Equation 261
11.1.3 Laplace's Equation 262
11.1.4 Beam Vibrations 264
11.2 Boundary Conditions 266
11.2.1 Heat and Diffusion Problems 266
11.2.2 Wave Equations 267
11.2.3 Laplace's Equation 267
11.3 Separation of Variables 268
11.3.1 Cooling of a Slab 268
11.3.2 Standing Wave Solutions 271
11.3.3 Steady State Heat Conduction 273
11.4 Problems 276

III Maple Application Topics 279

12 Introduction to Maple Applications 281

13 Plotting With Maple 283
13.1 Maple Plotting Structures 283
13.2 Remember Tables 284
13.3 Plotting Numerical Results 287
13.4 Plotting Vector Variables 288
13.5 Further Plotting 290

14 Maple and Laplace Transforms 297
14.1 Walking Maple Through Problems 298
14.2 Coding a Beam Problem Solver 301

15 Maple Linear Algebra Applications 307
15.1 Canonical Forms 307
15.1.1 Eigenvalues and Eigenvectors 307
15.1.2 Jordan Form Calculations 310
15.2 Matrix Exponentials 312
15.3 Stability Examples 315
15.4 Periodic Solutions with Maple 317

16 Runge–Kutta Designs 323
16.1 Orientation and Overview 323
16.2 Notational Issues 324
16.3 True Solution Derivatives 325
16.4 Runge–Kutta Derivatives 329
16.5 Traps and Tricky Bits 333
16.6 A Sample Run 334
16.7 Order Condition Code 338

17 Maple Packages 355
17.1 Introduction 355
17.2 Constructing Packages 356
17.3 A Control Design Package 357
17.3.1 Root Locus 358
17.3.2 Nyquist Locus 360
17.3.3 Bode Plots 363
17.3.4 Classical vs. State Space 364
17.3.5 LQR Utility Code 365
17.3.6 Least Squares Optimal Control 366
17.3.7 Control Equation Generation 371
17.3.8 LTI Realization Calculations 374
17.3.9 Initialization 376
17.4 Package Creation and Installation 376

17.5 How About Help? 377
17.5.1 Maple Libraries 377
17.5.2 Creating and Installing Help Files 378
17.6 Control Package Demo 380

IV Appendices **389**

A Review Problems **391**

B Laplace Transform Table **401**

References **403**

Index **405**

To Susan

DIFFERENTIAL EQUATIONS WITH MAPLE

AN INTERACTIVE APPROACH

Part I

Maple Use and Programming

1
Introduction to Maple

Maple is a symbolic calculation program, capable of dealing with both numeric and symbolic quantities. When properly harnessed, it can be a great help in understanding differential equations as well as many other areas of mathematics. Maple as distributed actually contains a large number of "cookbook" routines, both numerical and symbolic, that with suitable coaxing will produce "answers" to various standard problems.

While it is interesting (and even useful) that Maple has been programmed to recognize and make the required calculations for a variety of standard problems, the true usefulness of the program lies in its ability to make calculations that are "do-able in principle", but that are sufficiently burdensome so that one would be loathe to undertake them "by hand". Maple does not become more likely to lose minus signs in the small hours of the morning, and is oblivious to the fact that some intermediate calculation might run over several pages.

To take advantage of this aspect of Maple, some familiarity with use of Maple as a programming environment is what one desires.

Programming in Maple is not at all difficult, especially for anyone with previous experience in a programming language. It takes some investment of time to become familiar with the Maple syntax and data types, but the reward is that Maple can then be used to solve *your* problems, rather than just the predictable examples of introductory courses.

1.1 Calculator or What?

Maple is a software system primarily organized for doing symbolic mathematical calculations. That is, rather than producing strictly numerical output, Maple can generate results in what can be thought of as conventional mathematical notation, with a graphical user interface providing both user interaction and the resulting output. Figure 1.1 illustrates the appearance of the user interface.[1]

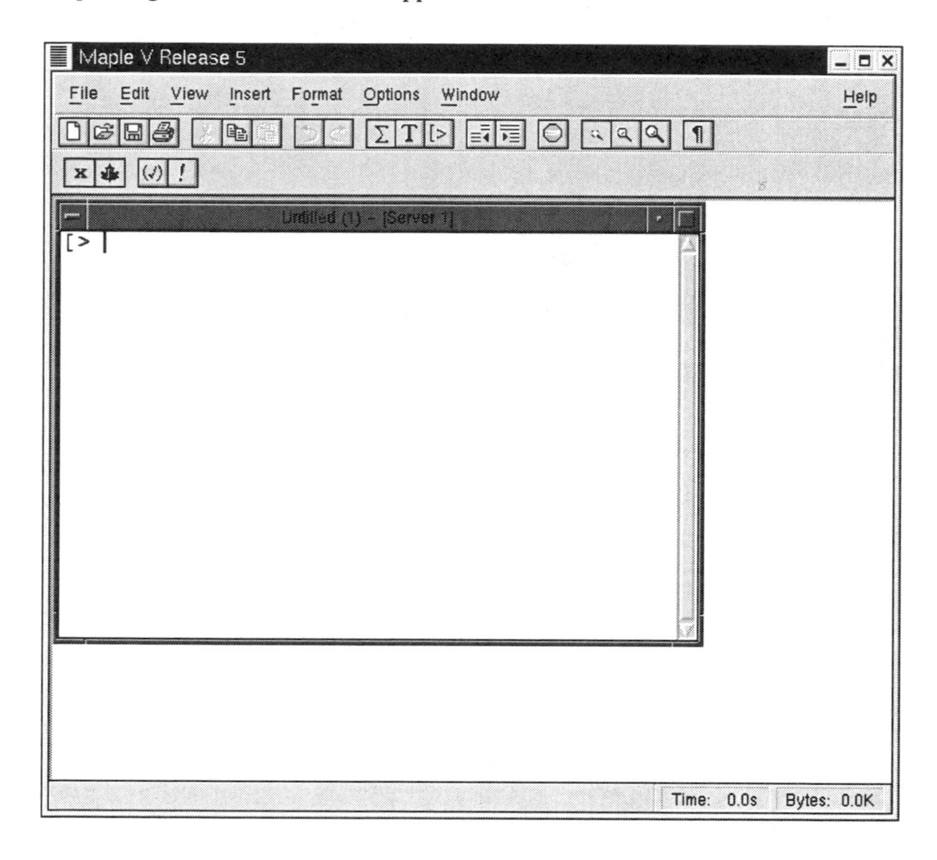

FIGURE 1.1. Maple user interface

Many people introduced to Maple come away with an initial impression of a huge, balky calculator with a vast number of built-in facilities, returning an assortment of hard to fathom error messages. The sheer size and variety of the library routines provided with the program make it an intimidating prospect for the uninitiated user.

[1]The Maple computational engine can also be run without the user interface, as a slave process. The details of this facility are platform dependent, while the graphical interface presents a largely consistent appearance across the supported platforms.

Many books that "use" Maple as an integral part of the text make no attempt to explain what the underlying basis and facilities of the program are, but rather provide a series of magic incantations which will cause the program to produce the solution to some particular problem. Maple has a large number of subroutine libraries which have been expressly written to provide ready solutions to standard homework problems of first and second year courses. An introduction to such facilities is available within Maple, and users can easily find online documentation for such facilities.

The reference for Maple programming facilities is also on line. Maple provides an extensive "help" facility which includes an index which is useful for getting oriented to the available facilities. The help menu provides a "new user tour", as well as information documenting changes between releases of the program.

1.1.1 First Contact

Exercise
If you have not used Maple before (or forgotten everything) select the "new user tour" from the help menu. This will lead to an interactive sampling of the Maple environment. An amazing amount of documentation is on line in Maple. Try the help index on the main menu bar, or by typing a question mark at the user prompt for commands you are willing to bet are included.

```
> ?snurglelurtz
> ?plot
> ?for
```

The program tries to guess what you might have wanted if `snurglelurtz` actually is not a valid command. Most interactive manipulation commands use conventional terminology, so that obtaining help online is relatively easy.

Exercise
Something like 90% of Maple is written in the Maple programming language. To learn how to program in Maple, the easiest place to start is to read some of the 15 megabytes or so of code that comes with the program. Like this:

```
> interface(verboseproc=2);
> print(sin);
> print(dsolve);
> .....
```

Exercise
That sure is a lot of code to debug, isn't it? Try

```
> sin(arcsin(x));
> sin(arcsin(2));
> exp(ln(x));
> ln(exp(x));
```

These results may appear surprising in that `sin(arcsin(2))` will produce a domain error on most hand calculators. The rationale for the Maple result lies in the area of complex variable theory, and principal branches of multivalued complex functions.

Exercise
The command which allows the printing of the built-in library routines is part of a set of user interface parameter settings. The available parameters are described in the help page for the `interface` procedure.

```
> ?interface
```

In addition to procedure printing, the interface variables control the use of "percent variables" as output abbreviations, and default formatting of plots. The names of the parameters are sometimes easy to forget, but they are all set by use of the `interface` command.

Exercise
Examples of standard calculations available include function plotting facilities. A command such as

```
> plot(sin(x), x= -1..1);
```

produces ready graphical output.

Exercise
Various types of differential equations have "cookbook" solution algorithms associated with them, and some Maple procedures are able to recognize the problem type and carry out the algorithm. An example of a first order differential equation problem solvable by a variety of methods is

$$\frac{dx}{dt} = x(t), \quad x(0) = 1,$$

which Maple manages to handle with aplomb:

```
> dsolve({diff(x(t), t) = x(t), x(0) = 1}, x(t));
```

$$x(t) = e^t$$

Exercise

Math discussion **Page:** 52

Maple is particularly useful for producing graphical output. The graphical depiction of the flow fields associated with a first order differential equation are particularly tedious to construct by hand. Maple has a dedicated command for producing the plots.

```
> with(DEtools);
> dfieldplot(diff(y(x),x)= y(x)*x^2, y(x), x=-5..5, y=-5..5);
```

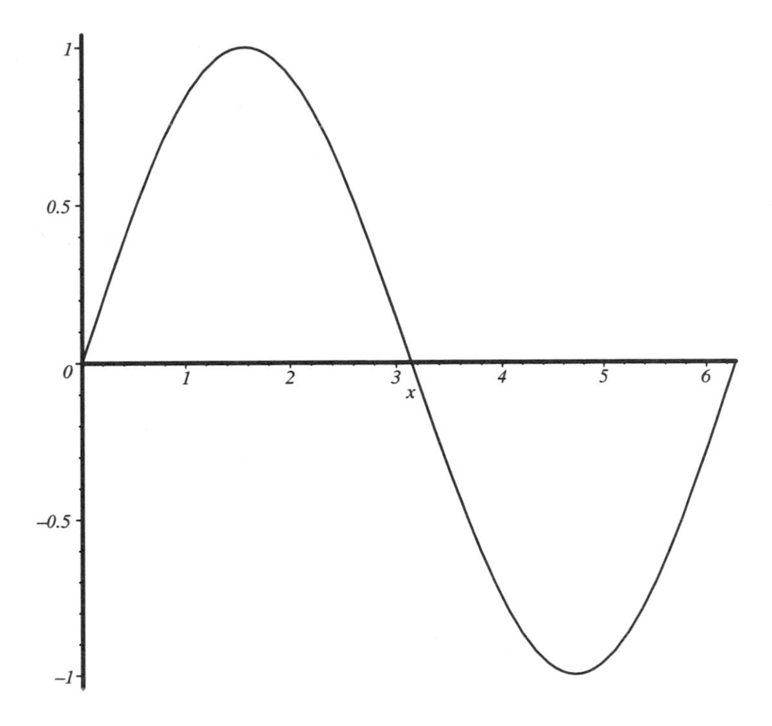

FIGURE 1.2. Maple plot of sin(x)

Exercise

Math discussion **Page:** 53

Solving separable differential equations results in the representation of the solution in the form of an implicit equation, or by another description, as a level curve of a function of two variables. If it is desired to visualize such a family of curves as a whole, the Maple built-in contour plotting routine can be invoked to produce a graph.

```
> with(plots);
> S := (x,y) -> ( 1/y + (x^2)/2);
> contourplot(S(x,y), x=-5..5, y=-5..5);
```

1.2 Programming in Maple

The fact that very high level interactive calculation facilities are available with Maple is not surprising, given that the system is designed to perform symbolic and numerical processing, and that the solution of such problems is largely a matter of matching known problem formats with successful solution algorithms.

The fact that Maple can with some coaxing recognize certain solvable problems does not undermine the fact that the ability to distinguish solvable from difficult or impossible mathematical problems is a basic skill required for making progress in either pure or applied areas of mathematics. There probably never were first or second year course instructors who thought that the mechanical solution of particular example problems was an end to be pursued for its own sake.

The real utility of Maple lies not in its ability to solve simple homework problems in closed form, but rather in the rich collection of mathematical manipulation routines available together with the programming language that allows one to write algorithms for solving problems using those provided subroutines.

The programming language that is part of Maple was designed to be friendly and familiar, in the sense that it is similar to other block structured programming languages. This was done on purpose, as earlier symbolic calculating programs used more arcane programming facilities, such as lisp. Any programming language has a declaration of syntax, so that the program doing the interpretation of the user's input can decode the input successfully. In the case of Maple programming language, the syntax is closer to PASCAL than to anything else. This refers to the choice of keywords (reserved words) and to the form of standard facilities like assignment statements and loop constructions.

The main differences between Maple and conventional programming languages like PASCAL and C are in the area of the types of data objects that are available to a program. Maple makes some different data types available in order to support its facilities for algebraic and symbolic manipulation. It also differs from conventional languages in that variables are not checked for type matches when code is compiled. Rather, many functions are written with the expectation that they may be invoked with arguments of different types. The burden is on the function to check the arguments at run time to make sure they are valid operands.

This mode of operation is the source of the obscure error messages that plague beginners in Maple use. It is probably impossible to write routines that completely check and eloquently report mangled arguments. This makes it inevitable that mangled arguments get passed down to lower level subroutines, and their complaints are likely to be couched in terms that make no sense to the casual user of the program. A way out of this dilemma is to know enough about Maple programming so that the error messages actually do indicate what is going wrong. If you have had some experience writing your own Maple procedures, the sorts of errors that typically occur are familiar and comprehensible.

1.3 Maple Variables

1.3.1 Variable Types

Maple variables are not declared as having some particular type like floating point number, integer, string, and so on. Instead, variables carry their types with them as they are passed around in programs. There are routines that access this information, and since Maple is an interactive system, examples can easily be tried.

```
> whattype(1.0);
```

float

```
> whattype(1/2);
```

fraction

```
> whattype(-2);
```

integer

```
> whattype(a);
```

string

1.3.2 Maple Expressions

Given the common currency of the phrase *mathematical expression*, it is not surprising that objects which may be thought of as *expressions* are widely used as variables in Maple. Symbolic expressions are used not only as user level variables to be manipulated by Maple, but internally by Maple to represent objects ranging from defined procedures to plotting data structures.

The simplest example is a function of a single argument.

```
> whattype(sin(x));
```

function

In computer programming, it is generally useful to have a graphical representation of the data structures that are being used. For expressions, a useful representation is as an "expression tree".

Trees consist of nodes interconnected by branches. The nodes generally represent functions or operations, and the branches leaving the node stand as place holders for the argument(s) of the node function. For a function of a single variable, the picture is the particularly simple one of Figure 1.3. Multiplication is an

FIGURE 1.3. Function expression

example of a binary operation (otherwise known as a function of two arguments.)

```
> whattype(x*y}
```

$*$

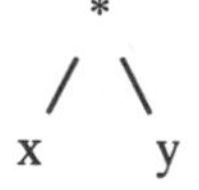

FIGURE 1.4. Multiplication expression

The expression tree corresponding to this is in Figure 1.4. Other binary expressions have similar expression trees. These include equations and ranges, as well as the usual arithmetic expressions.

```
> whattype(a=b);
```

$$=$$

```
> whattype(a..b);
```

$$range$$

```
> whattype(a+2);
```

$$+$$

One surprise is that Maple treats inversion as a unary operation, so that an expression that looks like a division actually is the product of one expression times the inverse of the other.

```
> whattype(a/b);
```

$$*$$

Maple expressions are naturally composite objects, and the language provides facilities for determining the expression structure and pulling apart the pieces. These manipulations are at the heart of symbolic manipulation where it is required to process the terms of a sum, examine the denominator of a fraction, test whether an argument is a sum or a single element, and so on.

Expressions are thought of as "functions" or (more in a computing frame of mind) "procedure calls". The graphical depiction is an *expression tree* as described above for simple functions of one and two arguments. The number of *operands* ("slots" the function has) is returned by `nops`.

```
> nops(1);
```

$$1$$

```
> nops(a/b);
```

$$2$$

```
> nops(sin(x));
```

$$1$$

The individual operands of a Maple expression are retrieved by number. Examples follow.

```
> op(1,1);
```

$$1$$

```
> op(1, a/b);
```

$$a$$

```
> op(2, a/b);
```

$$\frac{1}{b}$$

```
> op(0, a/b);
```

$$*$$

```
> op(0, sin(x));
```

$$\sin$$

For functions, the 0 numbered operand turns out to be the name of the function. For many Maple expressions, operand number 0 is used as the type of the expression.

A complex Maple expression can be defined as an example.

```
> symbolic_expr :=f(x,y, g( h(u,v+x), w));
```

$$symbolic\_expr := \mathrm{f}(x,\ y,\ \mathrm{g}(\mathrm{h}(u,\ v+x),\ w))$$

The graphical representation of this as an expression tree reflects the fact that expression trees are in fact recursive structures. Branches terminate in nodes that represent other expressions, and have argument branches of their own. The expression tree for `symbolic_expr` is in Figure 1.5. The component subexpressions can be assessed with the `op` commands.

```
> op(0, symbolic_expr);
```

$$f$$

```
> op(1, symbolic_expr);
```

$$x$$

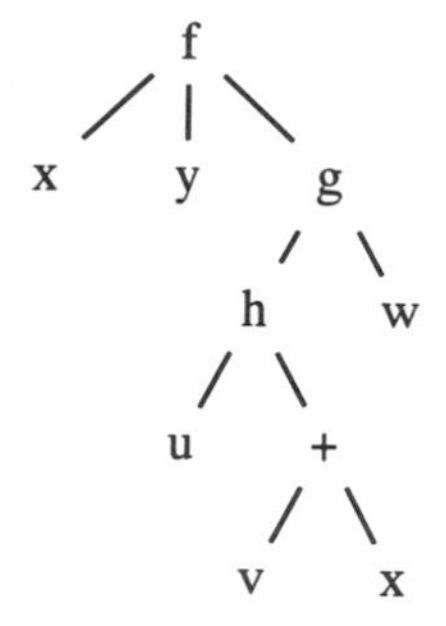

FIGURE 1.5. Composite expression tree

```
> op(3, symbolic_expr);
```

$$g(h(u, v+x), w)$$

```
> op(1, op(3, symbolic_expr));
```

$$h(u, v+x)$$

```
> op(2, op(1, op(3, symbolic_expr)));
```

$$v+x$$

```
> op(0, op(2, op(1, op(3, symbolic_expr))));
```

$$+$$

```
> op(1, op(2, op(1, op(3, symbolic_expr))));
```

$$v$$

An expression like the last one is totally opaque unless one has a picture of the variable structure in mind. The picture which makes the expression comprehensible is the tree of Figure 1.5.

1.3.3 Aggregate Types

Maple also has several aggregate types which conveniently represent such things as lists, sets, and ranges.

```
> whattype([a,b,3,sin(x)]);
```

list

```
> nops([a,b,3,sin(x)]);
```

$$4$$

```
> op(0,[a,b,3,sin(x)]);
```

$$list$$

Since lists are Maple expressions, their elements may be accessed by thinking of them as operands of the expression, as well as by using a subscript notation.

```
> mylist := [a,b,3,sin(x)];
>
> mylist[4];
```

$$sin(x)$$

```
> op(1, op(4, [a,b,3,sin(x)]));
```

$$x$$

Following conventional mathematical notation, sets are denoted by curly braces.

```
> whattype({a,b,c,c, sin(x)+2+exp(cow)});
```

$$set$$

Sets are also expressions, so that you can extract operands; the result of doing this is probably not what one expects, and so is not very useful.[2]

We have avoided making a blanket statement that "everything is an expression" in Maple. This seems nearly true in practice, but there is a ready counterexample to this thought. There are also expression sequences.

```
> whattype(a,b,c,c, sin(x)+2+exp(cow));
```

$$exprseq$$

```
> exseq := x,y,z;
```

$$exseq := x,\ y,\ z$$

```
> whattype(exseq);
```

$$exprseq$$

If this is a Maple expression, we ought to be able to access an operand.

```
> op(1, exseq);

Error, wrong number (or type) of parameters in function op
```

[2] Sets are stored internally by Maple using a hash table for access, and the operand order reflects this.

The workaround for this is to turn the "object" (evidently, we cannot call it an expression) into a list by surrounding it in square brackets. The arguments of a Maple procedure are actually an expression sequence, so that this becomes the idiom for accessing procedure arguments.[3]

```
> op(1, [exseq]);
```

$$x$$

1.3.4 Maple Tables

Maple is basically a symbolic manipulation engine, so that it has various facilities that support achieving this objective, as well as speeding it up. One of these is something that is common to other symbolic manipulation tools, where it masquerades as "association lists" or "associative arrays". Maple calls such objects "tables", and uses a notation (syntax) that makes expressions read as though one was dealing with some sort of an array which somehow manages to have symbolic rather than numerical subscripts.

```
> t := table();
```

$$t := \mathrm{table}([\\ ])$$

```
> t[1] := quail;
```

$$t_1 := quail$$

```
> t[rooster] := 2;
```

$$t_{rooster} := 2$$

```
> t[elephant] := `water buffalo`;
```

$$t_{elephant} := water\ buffalo$$

```
> t;
```

$$t$$

```
> whattype(t);
```

$$string$$

[3] Object oriented programming zealots may argue that some uniform base type should represent all Maple objects. The original design of Maple predates the rise of object orientation.

```
> eval(t);
```

table([
$elephant = water\ buffalo$
$1 = quail$
$rooster = 2$
])

```
> whattype(eval(t));
```

$$table$$

```
> nops(eval(t));
```

$$2$$

```
> op(0, eval(t));
Error, improper op or subscript selector
> op(1, eval(t));
>
> op(2, eval(t));
```

$$[elephant = water\ buffalo, 1 = quail, rooster = 2]$$

The examples above use the `eval` routine. This forces complete evaluation of an expression, and is a useful technique in Maple programs. Evaluation is a subtle issue in Maple, and is discussed at greater length below.

1.4 Maple Syntax

One can look at the official syntax definition of Maple, but unless one is familiar with the formalisms of programming language definitions, a collection of standard idioms probably is an easier way to learn. As far as language syntax goes, Maple is probably closest to PASCAL. An evident difference is that Maple is case sensitive, while PASCAL is not. The syntax rules are sufficiently close that replacing a PASCAL `begin` ..`end` with a Maple `do` .. `od` nearly translates a PASCAL routine into Maple. One catch in this scenario is that Maple `if` blocks have to be closed by a trailing `fi`, rather than being enclosed by block delimiters. Leaving off the Maple block termination statement is the primary source of compiler error messages for users familiar with other programming languages like PASCAL or C, where block delimiter pairs (BEGIN ... END or { ... }) are part of the language syntax. In Maple, the limitation of the governed block is part of the statement type, rather than a statement grouping mechanism. This fact is discussed further below, as it is a source of confusion for users familiar with PASCAL.

1.4.1 Function Declarations

Maple is a programming language in the same family (block structured languages) as C and PASCAL. PASCAL has notions of both procedures and functions, differentiated by whether or not they return a value. Maple follows C in using one formalism for both cases, with the idea that everything should be regarded as a function. The Maple equivalent of a PASCAL procedure is just a function that does not return a useful value to the calling entity.

Following C, Maple supports a `RETURN` statement to determine the value returned to the caller. If you do not use an explicit `RETURN` statement, what gets returned is the last thing that was evaluated. While this feature seems handy and is widely used, it will cause procedures that do not explicitly `RETURN(NULL);` to print garbage values when used interactively. Maple procedures are more readable and more maintainable if they use explicit `RETURN` statements. In Maple, procedures are valid variables like other types, and so the declaration syntax assigns the declaration block to a name.

The standard form of a Maple procedure declaration is the following:

```
.....
myfunction := proc( arguments )
local var1, var2, var3,..var;
options opt1, opt2...;
.....
.....
.....
RETURN( value );
end;
```

A procedure that squares its arguments is simple to define.[4]

```
> myproc := proc(x)
>    local y;
>    options remember, 'Copyright me';
>    y := x*x;
>    RETURN(y);
> end;
```

When a procedure is defined, it is available at the same level as all of the built-in Maple procedures, and hence is available for interactive use.

```
> myproc(2);
```

$$4$$

Maple procedures as defined have global scope. This parallels the structure of C language, rather than the block nesting arrangement used in PASCAL. There is also no notion of the file scope and separate linkage used in C. Unique procedure names must be chosen in Maple unless the intent is to override a previous definition.

---

[4]The procedure options, particularly "remember", are discussed below.

Procedure Expressions

Since a Maple procedure is constructed by name assignment in the same manner as other Maple objects, it is not surprising that it has the internal structure of an expression. The 0 operand of a procedure is the type.

```
> op(0, eval(myproc));
```

$$procedure$$

The first and second operands are the procedure arguments and local variables.

```
> op(1, eval(myproc));
```

$$x$$

```
> op(2, eval(myproc));
```

$$y$$

The third operand holds the option variables.

```
> op(3, eval(myproc));
```

$$remember, \, Copyright \; me$$

The fourth operand only has meaning for procedures declared with the `remember` option. It contains a Maple table, the entries of which are the previously computed argument-value pairs for the procedure. In this case the procedure has been invoked once to compute the square of 2. The remember table contains 2 as an index, with the result 4 as the corresponding value.

```
> op(4, eval(myproc));
```

```
table([
    2 = 4
    ])
```

The `remember` option is used as a mechanism to speed up computations. The idea is to store the results of previous (presumably laborious) computations in a table associated with the procedure. When the procedure is subsequently invoked, it can just return the previous result rather than having to redo the computation. This can be thought of as an instance of the time-storage space tradeoff that often occurs in problems of computational efficiency.

1.4.2 Assignments

The Maple syntax for assigning a value to a variable follows the PASCAL syntax convention. Users of C cause themselves trouble by forgetting this, and using an = sign. In Maple, = is a valid binary operator that can be used in expressions, and so the interpreter accepts and echoes an expression with an erroneously typed = sign without complaint. The response is typographically so close to what would have been echoed for the correct assignment statement that the user, now concentrating on the following statement syntax, never senses the error.[5] One way to combat this problem is to make a strong effort to read Maple assignment statements as " ... gets ... ", and never as " ... equals ... ".

```
.....
var := value;
```

1.4.3 Conditionals

The Maple syntax for conditionally executing a block of statements includes a trailing block delimiter.

```
.....
if test then
.....
.....
.....
fi
```

For a sequence of conditional tests, the syntax includes **elif** in spite of the tendency of most Maple beginners to type **elsif**. This is really a shorthand that avoids a proliferation of trailing `fi` statements.

```
.....
if test then
.....
elif test then
.....
else
.....
fi;
```

The conditional test constructs include the relations <, >= and <> (not equal).

1.4.4 Loops

If the aim of a programming language designer is to invent elegant, adaptable syntax, then the Maple looping construct syntax is notable. Every individual part

[5]The symptom of this is that subsequent use of the supposedly assigned variable appears mysteriously unevaluated. Typographic errors in variable names are also more of a problem in Maple than in most compiled computer languages, since Maple assumes "just another funny variable name".

of the loop controlling syntax is optional. If we follow the notational convention of enclosing optional parts in square brackets, the syntactical form is

```
.....
[[for var] [from start] [to finish] [by
step]
[while test]]
do
.....
if ...break; fi;
.....
if ... next; fi;
.....
od;
```

The structure allows a counted loop (using the **for**, governed by an additional termination condition (the **while** test), and allowing either an early exit (**break**) or loop short circuit (executing **next**).

It is hard to imagine the need for all of these options at once. A simple loop governed by a counter looks like the following.

```
.....
for var from start to finish
do
.....
.....
.....
od;
```

In this, the governing variable may be omitted, in which case Maple supplies an anonymous counter. The beginning counter value can also be omitted, and a default value of 1 is used. This idiom is quite common in Maple programs, and quite disconcerting for beginners.

The "indefinite" form of the Maple loop drops the counter syntax, and governs execution with a test condition.

```
.....
while test
do
.....
.....
.....
od;
```

1.4.5 Maple Evaluation

Users of Maple often proceed in program development along a natural path. Since the Maple program operates as an interactive interpreter, calculations can be tried out "piece by piece" before being embedded in a procedure. It is then a big surprise when phrases that seemed to work interactively suddenly return some of the procedure local variables instead of the expected calculations when the procedure runs. What is causing the difference is that interactive expressions are generally experienced fully evaluated, while within a procedure the evaluation only takes place (unless coded to behave differently) to a single level. This is a situation without a parallel in conventional procedural programming languages, and is driven by a need to control excessive computation in a symbolic computation environment.

Maple expressions entered into the worksheet environment have a structure that can be visualized in the same way as the symbolic expressions discussed earlier. This structure is the conceptual framework where the terms "fully evaluated" and "evaluated to one level" used loosely in the previous paragraph can be understood.

An example can be constructed from a set of symbolic variables, and a single procedure that multiplies together its arguments. [6]

```
> a:=b;
> b:=c;
> c:=d;
>
> mpy := proc()
>   local product, i;
>
>   product := 1;
>   for i from 1 to nargs do
>     product := product * args[i];
>   od;
> end;
>
> d:=mpy(e,f);
>
> f:=h;
> e:=i;
> i:=j;
>
> l:=mpy(m,n,o);
>
> h := l
```

These entries correspond to an evaluation tree in a natural way, identifying the tree branches with the statement assignments (and function arguments). The representation of the above situation is the following tree of Figure 1.6. Evaluating a interactively results in a full evaluation.

```
> a + 1
```

$$j\,m\,n\,o + 1$$

[6]The procedure illustrates that Maple procedures handle variable argument lists easily, using the built-in argument count, and the array of arguments.

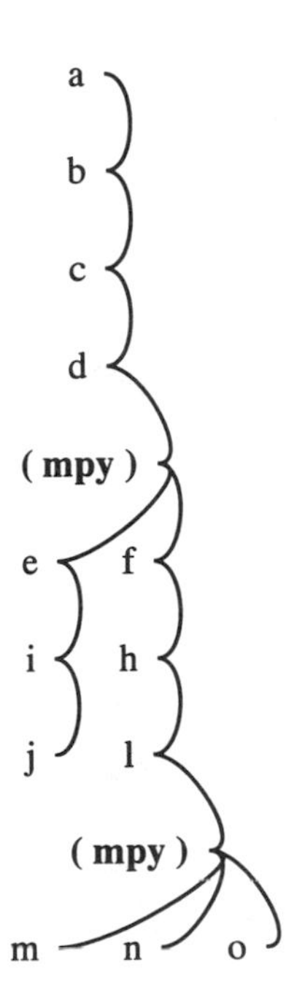

FIGURE 1.6. Runtime evaluation tree

Partial evaluation is obtained by using the `eval` command with an additional argument for the level of evaluation desired. Evaluating a to only one level corresponds to going down one level in the evaluation tree.

```
> eval(a,1);
```

$$b$$

The following results can all be understood by reference to the evaluation tree in Figure 1.6.

```
> eval(d,1);
```

$$e\,f$$

```
> eval(a,4) +1;
```

$$e\,f+1$$

```
> eval(d,2);
```

$$e\,h$$

```
> eval(a,5)+1;
```

$$e\,h+1$$

```
> a+1;
```

$$eh + 1$$

```
> eval(a,5) +1;
```

$$ih + 1$$

```
> eval(a,5) +1;
```

$$ih + 1$$

```
> eval(a,6) +1;
```

$$jh + 1$$

```
> eval(a,4) +1;
```

$$ef + 1$$

```
> eval(a,5)+1;
```

$$ih + 1$$

```
> eval(a,6)+1;
```

$$jl + 1$$

```
> eval(a,7) +1;
```

$$jmno + 1$$

Full evaluation of a in this case takes level 7. Yet "overdoing it" produces the hoped for full evaluation, and not an error message.

```
> eval(a,8)+1;
```

$$jmno + 1$$

Strategically inserted `eval` expressions in procedures usually solve such problems when they do occur. If efficiency is an issue, then determining what level of evaluation is actually required may be useful. This can be carried out interactively (as above) by experimenting with evaluation to various levels.

1.4.6 Details about Maple Procedures

In programming languages there are issues about how the arguments to a procedure are treated when the procedure is called. PASCAL uses keywords to make distinctions, while C and FORTRAN have flat conventions for what happens. The situation in Maple is that most arguments are fully evaluated as they are handed (by value) to a procedure, but there are exceptions to the rules. The exercises below illustrate how this works.

1.4.7 Procedure Arguments

Exercise

Maple routines sometimes have to modify variables that are defined back in the calling routine. PASCAL handles this by using a VAR declaration, and C uses pointer variables to get the effect. What is going on in Maple is not so obvious since Maple has no syntax that indicates the distinction between modifiable arguments, and those passed by value.

The default way arguments are passed in Maple is (for most types of arguments) by full evaluation. Try the following (without doing anything involving the variable cow first).

```
> f:=proc(x) x:= 5; end;
>
> cow;
> f(cow);
> f(cow);
>
> cow;
> f('cow');
>
> cow := 'cow';
> f(cow);
```

What is going on is that f is unhappy unless the argument evaluates to what is referred to as a *name*. When it is a *name*, the assignment inside f acts as though the argument were a Pascal VAR parameter. Any function argument that evaluates to something other than a name that causes an error in the assignment statement attempted within f, and so will result in the error report from f. Assigning a variable to its quoted self "clears" the variable back to the unassigned state, and restores the Maple type of the variable to *name*. When f is called again, Maple sees just a name, and so the procedure succeeds in assigning the value 5 to the variable cow.

Exercise

Maple supports tables which act like what are called associative arrays in other symbolic programming languages. What this means is you can use names as subscripts. Try these.

```
> a[donkey] := duck;
> a;
> eval(a);
> f(a[rubbish]);
> eval(a);
> a := 'a';
> eval(a);
> b:=table();
> eval(b);
```

Using something like a table defines it as a table. It is clearer to create it with `table()`, especially since you have to explicitly create matrices and vectors. WHEN YOU PASS TABLES OR ARRAYS TO A PROCEDURE IN MAPLE, THEY ARE PASSED AS THE NAME OF THE OBJECT AND NOT FULLY EVALUATED. This makes these variable types natural as VAR parameters for returning procedure results. Try this out.

Exercise
Tables are used internally in Maple to speed up computations, basically by cheating and returning the result from a previous invocation. The previous result is read from a record in a table stored within the procedure. This mechanism applies to user defined procedures that are built with a `remember` option. It is very useful for recording sequences of results associated with recursive calculations. Classics:

```
> myfact:= proc(n) option remember;
>
>   if n = 0 then 1 else n * myfact(n-1)
>   fi
> end;
>
> myfact(4);
>
> rabbits := proc(n) option remember;
>
> if n = 0 then 1
> elif n = 1 then 0
> else rabbits(n-1)+rabbits(n-2)
> fi
> end;
>
> seq(rabbits(n), n=1..10);
```

Exercise

Math background **Page:** 77

The existence theorem for ODE's involves a recursive procedure, generating a sequence of functions that in the limit converge to the solution of the differential equation. If the problem is to solve

$$\frac{dx}{dt} = f(x(t), t)$$

subject to the initial condition

$$x(0) = x_0,$$

the algorithm consists in plugging the previous approximate solution into the right hand side of the integral equation

$$x(t) = x_0 + \int_0^t f(x(t), t)dt.$$

For the equation $\frac{d}{dt}x(t) = x(t), x(0) = a$, the sequence of approximations can be generated by the recursive procedure listed below. Consider first

```
> int(1, t=0..t);
```

This produces an error message with some Maple releases, caused by the use of the same variable name both as the upper limit and the "dummy variable of integration". This means that any Maple code that carries out the successive integration procedure of the Picard iteration ought to take care to keep the upper limit and variable of integration distinct. The code below does this.

```
> iteration:= proc(n) option remember;
>
>  if n = 0 then RETURN(a);
>    else RETURN( a + subs(b =t, int(iteration(n-1), t=0..b)));
>  fi;
>
>  end;
>
> seq(iteration(n), n=1..5);
```

`plot` does not like free constants in its expressions, so that in order to plot the curves from the iteration, a must be assigned a value.

1.4.8 Programming Gotchas

As illustrated by the discussion above, Maple programming syntax has a lot of similarity to other block structured procedural languages, in particular to standard PASCAL.

The similarity cuts both ways however, since there are things that are different, and readily catch the unwary. The first thing is that Maple is case sensitive (like C and Java), while PASCAL is not.

The second has to do with the apparent parallel between the Maple `do .... od` and either the PASCAL `BEGIN .... END` , or C and Java `{ ... }`constructs. The connection is much more apparent than real, and inevitably traps programmers familiar with the other languages.

The `BEGIN ..... END` in PASCAL and braces in C and Java serve as generic block delimiters, used generally to group a sequence of statements. In contrast, the `do .... od` groups a sequence of Maple statements, but *only* a sequence governed by a `for` iteration or `while` test. The classic blunder writers of other block structured languages make is to follow an `if .. then` by a `do ... od`. Maple enters the `do ... od`, never finds an exit condition, and loops forever.

Virtually all former PASCAL and C, and Java programmers do this once, and learn this syntax point the hard way. It is also the way most programmers become acquainted with the Maple `debug` facilities and related debugging tools, since it is impossible to find where the code is hanging without tracing the execution path.

1.4.9 Maple Program Development

Our interest is in developing "serious" Maple programs, rather than one-off calculations or two page homework solutions. What this means is that the project will be developed as a series of procedures, and probably over a period of time.

Since a procedure definition in Maple syntactically is just an assignment to a variable that happens to denote a procedure, one could type in the whole series of required procedures, end to end, as it were.

In fact, this idea is found to not work very well in practice, even though all Maple novices attempt that tactic. The problem seems to be that the Maple user interface edit window is *modal*: it knows[7] when you are entering a procedure block, and when you are simply entering a statement to be directly executed by the Maple interpreter. A user inevitably decides that something has been left out, or can be copied from somewhere else, or . . . , and the user interface ends up out of synchronization with the user's idea of the mode. This is frustrating.

A more productive approach is to adopt the classical program development cycle: design, code, compile, test, diagnose, fix, compile, test . . . . This is readily done with the Maple user interface by using a separate text editor to write the Maple procedures.

The Maple `read` command then will enter the contents of the text file just as though they had been hand entered. Errors in the source code are reported, so effectively this compiles the procedures defined in the text file. The newly entered procedures can then be tested interactively in the Maple environment, and if an error is discovered, it is simply fixed in the independent text editor window. The `read` command then is re-executed, and testing continues.

With Maple code, it is even common to try out "pieces" of code interactively as the procedure is being written in order to verify that the memory of the syntax for some Maple library routine is accurate. That catches problems like reversed procedure arguments before they get cast in the code file.

---

[7]or at least thinks it does

Part II

Differential Equations

2 Introduction to Differential Equations

Differential equations are a major tool in a wide range of applied mathematical endeavors. The discussion of reasons for this leads ultimately to unanswerable metaphysical questions, but the proximate cause is the fact that most physical laws (at least from the time of Newton and Leibniz with the invention of calculus) are phrased as statements about the rates of change of some quantity of interest. A mathematical statement of such a law will consequently involve derivatives in its formulation, and hence represent a differential equation of some sort. The rough and ready definition of a differential equation is simply an equation involving derivatives of an unknown function whose solution is sought.

Physical problems in the general sense are not only the source of examples of differential equations, but provide invaluable guidance in the process of formulating "properly posed" mathematical problems for these equations. We will see that some problem statements lead to differential equations with what might be called under-determined solutions, or solutions involving free parameters in some sense. The "real problem" behind the mathematical formulation, and all the physical insight or common sense associated with it, is the best guide to what the natural "side conditions" that need to be specified might be.

2.1 Example Problems

In view of the central role that Isaac Newton's work played in convincing the world that "calculus formulations" were a valuable description of the natural universe, his *second law* almost has to be the first example of a differential equation. For a particle moving in such a way that the forces acting depend only on the position variable, this takes the form

$$\frac{d^2x}{dt^2} = \frac{F(x)}{M}.$$

This is an example of a second order differential equation. Experience with mechanics problems provides the insight that the above differential equation (i.e., the second law) is not in itself sufficient to determine the trajectory of the described particle. It is necessary to additionally specify an initial position and velocity in order to completely specify a solution.

Of course Newton's law is just one of many physical laws stated in terms of rates. A sampling of examples leading to differential equations is given below.

2.1.1 Simple Rate Equations

Many situations lead directly to first order differential equations. The radioactive decay of nuclear material is one such example. The observed law governing this phenomenon is that the percentage rate of decay of the radioactive species in question is a constant, characteristic of the material. That is,

$$\frac{\frac{dx}{dt}}{x} = -k,$$

or

$$\frac{dx}{dt} = -k\,x.$$

One may verify the solution

$$x(t) = x(0)\,exp(-k\,t),$$

where $x(0)$ is the quantity of the material present at the initial time. It is also easy to see that $x(t) = C\,exp(-kt)$ is a solution of the equation above no matter what the value of C might be. The choice of $C = x(0)$ picks out the solution that satisfies the correct initial condition.

What are described as flow problems (or mixing problems) also lead readily to first order equation models. The classic problem is a brine tank (constantly stirred), which has two different concentrations of brine flowing into it. One pipe supplies brine with a concentration of k_1 grams per liter, at a volume rate of R_1 liters per second. The second source delivers a concentration of k_2 at a rate of R_2. The problem is to find the concentration in the tank as a function of time, assuming the mixture is withdrawn at a rate that keeps the volume in the tank constant.

The key to this (and related) problems is to take the quantity of salt in the tank (rather than its concentration) as the quantity of immediate interest. If $x(t)$ denotes this quantity, the equation then follows from the indisputable conservation law

$$\frac{dx}{dt} = \text{rate of inflow} - \text{rate of outflow}$$

where the rates must be calculated in terms of grams per second of salt transferred. The rate of inflow is just

$$k_1 R_1 + k_2 R_2,$$

and the volume rate of outflow is $R_1 + R_2$ in order to keep the tank at a constant level. If the volume of the tank is V (liters), the exit concentration is $\frac{x(t)}{V}$, leading to the equation

$$\frac{dx}{dt} = k_1 R_1 + k_2 R_2 - \frac{R_1 + R_2}{V} x(t).$$

This is closely related to the radioactive decay problem, except that there is a source term present as a result of the input streams. Again, intuition about the problem suggests that it will be necessary to specify the initial amount of salt in the tank in order to completely specify the problem to be solved.

2.1.2 *Mechanical Problems*

The harmonic oscillator equation

$$\frac{d^2x(t)}{dt^2} = -\frac{k\,x(t)}{M}$$

is familiar as the model of a single spring mass system. The physical arrangement leading to this conclusion is given in Figure 2.1. As indicated in the diagram, the

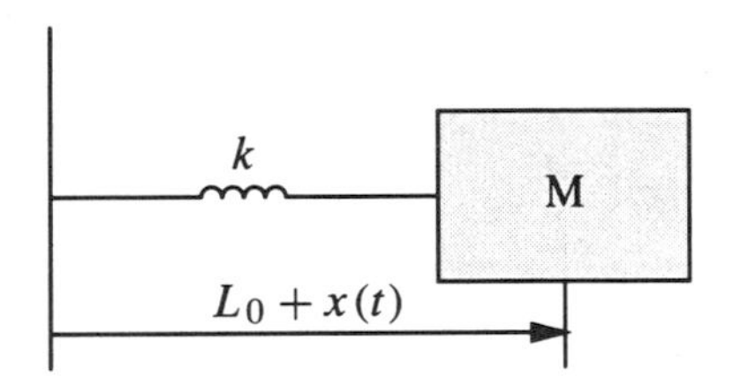

FIGURE 2.1. A truthful spring mass system

variable $x(t)$ represents the deviation of the center of mass of a block from the "rest position" of the block, at a distance L_0 from the wall attachment point. The equation given above then follows from the observations that

$$\frac{d^2}{dt^2}x(t) = \frac{d^2}{dt^2}(x(t) + L_0),$$

while the restoring force acting on the block is just

$$-k\,(L_0 + x(t) - L_0) = -k\,x(t),$$

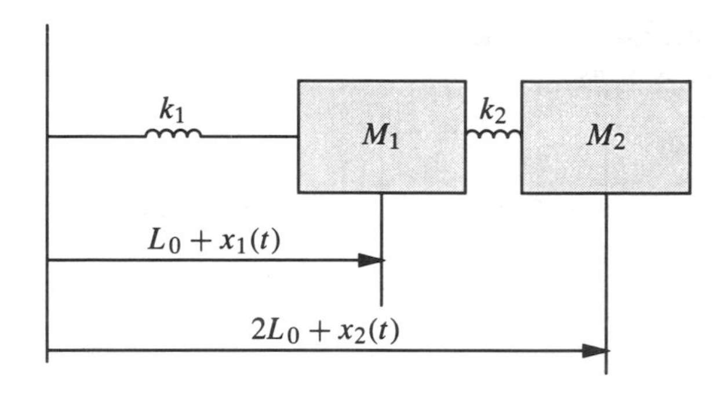

FIGURE 2.2. Dual spring mass system

since the spring force is generated by the stretch of the spring relative to its equi lib-rium length.

If two masses are interconnected as shown in Figure 2.2, then a more complicated system results. Again, the variables in the problem represent the deviation of the centers of mass from their equilibrium positions.[1] The "outboard" mass is governed by

$$M_2 \frac{d^2 x_2(t)}{dt^2} = -k_2 \left(x_2(t) - x_1(t)\right)$$

since the only force acting is as a result of the extension of the spring between the two masses. The inboard mass has two springs acting, resulting in

$$M_1 \frac{d^2 x_1(t)}{dt^2} = -k_1 x_1(t) + k_2 \left(x_2(t) - x_1(t)\right).$$

This represents a coupled pair of second order equations. When (if ??) this is solved, one may anticipate that four pieces of additional information, namely, initial positions and velocities of both masses will have to be specified to unambiguously predict the future motions of the system.

2.1.3 *Electrical Models*

Solving voltage-current problems for electrical circuits consisting solely of (linear) resistors boils down to solving a system of (linear) equations for the unknown voltages and currents. Differential equations arise when energy storage elements are introduced into the circuits. The fundamental models are of capacitors and inductors.

The replacement for Ohm's Law of resistive circuit fame that applies to these elements is a derivative relation between the voltage across and the current through

[1]The diagram assumes the equilibrium distances are equal, although the same equations result even if they are assumed to be different. Convince yourself of this.

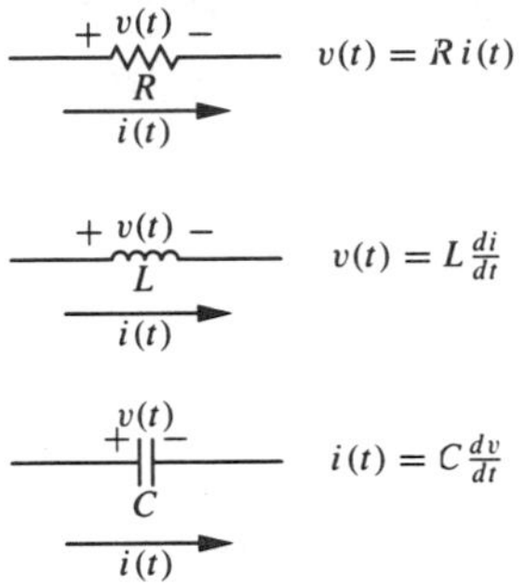

+ $v(t)$ −
R
$i(t)$
$v(t) = R\,i(t)$

+ $v(t)$ −
L
$i(t)$
$v(t) = L\frac{di}{dt}$

$v(t)$
+ −
C
$i(t)$
$i(t) = C\frac{dv}{dt}$

FIGURE 2.3. Circuit elements

the element. These are (see Figure 2.3 for the sign conventions)

$$L\frac{di(t)}{dt} = v(t),$$

and

$$C\frac{dv(t)}{dt} = i(t)$$

for the inductor and capacitor respectively. The L and C are constants specific to each instance of the device, called *inductance* and *capacitance* respectively.

A simple first order differential equation arises from a circuit containing a source, resistor, and capacitor, as illustrated in Figure 2.4. The differential equation arises

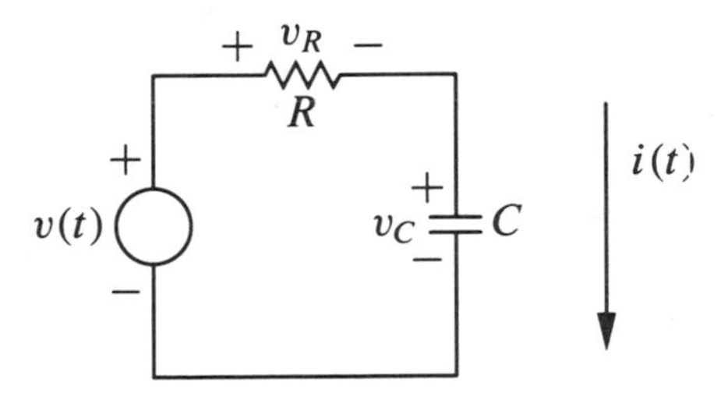

FIGURE 2.4. RC circuit

from noting that the current flowing through the capacitor is the same as that flowing through the resistor. Hence,

$$C\frac{dv_C(t)}{dt} = \frac{(v(t) - v_C(t))}{R}.$$

The problem of deriving the governing equations of circuits has been heavily studied (as one might expect.) One of the insights that have emerged is that capacitor voltages together with inductor currents (barring circuit degeneracies) are the appropriate variables with which to formulate the governing equations. This can be understood on the basis that these constitute the energy storage elements in the circuit. An example of a simple circuit is in Figure 2.5.

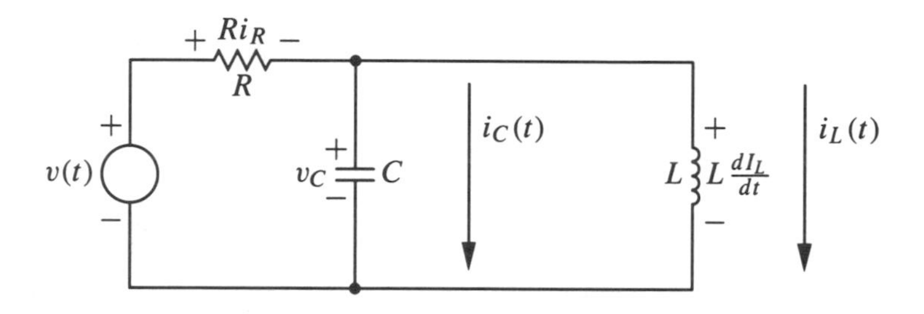

FIGURE 2.5. RLC circuit

The governing differential equations follow from the observations that the capacitor voltage $v_C(t)$ appears directly across the inductor, and the current flowing into the capacitor is whatever flows through the resistor diminished by the flow down through the inductor. The resulting system is

$$L\,\frac{di_L(t)}{dt} = v_C(t)$$

$$C\,\frac{dv_C(t)}{dt} = \frac{(v(t) - v_C(t))}{R} - i_L(t).$$

Intuition (now electrical) suggests that specification of initial values $i_L(0)$ and $v_C(0)$ should be exactly what is required to determine the evolution of the above set of equations.

2.1.4 *Kinetics*

Chemical kinetics studies the behavior of chemical reactions as they evolve over time, as contrasted with the calculations of simple product amounts associated with mass balance exercises. Chemical reactions give rise to interesting behavior. Some reactions can go into an oscillatory motion over time, and visible traveling reaction waves can be seen in particular cases (see [5]).

Chemical reaction problems are also somewhat notorious as examples of problems that are difficult to solve numerically in an accurate fashion. The problem arises because of a mixture of fast and slow time scales in the same reaction; the resulting numerical problems are termed “stiff”.

The rate equations for chemical reactions follow from the law of mass action, which allows the governing equations to be written directly from the reaction balance equation. As an example an enzyme reaction can be considered. The equation of balance is of the form

$$n\,S + E \underset{k_2}{\overset{k_1}{\rightleftarrows}} I \overset{k}{\rightarrow} P + E.$$

The interpretation is that n molecules of the "substrate" S combine with one of the "enzyme" E to form an intermediate I, with forward and backward reaction rates of k_1 and k_2 respectively. The intermediate decays to produce a product P, liberating the enzyme for reuse in the process. This latter reaction is treated as unidirectional with rate k. This description is centering on the major mechanism of the situation, while in actuality there are relatively high concentrations of other species present in the reaction vessel, but treated as constant as far as the above is concerned.

The equations of motion for the concentrations become

$$\frac{d[P]}{dt} = k\,[I]$$

$$\frac{d[E]}{dt} = -k_1[E]\,[S]^n + k_2[I] + k[I]$$

$$\frac{d[I]}{dt} - \frac{d[E]}{dt}$$

$$\frac{d[S]}{dt} = -k_1[E]\,[S]^n + nk_2\,[I].$$

The powers in the above equations come from the mass action rule, while the factor of n in the last equation reflects the fact that the back reaction of one molecule of I produces n molecules of the substrate species.

2.1.5 Beam Models

Everyone is aware that overloaded structural elements can fail. At one time the mechanisms of failure were not well understood, and experience was the main guide for the design of structural beams.

The analysis of the deformations and force distributions of continuous materials is properly part of solid mechanics. Solid mechanics in general form is based on partial differential equation models. A simplified analysis suffices for many structural beams, and this approach (called the Euler–Bernoulli theory) is based on ordinary differential equation models.

Concepts and Sign Conventions

The object of study is a transversally loaded beam. A naturally occurring example is a dwelling floor joist, or roof joist. The duty of such a structural element is to support the imposed loads, without excessive deflection, or material failure. What is required is a model whose variables will provide (estimate, since all models are approximate) the deflection and internal forces resulting from a given load distribution. The basic picture is of a beam loaded "from above", supported at the ends, and sagging under the load. This is illustrated schematically in Figure 2.6. The diagram also illustrates the basic problem coordinates. The x measures

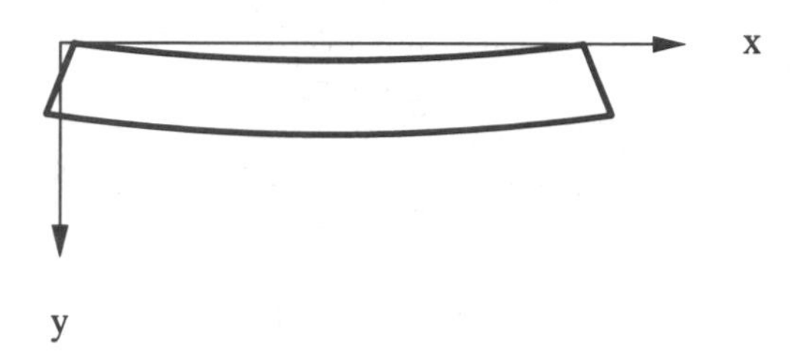

FIGURE 2.6. Basic beam coordinates

position along the beam, and y is used for the resulting deflection.[2]

If the beam is held bent as the diagram indicates, the material at the top of a cross section through the beam will be under compression, while a state of tension will prevail through the lower portions. This situation can be visualized by imagining a "free body section" extracted from the deformed beam. The "free body section" can be kept in equilibrium by applying to the "cut surface" the forces which were applied by the adjacent section of deformed beam material. The force varies is both magnitude and direction across the cut face. The vector which represents the force per unit cross sectional area is called the *stress vector* acting on the cross section. A diagram illustrating the equilibrium situation is in Figure 2.7. The stress vector at each point of the cross section can be resolved into components normal and parallel to the cross section surface. The *normal stress* distribution is shown in Figure 2.8. Because of the variation of the normal stress[3] a couple, or moment is generated by the bending of the beam. There are also components of stress parallel to the cross section face, referred to as *shear stress.*

The Euler–Bernoulli model considers not so much the distribution of stresses across the section, but the net resulting moment (due to the normal stress distribution) and shear (thought of as the total transverse force acting across the cross

---

[2]Notice that the choice of coordinate direction will make the deflection positive in the usual structural situation. There are differing sign conventions between structural texts, and some mechanical treatments.

[3]The assumption that the normal stress varies linearly across the cross section, as suggested by the Figure, is the basis of Euler–Bernoulli beam theory.

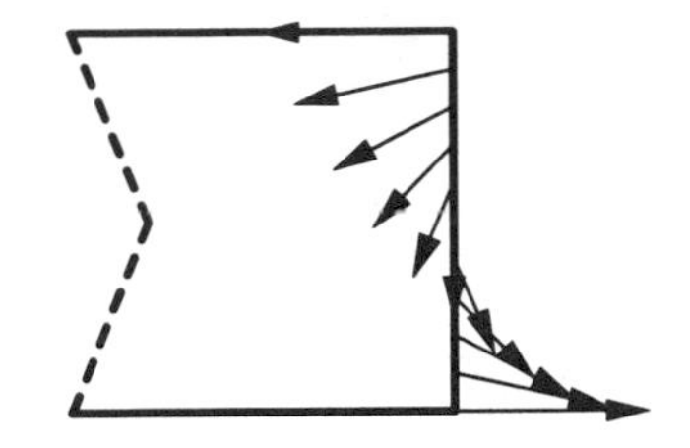

FIGURE 2.7. Free body stress distribution

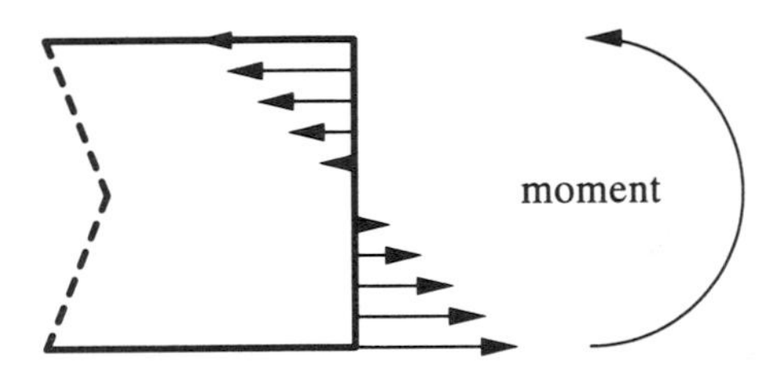

FIGURE 2.8. Normal stress and resulting moment

section). The net moment and shear must be taken to vary according to position along the beam, since the load carried by the beam varies throughout the length. The problem moment and shear variables are pictured in Figure 2.9 for a "small" piece of beam of length Δx.[4]

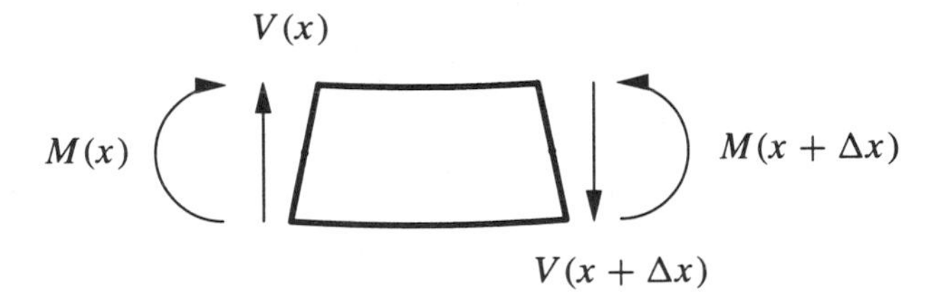

FIGURE 2.9. Shear and bending moment conventions

Shear and Bending Moment Equations

The differential equations describing this problem come from the conditions that the deformed beam must satisfy in order to be in mechanical equilibrium. By taking moments about the center point on the left end of the "beam slice" of Figure 2.9

[4] There are again sign conventions associated with this diagram. A positive moment corresponds to conventional loading from above, and positive shear is as indicated.

we get

$$\begin{aligned} M(x) - M(x + \Delta x) + V(x + \Delta x)\,\Delta x &= 0, \\ \frac{M(x) - M(x + \Delta x)}{\Delta x} &= -V(x + \Delta x). \end{aligned}$$

As $\Delta x \to 0$, this becomes the derivative relation

$$\frac{dM}{dx} = V(x)$$

relating the moment and shear variables. We are assuming that the beam is loaded by weights that provide a downward force of $w(x)$ units [5] per unit length. The forces acting in the vertical direction are illustrated in Figure 2.10, and must also be in equilibrium. Hence,

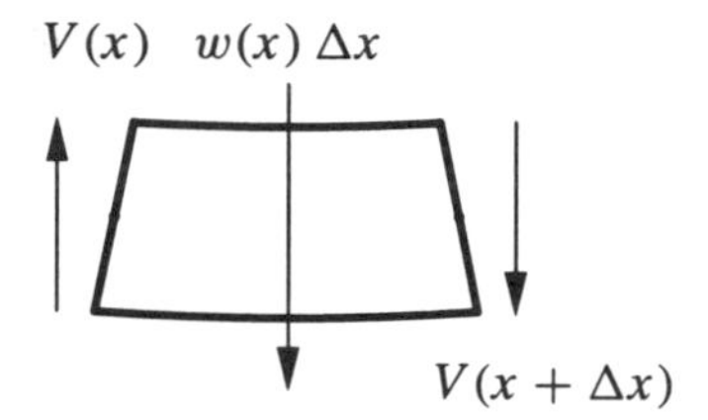

FIGURE 2.10. Vertical force balance

$$V(x) - w(x)\,\Delta x - V(x + \Delta x) = 0.$$

As $\Delta x \to 0$, the shear and load are related by

$$\frac{dV}{dx} = -w(x).$$

The shear and bending moment relations above are a matter of definition of the quantities involved and the attendant sign conventions. The mechanical content of the beam equations lies in the connections between the geometrical deformation of the material and the resulting shears and moments.

The deformations are described in terms of the "neutral axis" of the beam. It was mentioned before that generally one side of the deformed beam is under compression, while on the other side the beam material is in a state of tension. This is indicated on the cross sectional view of Figure 2.11.

The assumption of Euler–Bernoulli beam theory is that the normal component of stress varies *linearly* across the depth of the beam. This means that the assumed form is simply

$$K(y - m),$$

[5] Newtons, or pounds (force)

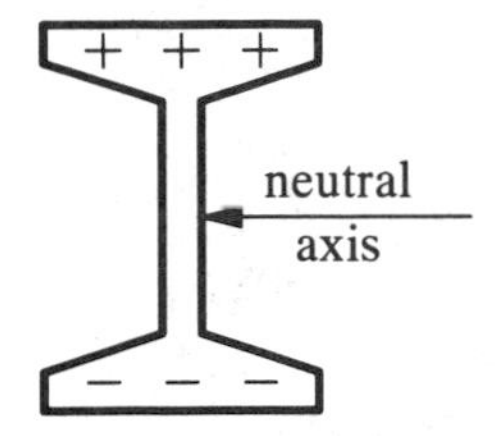

FIGURE 2.11. Axial stress and neutral axis

for some constants A, m. The location of the "neutral axis" can be determined by the condition that the total force acting on a cross sectional face must vanish. Hence,

$$\int_{\text{cross section area } A} K(y-m)\,dA = 0.$$

This condition shows that the constant m is actually the *y axis centroid* of the beam cross section. The location of the neutral axis can be determined by carrying out this geometrical calculation.

The constant K is evaluated by reference to Hooke's law: stress is proportional to the *strain* in the loaded material. It is the "relative displacement rate" of adjacent material elements that is the source of the stress. If the body is translated sideways "as a whole" with no relative deformation, no stress results.

Discussing *strain* carefully requires introducing the *displacement field* of elasticity theory, but the quantity can be calculated explicitly for the model of elementary beam theory.

The assumption that stress varies linearly across the beam face can be stated in the form that planar cross sections remain planar after the deformation. We consider two "close" cross sections (separated by Δx), and the positions of two points at a distance h from the neutral axis (the centroid of the cross section). The situation after deformation is illustrated in Figure 2.12. In the diagram, $y(x)$ plays

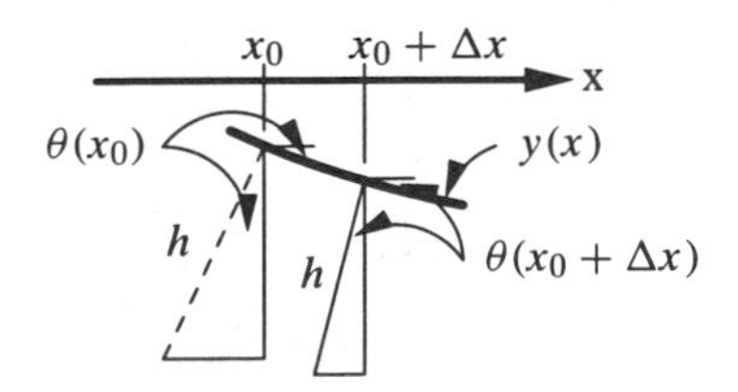

FIGURE 2.12. Material point displacements

the role of the deformed neutral axis.[6] The object of the calculation is the rate of change of the axial displacements (of points originally at a distance h from the neutral axis) with respect to x, distance along the beam.

The displacements are calculated by using using triangles, whose angle is the angle of the curve from the horizontal, with h as the hypotenuse. This leads directly to the calculation of the relative displacement of the points in the axial direction.

The change in displacement between the deformed and undeformed positions is

$$h \sin(\theta(x_0)) + \Delta x - h \sin(\theta(x_0 + \Delta x)) - \Delta x.$$

One of the assumptions of Euler beam theory is that the length of the beam is "essentially the same" after deformation as before. In view of the arc length formula

$$ds = \sqrt{1 + y'(x)^2}\, dx,$$

this assumption requires that $y'(x)$ be a small quantity throughout. Since

$$\tan(\theta(x)) = y'(x),$$

and $\tan(\theta) \approx \sin(\theta)$ for small angles, we might as well use the approximation

$$\sin(\theta(x)) = y'(x)$$

in the calculation of the relative displacement. Using this approximation, the axial strain is computed as

$$\lim_{\Delta x \to 0} \frac{h \sin(\theta(x_0)) - h \sin(\theta(x_0 + \Delta x))}{\Delta x} = -h\, y''(x_0).$$

The strain at a point in the deformed beam is proportional to both the distance of the point from the neutral axis, and the second derivative (curvature) of the axis.

Displacement Equations

The relationship between stress in a material body (such as a beam) and the strain the material is undergoing is a linear one. The constant of proportionality in the beam case is called Young's modulus and is denoted by E. If y_m is used for the y coordinate centroid of the cross section, the expression for the normal stress in the beam at location y will just be

$$-(y - y_m) E y''(x),$$

since the neutral axis is at the centroid of the cross section, and the distance from the centroid plays the role of the h in the calculation of the strain. The stress on a cross section thus has the linear distribution illustrated in Figure 2.7.

---

[6] The orientation is chosen is accord with the structural convention that the deformation axis increases "downwards".

The bending moment set up in the deformed beam is the result of this stress distribution. The total moment may be calculated by multiplying this (force per unit cross sectional area) by the moment arm through which it acts. Taking the moments about the centroid of the cross section, the element of moment acting on a cross sectional strip of area dA is

$$dM = -(y - y_m)^2 E y''(x)\, dA.$$

The total comes from integrating this over the area of the cross section, which gives the moment expression

$$\begin{aligned} M(x) &= \int_{\text{cross section area } A} -(y - y_m)^2\, E\, y''(x)\, dA \\ &= -E\, I_{yy}\, y''(x), \end{aligned}$$

where I_{yy} is the moment of inertia of the cross section.[7]

Although some beam calculations (support reaction forces) can be made working only with the moment and shear expressions, one can also consider the displacement $y(x)$ as the primary variable, and compute the other quantities in terms of the displacement function. Since the shear $V(x)$ is related to the load function $w(x)$,

$$\frac{dV}{dx} = -w(x),$$

while

$$\frac{dM}{dx} = V(x),$$

we have

$$\begin{aligned} V(x) &= -E\, I_{yy}\, y'''(x), \\ -w(x) &= -E\, I_{yy}\, y''''(x), \end{aligned}$$

so that the beam displacement satisfies the fourth order differential equation

$$E\, I_{yy} \frac{d^4}{dx^4} y(x) = w(x).$$

As a differential equation, this turns out to be a simple one to deal with. With y solved for, the bending moment $M(x)$ and shear $V(x)$ can be evaluated by differentiation. It is also possible to solve Euler beam problems by successive anti-differentiation, computing in turn the shear V, the bending moment M, and finally the displacement y. Carrying this approach out graphically results in so-called shear and bending moment diagrams.

[7] The sign of this expression should be noted. With the displacement axis oriented "downwards", a beam section with a conventionally positive bending moment acting will have a negative displacement second derivative.

End Conditions

The most common model is a pin joint. This often corresponds physically to a simple beam end supported without restraint from below, although some bridge beams are constructed with actual hinge pins at the ends. A symbolic representation of this is in Figure 2.13. The various physical end conditions correspond to different

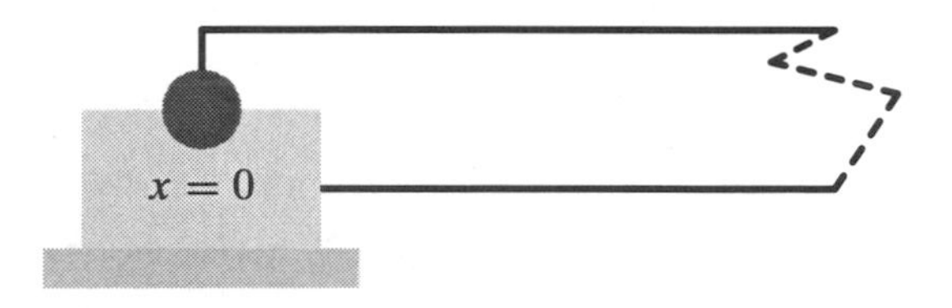

FIGURE 2.13. A pin joint boundary

sets of initial conditions for the differential equation modeling the beam deflection. For a pin joint attachment, the displacement must be 0 at the corresponding end. A pin joint is free to rotate, and cannot exert any bending moment on the end of the beam. The boundary conditions for a pin joint at $x = 0$ are hence

$$y(0) = 0, \quad -E I_{yy} \frac{d^2 y}{dx^2}(0) = 0.$$

Some beam ends are embedded in masonry, or are otherwise restrained (by welding, or bolting) from either moving vertically, or rotating at the built in end. The situation is as illustrated in Figure 2.14. The boundary condition that enforces

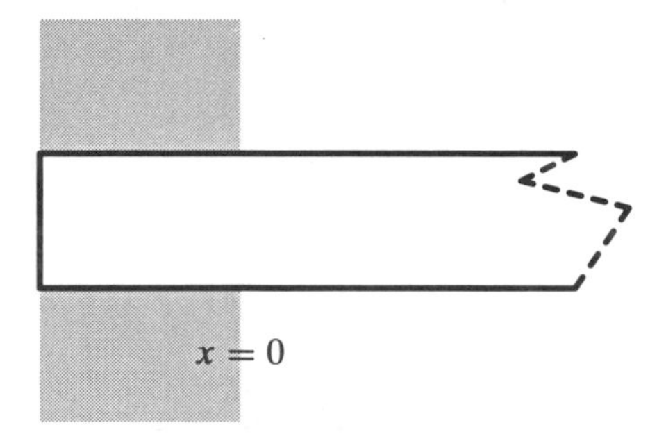

FIGURE 2.14. A built-in end condition

the condition of no rotation is that the slope of the deflection at the built in end must vanish, as well as the deflection itself. Hence

$$y(0) = 0, \quad \frac{dy}{dx}(0) = 0,$$

for a built in end at $x = 0$.

Cantilever structures have no restraint support on one end of the beam.[8] The

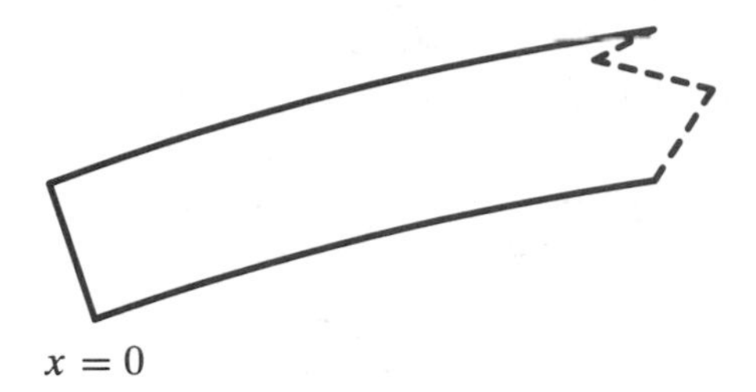

FIGURE 2.15. A free beam end

configuration appears in Figure 2.15, from which it is clear that both nonzero displacement and nonzero slope are to be expected here. The boundary conditions that arise come from the fact that there are no external forces (or torques) from the external environment acting on the free end of the beam. Both the shear and moment expressions must vanish at a free end. If the free end is located at the beam end $x = L$, then

$$-EI_{yy}\frac{d^2y}{dx^2}(L) = 0, \quad -EI_{yy}\frac{d^3y}{dx^3}(L) = 0$$

are the boundary conditions that must apply in this case.

With three choices of boundary conditions on each end of a beam, there are a lot of basic configurations, leaving aside variations in the applied load pattern $w(x)$. Maple makes it possible to easily create a beam solver with symbolic problem description input.

Maple connection **Page:** 301

2.2 Order, Dimension and Type

Differential equations are classified into several type categories, with the aim of differentiating between the more or less tractable and intractable cases. The simplest problems are the first order ones, in which the problem is represented by a single equation for an unknown function. The generic representation of such a problem is

$$\frac{dy}{dx} = F(y, x)$$

[8]For the problem to make physical sense, the other end of the beam must be a built-in one.

where $F(y, x)$ represents the slope of the unknown function $y = y(x)$.

First order equations of this sort have an easy geometric interpretation arising from the slope specification idea. One can visualize $F(y, x)$ as defining a "flow field" on the plane R^2 where the solution curve may be visualized. If the flow field is graphed by drawing short tangent lines throughout the plane, the solution may be thought of as being constructed by traveling tangent to these lines. See Figure 2.16. These diagrams show graphically that a point on the curve must be specified

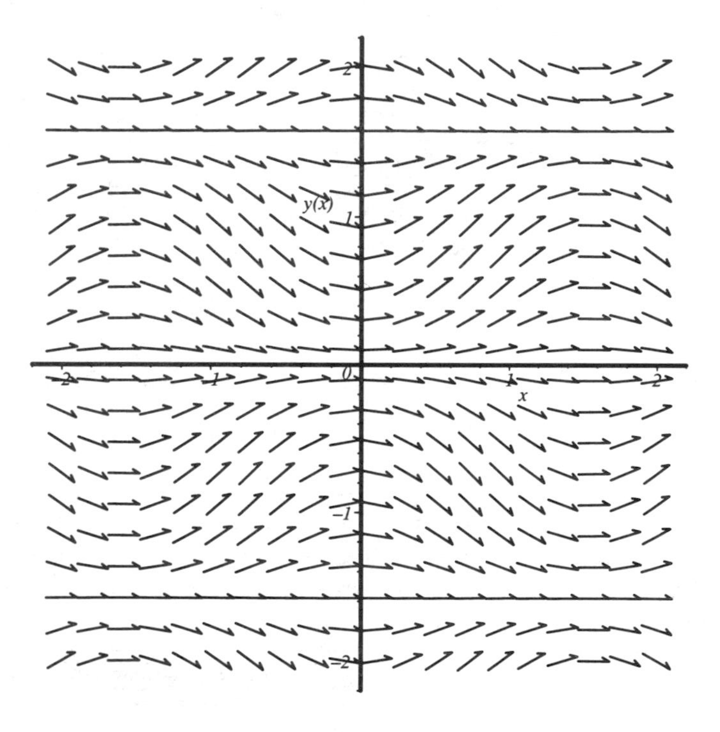

FIGURE 2.16. A flow field

in order to obtain a unique solution to such a problem. The equation above alone represents a family of solutions, and specifying a different point will result in a different solution curve.

Several of the examples above involve derivatives of order higher than one. The mechanical examples naturally arrive with second order derivatives and are classified as second order equations. The order of a differential equation may hence be taken as the order of the highest derivative appearing in it.

The models above differ in more ways than the order of the equations. Some involve a single scalar function, while others are in reality simultaneous differential equations for several unknown functions. It might be thought useful to "eliminate variables" in order to obtain a single equation, perhaps of higher order. For example, one might differentiate the second of the equations for the coupled spring mass system twice, and then use the other equation to eliminate the "outboard" variable. This procedure is atypical, however, as there is no apriori reason to believe that such elimination of variables is possible. There actually is no reason to think it desirable either, as there are compelling reasons to consider a coupled system of first order equations (such as the electrical or kinetic examples above) as the natural format.

Perhaps the most important classification of differential equations is in the linear and nonlinear camps. Of the examples above, all except the kinetics problem have equations that are linear in the unknown function(s), and hence are classified as linear problems. It will become evident below that linear differential problems are better understood than the nonlinear case by an embarrassingly wide margin. The well-developed machinery of linear algebra can be brought to bear on linear problems with effect, while the conquest of even a single nonlinear member of the genre is cause for celebration.

2.2.1 *First and Higher Order*

Since both "order" and systems of equations have been introduced in the above discussion, this is an opportunity to point out that the two ideas are not independent. In a real sense, one can trade the former for the latter, and this leads to the idea that systems of first order equations are the "natural thing".

This issue is easiest to approach by considering a concrete example, say the spring mass harmonic oscillator problem. The governing equation in its initial format is the second order linear model

$$\frac{d^2x(t)}{dt^2} = -\frac{k\,x(t)}{M}.$$

This can be represented as a first order system by introducing the position and velocity of the particle as simultaneous (related) variables. The relationships are

$$\frac{d}{dt}x = \frac{dx}{dt}$$

and

$$\frac{d}{dt}\frac{dx}{dt} = -\frac{k}{M}x.$$

The above is a system of first order differential equations for the variables x and $\frac{dx}{dt}$. It might be easier to see this if v is used as the "name" of the velocity variable. Then

$$\frac{d}{dt}x = v$$

and

$$\frac{d}{dt}v = -\frac{k}{M}x$$

is the simultaneous system. The single second order system has been replaced by a pair of first order equations.

It is even helpful to write the above equation in a vector notation, as in

$$\frac{d}{dt}\begin{bmatrix} x \\ v \end{bmatrix} = \begin{bmatrix} v \\ -\frac{k}{M}x \end{bmatrix},$$

$$\frac{d}{dt}\begin{bmatrix} x \\ v \end{bmatrix} = \begin{bmatrix} 0 & 1 \\ -\frac{k}{M} & 0 \end{bmatrix}\begin{bmatrix} x \\ v \end{bmatrix},$$

which might be taken as the standard form for such a problem.

2.2.2 *Partial Derivative Problems*

The models and equations above involve only a single independent variable, and are called "ordinary differential equations" as a consequence. The contrast is with problems that involve derivatives with respect to several independent variables, consequently partial derivatives. One might argue that everything in the universe involves manifold complications, and hence that partial differential equations ought to be the modeling tool of discerning modelers. The tractability gulf between partial and ordinary differential equations stifles this line of argument.

The motions of a vibrating string are governed by

$$\frac{1}{c^2}\frac{\partial^2 u}{\partial t^2} = \frac{\partial^2 u}{\partial x^2}$$

where $u(x,t)$ represents the vertical displacement of the string as a function of distance along the string as well as time. This ranks as one of the most tractable of partial differential equations, with a solution of the form

$$\Sigma_{n=1}^{\infty} \sin\left(\frac{n\pi x}{L}\right)(a_n \cos(n\omega_0 t) + b_n \sin(n\omega_0 t)).$$

Solution methods for partial differential equations are built on the methods and results of the ordinary differential equation case.

2.3 Problems

1. Derive the all encompassing model of the universe.
2. If the differential equation in question is

$$\frac{dy}{dx} = x\,y,$$

sketch the associated flow field.

3. In the spring mass example discussed above, the "near" spring is attached to an immovable wall.Suppose that the wall is replaced by another mass. Determine the equations governing the motion of three masses interconnected by two springs. The system is illustrated in Figurc 2.17.

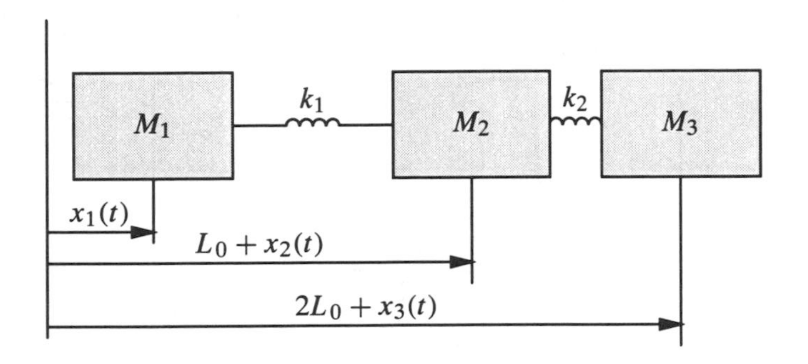

FIGURE 2.17. A three mass system

4. A stream carrying a sediment at a concentration of .5 grams per liter is flowing into a pristine mountain lake with a surface area of 250,000 square meters, and average depth of 20 meters. Find a differential equation whose solution would give the evolution of the amount of sediment in the lake as a function of time. You may assume that water evaporates from the surface at a rate that balances the inflow of water.

5. Consider the RLC series circuit in Figure 2.18.

   (a) First derive a set of first order equations governing the circuit.

   (b) Now suppose the inductor in the RLC circuit of Figure 2.18 is replaced by a short circuit. What does the governing equation become in this case? As solution methods for differential equations have yet to be discussed, can you *guess* a solution of this simpler equation for the RC circuit, for the case where the source voltage $v(t) = 0$?

6. How long does it take for a radioactive substance to decay to half of its original quantity? Assume the "rate constant" is α.

7. A simple model for heat transfer[9] considers the transfer between an ambient environment at temperature T_a, and a material body assumed to be at a spatially uniform temperature of T. The assumption is that the rate of transfer

[9]More complex models account for spatial variations of temperature and result in partial differential equations.

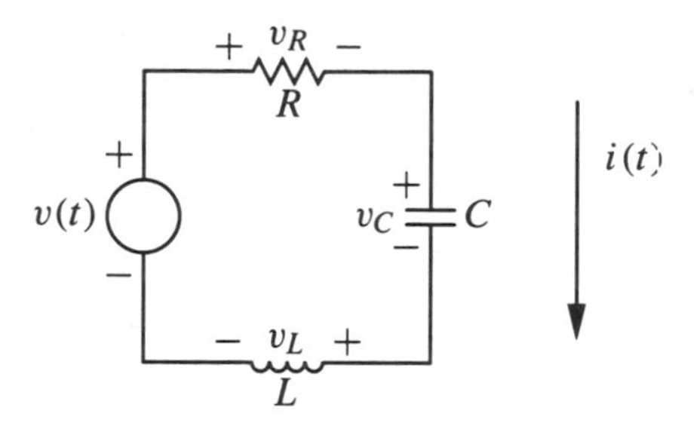

FIGURE 2.18. RLC series circuit

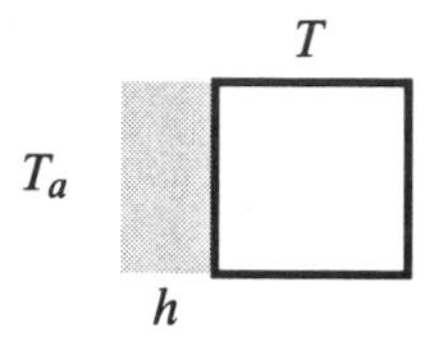

FIGURE 2.19. Basic heat transfer model

of heat energy from the body and the environment (per unit of surface area) is

$$h\,(T - T_a),$$

where h is the heat transfer coefficient. The problem is illustrated in Figure 2.19.

Using the fact that the stored heat energy in the material body is

$$M\,c\,T,$$

where M is the body mass, and c the specific heat, show that the differential equation governing the body temperature is

$$\frac{dT}{dt} = -\frac{hA}{Mc}\,(T - T_a)\,.$$

Here A represents the cross sectional area of the slab.

8. Limits on the switching speed of digital integrated circuits arise from the physical nature of the transistors used to build the switching devices. MOS (metal oxide semiconductor) transistors have a controlling terminal called a *gate*. The gate is modeled as being a capacitor, which must be charged to a certain level in order to switch the device. On the circuit, the gates are interconnected by polysilicon, which introduces a resistance in series leading to and between the gates. A three gate model is given in Figure 2.20.

   Find the differential equations for the gate (i.e., capacitor) voltages v_1, v_2, v_3 of the three gate model.

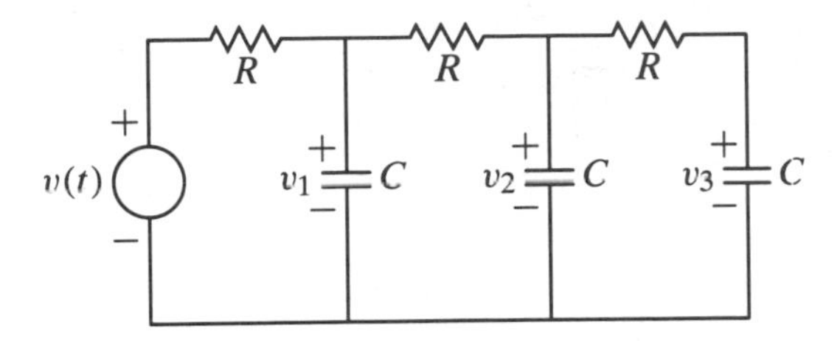

FIGURE 2.20. Gate capacitance model

9. In an Euler beam model, the shear stress is given by

$$V(x) = -E\, I_{yy} \frac{d^3 y}{dx^3},$$

where $y(x)$ is the beam displacement.

For conventional dimension wood timber beams, failure tends to occur when the shear stress exceeds the limit for the material (the fibrous structure of wood makes it relatively strong in tension/compression). Such beams are often installed (say as floor joists) simply supported at the ends by foundation walls. Such beams support their own weight w_0 as well as an applied load $l(x)$, so that the total vertical load

$$w(x) = w_0 + l(x)$$

is actually positive throughout.

Using elementary calculus, show that for such a beam the maximum shear stress must occur at one of the beam ends, so that the beam should be inspected for shear failure (splits) at the support points. Anyone for a tour of the basement?

10. Suppose that two mixing tanks of volumes V_1 and V_2 are used for combining "salt streams". The first tank receives two fluid streams at rates R_1 and R_2, with corresponding concentrations k_1 and k_2. This vessel is continuously stirred, so that the concentration within the tank is uniform throughout.

The overflow from the first tank is combined with an input of rate R_3 and concentration k_3 in the second (continuously stirred) tank.

Find the differential equations governing the evolution of the concentrations in the two tanks. You may, of course, assume that R_1, R_2, R_3 and k_1, k_2, k_3 are specified in a compatible system of units even though that is unlikely to happen in the wild.

11. A coupled heat transfer problem is illustrated in Figure 2.21. The model consists of a sequence of material layers of "high thermal conductivity" so that they may be considered to be at a spatially uniform temperature

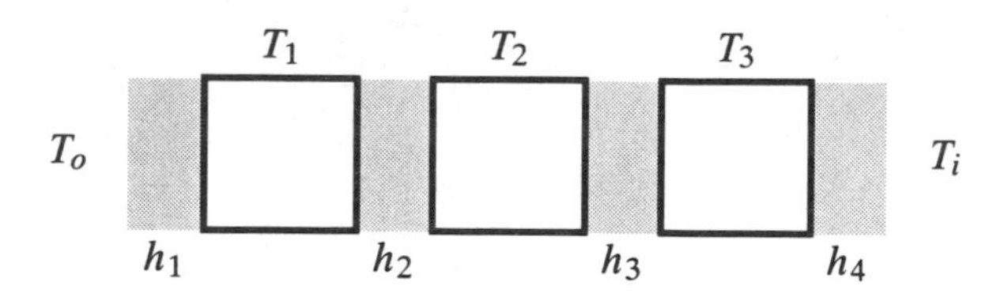

FIGURE 2.21. Heat conduction layers

throughout their (individual) physical extents. The layers are separated by insulation layers, each of which has a heat transfer coefficient (see problem 7) of h_i. The layers are also in thermal contact with exterior and interior ambient temperatures of T_o and T_i respectively.

Assuming that the material layers all have the same mass M, specific heat c and cross sectional area A, find the set of simultaneous differential equations satisfied by the temperatures T_1, T_2, T_3 of Figure 2.21.

12. Partial differential equations are generally much more difficult to solve than ordinary differential equations. The solutions in general are required to satisfy "spatial" boundary conditions in addition to satisfying the governing given equation. Find constants λ, k such that $u(x, t) = e^{\lambda t} \sin(k x)$ satisfies

$$\frac{\partial u}{\partial t} = \kappa \frac{\partial^2 u}{\partial x^2},$$

and also meets the boundary conditions $u(0, t) = u(L, t) = 0$.

13. Suppose that f and g are twice differential functions (of a single variable). Show that

$$u(x, t) = f(x - ct) + g(x + ct)$$

actually satisfies the partial differential equation

$$\frac{1}{c^2} \frac{\partial^2}{\partial t^2} u = \frac{\partial^2}{\partial x^2} u.$$

3
First Order Equations

First order scalar differential equations are the natural starting point in any study of the topic. The solutions can be easily visualized as curves in the plane, and there are many examples for which explicit solutions are possible. Beyond the fact that the problem leading to the equation may be of interest, such exact solutions are a useful aid when numerical solution methods are considered. If one has exact solutions for sample problems, it is easy to examine the errors of any numerical scheme by subtracting the exact solution values from the numerical approximate results. Estimating errors becomes much more difficult when exact solutions are not available for comparison.

The first order scalar problems also stand out as sources of nonlinear problems for which exact solutions can be calculated. They can then be used as a starting point for studying the differences between linear and nonlinear models.

The form of problem under discussion is

$$\frac{dy}{dx} = F(y(x), x),$$

which can be thought of as specifying the tangent lines to a family of curves in the plane. The easiest pictures of this interpretation can be constructed using routines in the Maple `DEtools` package devoted to exactly drawing the direction field associated with a differential equation of first order. For the differential equation

$$\frac{dy}{dx} = y x^2,$$

the Maple commands to produce the appropriate plot are discussed in the Maple exercises.

Maple reference Page: 6

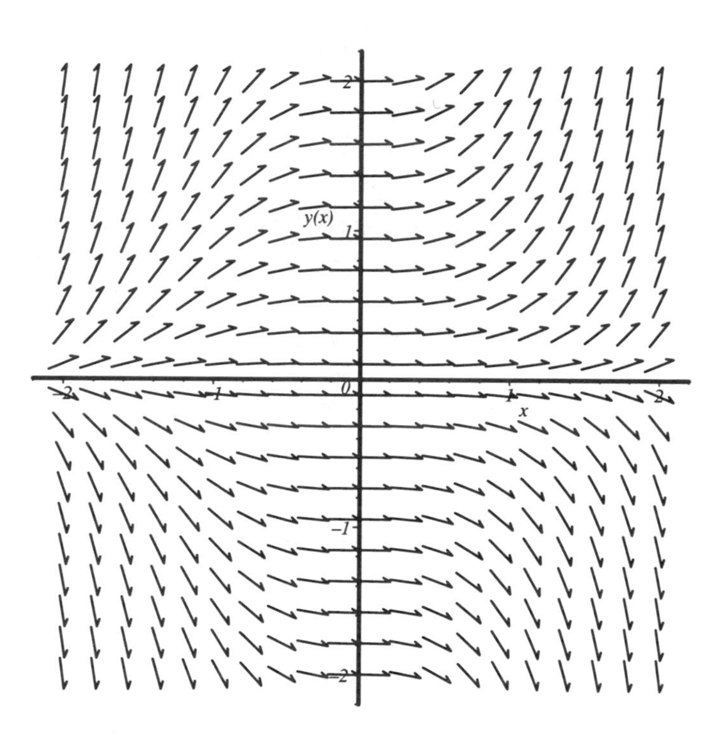

FIGURE 3.1. A dfield plot

Following that example should cause a plot like the one in Figure 3.1 to appear; the paths of solution trajectories are evident in the result. The appearance also is the source of the "flow field" terminology since the arrows resemble motion paths of particles moving with a fluid flow.

Similar plots can be produced for any first order scalar equation, and can be a useful starting point to get a qualitative view of behavior of the system under study.

3.1 Separable Problems

In some sense the process of solving a differential equation consists of "integrating" it in order to "undo" the derivatives in the equation. While it is generally accurate to think metaphorically of this process, separable differential equations are of such a form that direct integration produces solutions. The example

$$\frac{dy}{dx} = x\,\frac{(1+y^2)^2}{2y},$$

is actually a separable equation. It is solved by the formal steps

$$2y\,dy = x\,(1+y^2)^2\,dx$$

$$\frac{2y\,dy}{(1+y^2)^2} = x\,dx$$

$$\int \frac{2y\,dy}{(1+y^2)^2} = \int x\,dx$$

$$-\frac{1}{1+y^2} = \frac{x^2}{2} + C.$$

This last expression describes a family of curves in the plane, with different curves corresponding to different values of C. To pick out a particular curve, a point on the curve must be specified. For example,

$$y(0) = 1$$

leads to

$$-\frac{1}{2} = 0 + C,$$

and the curve

$$-\frac{1}{1+y^2} = \frac{x^2}{2} - \frac{1}{2}.$$

Maple can also produce plots of level curves, and so can illustrate the solutions of a differential equation such as the above. A plot of the level curves is given in Figure 3.2.

Maple reference **Page: 7**

A general separable first order equation is (by definition) one which may be written in the form

$$\frac{dy}{dx} = \frac{G(x)}{F(y)}$$

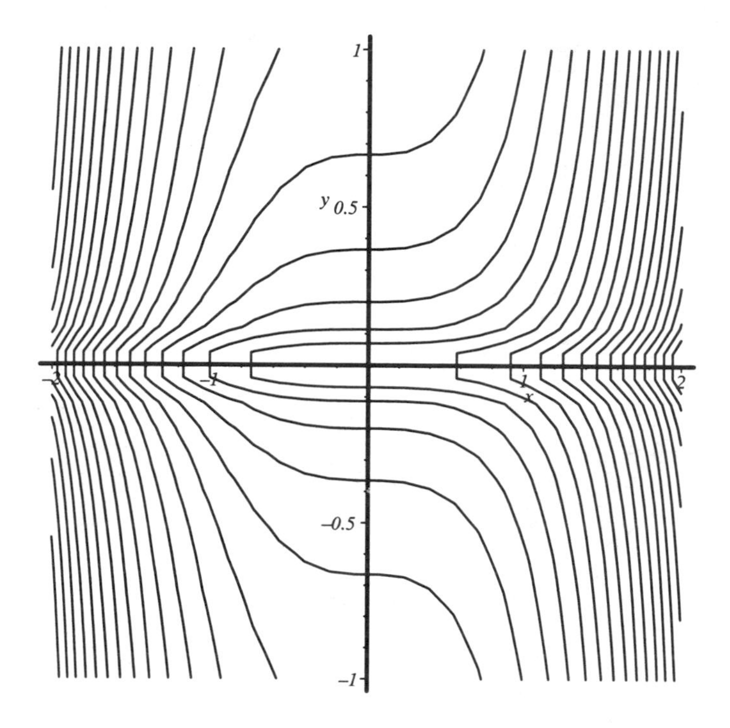

FIGURE 3.2. Solution contours

for some functions (of a single variable) $G(\cdot)$ and $F(\cdot)$. The formal solution procedure proceeds down the track

$$F(y)\,dy = G(x)\,dx,$$

$$\int F(y)\,dy = \int G(x)\,dx + C.$$

The procedure above amounts to "cross multiplying" the differential expressions dx and dy, and as such might appear to be on shaky logical ground. However, the procedure can be justified in the following manner. Cross multiply the $F(y)$ in the above to obtain

$$F(y)\frac{dy}{dx} = G(x),$$

and then integrate (anti-differentiate) both sides with respect to x, resulting in the equality (modulo a constant, since they are anti-derivatives)

$$\int F(y)\frac{dy}{dx}\,dx = \int G(x)\,dx.$$

The integral on the left really is the representation for an integral done with a change of variable. The formalism associated with the substitution is

$$u = y(x),$$

$$du = \frac{dy}{dx}\,dx,$$

with the substituted form of integral just

$$\int F(u)\,du.$$

Since the y variable must be substituted for u at the end of the substitution process to obtain the answer in terms of the original variables, this last integral may as well be calculated as

$$\int F(y)\,dy,$$

which is just exactly what the formal manipulation leads to.

3.1.1 Separable Examples

Example
Conic sections in standard form arise from separable differential equations. The equation

$$y\,\frac{dy}{dx} = x$$

is solved by the procedure

$$\begin{aligned} x\,dx &= y\,dy \\ \frac{1}{2}x^2 &= \frac{1}{2}y^2 + C, \end{aligned}$$

and so represents a hyperbola.

Example
The equation[1]

$$\frac{dx}{dy} = \frac{y}{\sqrt{1-x^2}}$$

[1] For curves in the plane, which variable is to be regarded as the independent one is often negotiable.

is easily noticed to be exact. Separating the variables and integrating gives an implicit form of the solution curve

$$\begin{aligned}\sqrt{1-x^2}\,dx &= y\,dy,\\ 1/2x\sqrt{1-x^2}+1/2\arcsin(x) &= \frac{1}{2}y^2+C.\end{aligned}$$

While there are as many silly explicitly solvable separable equations as there are pairs of anti-derivatives, some separable equations arise as models for problems of interest.

Example

An example that has wider ramifications than might first appear arises from considering a time variable decay rate problem in the form of the model equation

$$\frac{dx}{dt}=a(t)x(t).$$

This is a prototype problem for the general first order linear equation

$$\frac{dx}{dt}=a(t)x(t)+f(t),$$

which might be thought of as a general time-variable growth process, with an autonomous input rate given by $f(t)$. While this problem is of enough interest to deserve a treatment on its own, the above homogeneous version can be solved (with a bit of care) by separation of variables.

The separated equation is

$$\frac{dx}{x}=a(t)\,dt,$$

to be solved subject to the initial condition $x(0)=x_0$.

To solve the problem incorporating the initial conditions, integrate the separated equation between 0 and the current time t. That gives

$$\begin{aligned}\int_{x_0}^{x}\frac{dx}{x}\,dx &= \int_0^t a(t)\,dt\\ \ln(|x(t)|)-\ln(|x_0|) &= \int_0^t a(\tau)\,d\tau\\ |x(t)| &= |x_0|\exp\left(\int_0^t a(\tau)\,d\tau\right).\end{aligned}$$

To get rid of the absolute values, cases have to be considered.

The quantity $x(t)$ is either positive or negative. If $x(t)>0$, then $|x(t)|=x(t)$, and the solution formula becomes

$$x(t)=|x(t)|=|x_0|\exp\left(\int_0^t a(\tau)\,d\tau\right).$$

Evaluating this equation at $t = 0$, the exponential term becomes 1, and the equation reduces to

$$x(0) = |x_0| = |x(0)|$$

which requires that the initial condition x_0 is positive in this case.

If we assume that $x(t) < 0$, then

$$x(t) = -|x(t)| = -|x_0| \exp\left(\int_0^t a(\tau)\, d\tau\right),$$

at $t = 0$ leading to

$$x(0) = -|x_0| = x_0$$

and revealing that it must have been true that $x_0 < 0$ in this case.

The two cases then result in the expression

$$x(t) = x_0 \exp\left(\int_0^t a(\tau)\, d\tau\right),$$

which holds for the two cases, whether $x_0 > 0$ or $x_0 < 0$.[2]

Example

A classic example of a serious problem analyzable because it separates is a model for predator-prey interaction originally due to Volterra, but still basic in ecological dynamics.

The model assumes two interacting species, a prey with population $x(t)$, and a predator with population $y(t)$. The assumptions in this basic model are that the prey are so prolific that they would grow exponentially in the absence of predation, while the predators require the food supplied by the prey to survive, and would exponentially decay without their (one and only) source of nutrition. The governing equations are

$$\begin{aligned} \frac{dx}{dt} &= gx - pxy \\ \frac{dy}{dt} &= -dy + exy. \end{aligned}$$

Here g is the intrinsic growth rate of the prey, while d is the natural death rate for the predators. The p is the predation rate, effectively a death rate for the prey proportional to the predator density. Conversely, e represents a birth rate for predators proportional to the ambient food density.

The x and y variables represent populations (or densities) and so only non-negative ranges of the variables are of interest. It turns out that solutions started in the positive quadrant remain there for all time.

[2]For completeness, the formula also works when the initial condition vanishes.

It is useful to consider "boundary cases" for special solutions of the equations. If the predator population is initially zero, and the prey is not, then we obtain solutions

$$\begin{aligned} x(t) &= x(0)\, e^{gt} \\ y(t) &= 0, \end{aligned}$$

and in the absence of prey

$$\begin{aligned} x(t) &= 0 \\ y(t) &= y(0)\, e^{-dt}. \end{aligned}$$

The model also has an equilibrium state, obtained by setting the time derivatives to zero and solving for values of x and y which stay constant. The result is the equilibrium point

$$\begin{aligned} y &= \frac{g}{p} \\ x &= \frac{d}{e}, \end{aligned}$$

which certainly increases belief in the "food web".

The picture of the system behavior at this point is that initial points along the y axis head for the origin, and initial points along the x axis give a prey population explosion. In addition, there is a single point in the region $x > 0,\ y > 0$ where the populations are in equilibrium. The question is what happens to the rest of the points in the region of interest.

A representation of the predator-prey trajectories can be obtained by eliminating time as the variable between the two equations (effectively, just divide) to get

$$\frac{dy}{dx} = \frac{(-d + ex)\, y}{(g - py)\, x}.$$

This separates into the easily integrated form

$$\left(\frac{g}{y} - p\right) dy = \left(\frac{-d}{x} + e\right) dx,$$

which gives the implicit form of the trajectories as

$$g \ln\left(\frac{y}{y_0}\right) + d \ln\left(\frac{x}{x_0}\right) = e\,(x - x_0) + p\,(y - y_0).$$

These represent closed loops around the equilibrium point (see Figure 3.3).

One of the biologically interesting aspects of these solutions is that they show wild population swings for initial conditions far from the equilibrium point, and that the populations actually cycle rather than tending to the equilibrium.

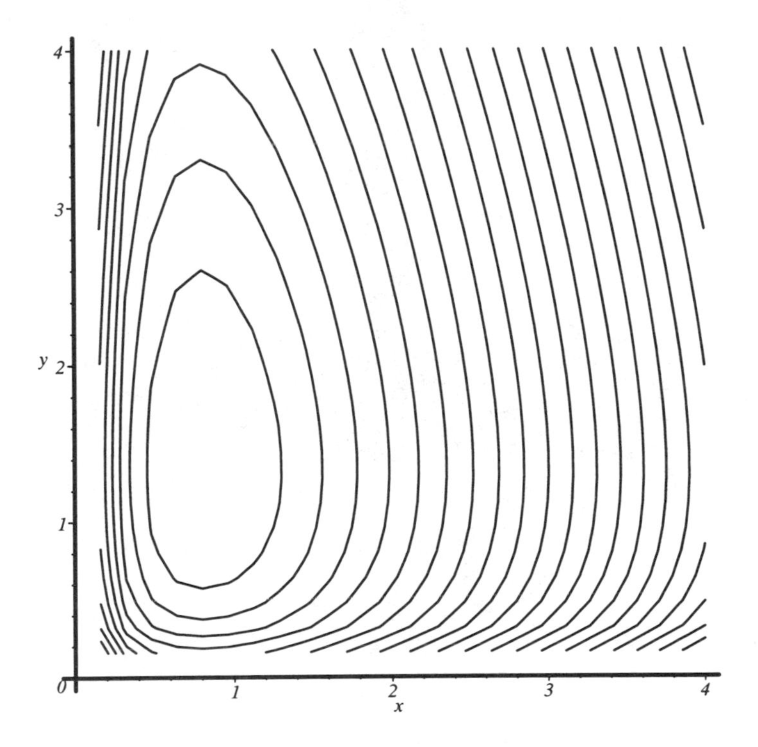

FIGURE 3.3. Species trajectories

3.2 Level Sets and Exact Equations

The separable equations lead to a picture of the solution curves as contour plots (or level curves in other terminology) of function of two variables. The function of two variables is rather special, in that the variables (due to the separability of the equation) are separated in the resulting function. One can obtain a differential equation description for more general level curves.

If the starting point is a (smooth) function $\phi(x, y)$ of two variables, then one can imagine that $\phi(x, y)$ is such that (at least locally) the implicit equation

$$\phi(x, y) = C$$

can be solved to define $y(x)$. The visualization of such a surface and the associated set of level curves in the plane are given in Figure 3.4.

To assume that the implicit equation can be solved for y in terms of x means that the resulting "expression" satisfies identically

$$\phi(x, y(x)) = C.$$

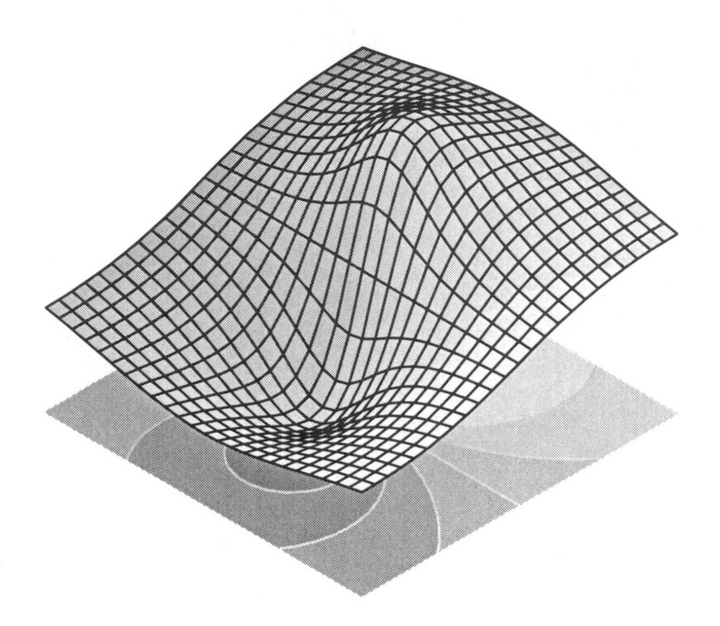

FIGURE 3.4. A surface and level curves

The slope of the resulting curve $x \mapsto y(x)$ can then can be calculated by differentiating the above implicit equation using the chain rule. This gives

$$\frac{\partial\phi}{\partial x}\frac{dx}{dx} + \frac{\partial\phi}{\partial y}\frac{dy}{dx} = 0.$$

Solving this for the slope of the level curve,

$$\frac{dy}{dx} = -\frac{\frac{\partial\phi}{\partial x}}{\frac{\partial\phi}{\partial y}}.$$

To make this closer to the formalism associated with separable problems, it is possible to use the notation

$$\frac{\partial\phi}{\partial x}\,dx + \frac{\partial\phi}{\partial y}\,dy = 0$$

for the above equation. Note that this is a formalism, and the meaning of the expression in terms of differential equations associated with it is just the original differential equation immediately above. This expression also occurs in the context of vector calculus, where it is interpreted as the equation of the tangent plane at a point on the surface. One could interpret the solution of the differential equation as the problem of finding the level surface, given the equation of the tangent plane.

Presumably, if one knew that a differential equation came from the level curves of some particular function, then a solution could be written down immediately, at least in implicit form. If the function in question is known to be $\phi(x, y)$, then the implicit solution is just

$$\phi(x, y(x)) = C,$$

and so this is a form of the solution, naturally appearing in an implicit form. Of course, this is a somewhat circular discussion, since we obtained the governing differential equation in the first place by assuming that $y(x)$ was determined by such an implicit equation.

The realistic version of the problem is to determine whether a differential equation presented in the form[3]

$$M(x, y)\,dx + N(x, y)\,dy = 0,$$

without a hint of its origin, actually represents a level curve of some function of two variables.

The way to tell this is to make use of the fact that the cross second partial derivatives of a smooth function of two variables are actually equal. That is,

$$\frac{\partial^2 \phi}{\partial x\,\partial y} = \frac{\partial^2 \phi}{\partial y\,\partial x},$$

as long as the indicated partial derivatives are continuous.

If the proffered differential equation really arises as a level curve, then the coefficients $M(x, y)$ and $N(x, y)$ must come from a ϕ as above, so that we can identify the equation coefficients with the appropriate partial derivatives,

$$\frac{\partial \phi}{\partial x} = M(x, y),$$

and

$$\frac{\partial \phi}{\partial y} = N(x, y).$$

The coefficient functions then will have to satisfy

$$\frac{\partial M(x, y)}{\partial y} = \frac{\partial N(x, y)}{\partial x}$$

in order to meet the cross partial condition. If the given equation satisfies this condition, the equation is said to be an *exact differential equation*.

[3]That is, the differential equation $\frac{dy}{dx} = -\frac{M(x,y)}{N(x,y)}$

To see how this all might work, consider the following contrived example. The function ϕ is defined as

$$\phi(x, y) = x^2 + x + y^2 + 3y,$$

giving the partials

$$\frac{\partial \phi}{\partial x} = 2x + 1,$$

$$\frac{\partial \phi}{\partial y} = 2y + 3.$$

The equation in what might be called neutral form is then

$$(2x + 1)\, dx + (2y + 3)\, dy = 0.$$

This is easily seen to be exact because the cross partials vanish identically in this particular case.

The general problem will be to reconstruct the function ϕ from the neutral form, so the starting point has to be identifying the coefficient of the x differential with the x partial of ϕ. Hence (pretending we do not already know ϕ)

$$\frac{\partial \phi}{\partial x} = 2x + 1,$$

from which it follows that ϕ has the form

$$\phi = (x^2 + x) + f(y).$$

The "constant of integration" must be allowed to depend on y, since that would disappear when the x partial derivative is computed. The unknown function f is determined by imposing the requirement that the y partial of the ϕ just calculated must match the dy coefficient of the original equation. Differentiating the above representation for ϕ with respect to y, we get

$$\frac{\partial \phi}{\partial y} = 0 + f'(y),$$

and hence (matching the value of $\frac{\partial \phi}{\partial y}$ from the original equation)

$$f'(y) = (2y + 3),$$

$$f(y) = y^2 + 3y + C.$$

Once we know $f(y)$, we obtain

$$\phi(x, y) = x^2 + x + y^2 + 3y + C = 0$$

as the implicit form of the solution curves.

The discussion above really only shows that the condition for exactness is a necessary condition for (smooth) level curves. In fact, exactness is also sufficient. Seeing this is just a matter of writing out the procedure used for the example in the general case, and verifying that it must work there as well. This argument is carried out in the following section.

3.2.1 Solving Exact Problems

We start with

$$M(x, y)\,dx + N(x, y)\,dy = 0$$

and assume that the coefficient functions are such that the equation is exact, meaning that they satisfy (the cross partial condition)

$$\frac{\partial M(x, y)}{\partial y} = \frac{\partial N(x, y)}{\partial x}.$$

Following the indicated procedure, we would compute in turn

$$\frac{\partial \phi}{\partial x} = M(x, y),$$

$$\phi = \int_a^x M(x, y)\,dx + f(y),$$

where the antiderivative must be calculated treating y as a constant. Differentiating the above with respect to y, we get

$$\frac{\partial \phi}{\partial y} = \int_a^x \frac{\partial M}{\partial y}\,dx + f'(y).$$

Identifying the $\frac{\partial \phi}{\partial y}$ partial with N and solving for the term $f'(y)$

$$f'(y) = N(x, y) - \int_a^x \frac{\partial M}{\partial y}\,dx.$$

At this point we are supposed to integrate to obtain f. But is the right hand side of that equation really a function only of y? The assumption of exactness is what is needed to verify this. If the expression actually only depends on y in spite of outward appearances, then its partial with respect to x must vanish. But

$$\frac{\partial}{\partial x}\left(N(x, y) - \int_a^x \frac{\partial M}{\partial y}\,dx\right) = \frac{\partial N}{\partial x} - \frac{\partial M}{\partial y},$$

and this expression vanishes since the equation is assumed exact. The antiderivative problem for f will therefore actually make sense and be solvable.

One may verify that starting by identifying the "other" partial of ϕ with $N(x, y)$ is an equally valid procedure. This of course masks the fact that one approach may lead to a simpler anti-derivative problem than the other beginning produces.

3.2.2 Exact Equation Examples

Example
Given the equation

$$(e^{2y} - y\ \cos(xy))\,dx + (2x\,e^{2y} - x\ \cos(xy) + 2y)\,dy = 0$$

we identify

$$\begin{aligned} M(x, y) &= e^{2y} - y\cos(xy) \\ N(x, y) &= 2x\,e^{2y} - x\cos(xy) + 2y \end{aligned}$$

and verify that

$$\frac{\partial M}{\partial y} = 2\,e^{2y} + xy\ \sin(xy) - \cos(xy) = \frac{\partial N}{\partial x}$$

so the equation is exact. Start with

$$\begin{aligned} \frac{\partial \phi}{\partial y} &= 2x\,e^{2y} - x\ \cos(xy) + 2y \\ \phi &= 2x \int e^{2y}\,dy - x \int \cos(xy)\,dy + 2 \int y\,dy \\ &= xe^{2y} - \sin(xy) + y^2 + h(x) \\ \frac{\partial \phi}{\partial x} &= e^{2y} - y\cos(xy) + h'(x) \\ &= M(x, y) \\ &= e^{2y} - y\cos(xy) \end{aligned}$$

so that $h' = 0$, and h really is constant. The implicit form of the solution curves then is

$$x\,e^{2y} - \sin(xy) + y^2 + C = 0.$$

Example
Another example is

$$(\cos(x)\sin(x) - xy^2)\,dx + y(1 - x^2)\,dy = 0.$$

This too is exact because

$$\frac{\partial M}{\partial y} = -2xy = \frac{\partial N}{\partial x}.$$

Picking $\frac{\partial \phi}{\partial y} = N$ appears to lead to simpler anti-derivatives. Hence the solution proceeds

$$
\begin{aligned}
\frac{\partial \phi}{\partial y} &= N(x, y) \\
&= y(1 - x^2) \\
\phi &= \frac{y^2}{2}(1 - x^2) + h(x) \\
\frac{\partial \phi}{\partial x} &= -xy^2 + h'(x) \\
&= -xy^2 + \frac{1}{2}\sin(2x) \\
h(x) &= -\frac{1}{4}\cos(2x) + C
\end{aligned}
$$

so that the level curve is

$$\frac{y^2}{2}(1 - x^2) - \frac{1}{4}\cos(2x) + C = 0.$$

3.3 Linear Equations

Linear differential equations hold a central place in the universe due to the large amount of knowledge available about the behavior of linear systems. Many engineering systems are designed on the basis of linear models, assuming (or better, verifying) that the "true" system has behavior sufficiently close to the linear idealization. All of this utility is a consequence of the fact that explicit solution representations (i.e., formulas) are available for linear systems. The base of this giant pyramid scheme is the analysis of a single first order differential equation, to which we now turn.

The model under analysis is

$$\frac{dx}{dt} = a(t)\,x(t) + f(t),$$

with the initial condition $x(0) = x_0$. We recall that in the absence of the forcing function $f(t)$, the solution of the above takes the form

$$x(t) = x_0\, e^{\int_0^t a(\tau)\,d\tau}.$$

This generalized the solution

$$x(t) = x_0\, e^{-ct}$$

for the constant decay rate problem

$$\frac{dx}{dt} = -c\,x(t),$$

effectively replacing the constant growth/decay rate with the running average rate. This interpretation arises from writing the time varying case solution as

$$x(t) = x_0\, e^{t\left(\frac{1}{t}\int_0^t a(\tau)\,d\tau\right)}.$$

This beguiling simplification unfortunately does not usually apply to the more complex problems where vector valued unknowns are involved, but it at least makes the general formula understandable in the scalar case.

We earlier solved the homogeneous form of the problem by separation of variables, but with a view toward solving the more complicated forced case above, it is useful to look at another approach to the original case.

Suppose one looks for a solution of

$$\frac{dx}{dt} = a(t)\,x(t)$$

in the form of a product

$$x(t) = z(t)\, e^{\int_0^t a(\tau)\,d\tau}.$$

If this assumed form is substituted into the equation (with the intent of obtaining an equation for the new variable $z(t)$) what results is

$$\begin{aligned}
\frac{d}{dt}\left(z(t)\, e^{\int_0^t a(\tau)\,d\tau}\right) &= a(t)\left(z(t)\, e^{\int_0^t a(\tau)\,d\tau}\right),\\
\frac{dz}{dt}\, e^{\int_0^t a(\tau)\,d\tau} + z(t)\,\frac{d}{dt}\, e^{\int_0^t a(\tau)\,d\tau} &= a(t)\,z(t)\, e^{\int_0^t a(\tau)\,d\tau},\\
\frac{dz}{dt}\, e^{\int_0^t a(\tau)\,d\tau} + a(t)\,z(t)\, e^{\int_0^t a(\tau)\,d\tau} &= a(t)\,z(t)\, e^{\int_0^t a(\tau)\,d\tau},\\
\frac{dz}{dt}\, e^{\int_0^t a(\tau)\,d\tau} &= 0,\\
\frac{dz}{dt} &= 0.
\end{aligned}$$

This is not very surprising, since in the original problem solution the $z(t)$ is just x_0, which is in fact a constant. What is surprising is that the same substitution applied to the problem including the forcing term results in a formula for the solution of the more complicated situation. The calculations are nearly the same, except for the added forcing term $f(t)$ in the inhomogeneous equation.

$$\begin{aligned}
\frac{d}{dt}(z(t)\, e^{\int_0^t a(\tau)\,d\tau}) &= a(t)\, z(t)\, e^{\int_0^t a(\tau)\,d\tau} + f(t), \\
\frac{dz}{dt}\, e^{\int_0^t a(\tau)\,d\tau} + a(t)\, z(t)\, e^{\int_0^t a(\tau)\,d\tau} &= a(t)\, z(t)\, e^{\int_0^t a(\tau)\,d\tau} + f(t), \\
\frac{dz}{dt}\, e^{\int_0^t a(\tau)\,d\tau} &= f(t), \\
\frac{dz}{dt} &= e^{-\int_0^t a(\tau)\,d\tau}\, f(t) \\
z(t) &= z_0 + \int_0^t e^{-\int_0^\sigma a(\tau)\,d\tau}\, f(\sigma)\, d\sigma.
\end{aligned}$$

The solution formula follows from this just by multiplying through by

$$e^{\int_0^t a(\tau)\,d\tau}$$

to get $x(t)$ on the left, noticing that $x_0 = z_0$, and combining the exponentials into a single integral. This gives

$$\begin{aligned}
x(t) &= z(t)\, e^{\int_0^t a(\tau)\,d\tau}, \\
x(t) &= z_0\, e^{\int_0^t a(\tau)\,d\tau} + \int_0^t e^{-\int_0^\sigma a(\tau)\,d\tau + \int_0^t a(\tau)\,d\tau}\, f(\sigma)\, d\sigma, \\
x(t) &= x_0\, e^{\int_0^t a(\tau)\,d\tau} + \int_0^t e^{\int_\sigma^t a(\tau)\,d\tau}\, f(\sigma)\, d\sigma,
\end{aligned}$$

known widely as the "variation of parameters formula". The name comes from the solution method: a variable $z(t)$ was introduced in place of the original (constant parameter) x_0.

Integrating Factors

A variant on the above solution method introduces an "integrating factor".[4]

Definition
The integrating factor for the linear differential equation

$$\frac{dx}{dt} = a(t)\, x(t) + f(t)$$

[4] Integrating factors are sometimes defined without specifying a lower limit. Due to the ambiguity of anti-derivatives (up to an additive constant), this amounts to leaving an unspecified scale factor in the definition of the integrating factor μ.

is defined by

$$\mu(t) = e^{-\int_0^t a(\tau)\,d\tau}.$$

Solving the problem by means of the integrating factor proceeds by multiplying the original equation through by the integrating factor. This puts the equation in the form

$$e^{-\int_0^t a(\tau)\,d\tau}\left(\frac{dx}{dt} - a(t)\,x(t)\right) = e^{-\int_0^t a(\tau)\,d\tau}\, f(t).$$

The left side of the equation then is in the form of the derivative of a product

$$\frac{d}{dt}\left(x(t)\, e^{-\int_0^t a(\tau)\,d\tau}\right) = e^{-\int_0^t a(\tau)\,d\tau}\, f(t).$$

This expression can be "solved" by simply integrating both sides (as a definite integral) over the range $[0, t]$. The result is

$$x(t)\, e^{-\int_0^t a(\tau)\,d\tau} - x_0 = \int_0^t e^{-\int_0^\sigma a(\tau)\,d\tau}\, f(\sigma)\,d\sigma,$$

and solving this for $x(t)$ gives the previously obtained expression. In view of the relationship between x and z, this procedure effectively introduces z into the problem without naming it.

A variant of the "integrating factor method" proceeds using anti-derivatives instead of definite integrals. The formalism gives (using $\mu(t)$ to hide the fact that it involves an exponential of an integral)

$$\frac{d}{dt}(x(t)\mu(t)) = \mu(t)\, f(t),$$

$$x(t)\mu(t) = \int^t \mu(\sigma) f(\sigma) d\sigma + C,$$

$$x(t) = \frac{1}{\mu(t)} \int^t \mu(\sigma) f(\sigma) d\sigma + C\,\frac{1}{\mu(t)}.$$

The constant of integration C must now be evaluated in terms of any given initial condition. An instructive exercise is to evaluate C for the case in which the initial condition is $x(0) = x_0$, and verify that the result is just the "variation of parameters" formula again.

Committing the integrating factor recipe to memory is no less work than is required to know the "whole formula" associated with the general solution. The difference in what has to be remembered amounts to a minus sign in the exponent. In addition, the general formula has physical interpretations that bear on the general idea of linear system models. This topic is discussed next.

Weighting Patterns

The first thing that should be mentioned is that in the discussion of the linear equation

$$\frac{dx}{dt} = a(t)\,x(t) + f(t),$$

there is really nothing special about the choice of $t = 0$ as the initial time at which the solution is specified. The choice was made simply to avoid another parameter complicating the derivation. If the initial condition were specified instead as

$$x(t_0) = x_{t_0},$$

and all of the integrals were calculated over $[t_0, t]$ instead of $[0, t]$, then the resulting solution expression would end up as

$$x(t) = x_{t_0}\, e^{\int_{t_0}^{t} a(\tau)\,d\tau} + \int_{t_0}^{t} e^{\int_{\sigma}^{t} a(\tau)\,d\tau}\, f(\sigma)\,d\sigma.$$

This is what should be taken as the general solution of the scalar linear differential equation. This expression appears complicated, but that is largely because of the "integrals in exponents" effect. If we invent the term *weighting pattern*, the solution can be written in a more appealing form.

Definition
The *weighting pattern* of a linear differential equation is a function of two variables given by

$$w(t, \sigma) = e^{\int_{\sigma}^{t} a(\tau)\,d\tau}.$$

In terms of the weighting pattern the linear differential equation solution can be written in the easily remembered form

$$x(t) = w(t, t_0)\, x_{t_0} + \int_{t_0}^{t} w(t, \sigma)\, f(\sigma)\,d\sigma.$$

This expression represents the solution as a sum of two terms. The first is an *initial condition response*

$$w(t, t_0)\, x_{t_0},$$

which represents the part of the solution due to the initial condition. If the forcing function $f(t)$ in the governing equation vanishes, then this represents the entire solution.

When the forcing function is present, the solution contains a *forced response*

$$\int_{t_0}^{t} w(t, \sigma)\, f(\sigma)\,d\sigma,$$

and this expression is the source of the *weighting pattern* terminology. The forced response is an integral over a time interval from the initial time t_0 up to the current time t. One of the factors of the integrand is $f(\sigma)$, which can be thought of as a *past input*, since the σ variable ranges over $[t_0, t]$ as the integral is evaluated.

The forced response integral "adds up" the effect of the past inputs, and it does so giving different weight to input values occurring at different times in the past. The weight associated now (at time t) with the past input (at time σ) is simply

$$w(t, \sigma),$$

the value of the weighting pattern function. The weighting pattern depends on two variables because both the time of observation, and the time of the input occurrence matter in the response.

3.3.1 Linear Examples

Example

The simplest first order linear examples involve constant coefficients. For example

$$\frac{dy}{dx} = -y + e^{-2x}$$

has the solution

$$y(x) = e^{-x}\, y(0) + \int_0^x e^{-(x-\zeta)} e^{-2\zeta}\, d\zeta,$$

$$y(x) = e^{-x}\, y(0) + e^{-x}\,(-1 + e^{-x}) = e^{-x}\, y(0) + e^{-2x} - e^{-x}.$$

Example

The equation of motion of a permanent magnet type servo motor is

$$\frac{d^2\theta}{dt^2} + a\frac{d\theta}{dt} = e(t).$$

Here θ is the angular displacement of the motor, and $e(t)$ is the terminal voltage. In spite of the appearance of a second derivative in the above equation, the equation is actually a first order linear (constant coefficient, even) example, as long as it is regarded as an equation for the angular velocity. The solution then is

$$\frac{d\theta}{dt} = e^{-at}\frac{d\theta}{dt}(0) + \int_0^t e^{-a(t-\tau)} e(\tau)\, d\tau.$$

If the displacement is required, it may be obtained by simply integrating the angular velocity result.

3.4 Tricky Substitutions

While linear, separable, and exact equations have ready solution methods, other equations of interest may not fall into one of these categories. In this case the alternatives are either to settle for numerical approaches, or make an inspired change of variable that reduces the problem to one of the "easy" cases. In the early days of differential equations (which is to say, shortly after the invention of calculus) there was interest in such possibilities, and the names of certain well-known mathematicians are associated with some equations that can be solved by these methods. It is worth checking whether your problem is one of the well-known types susceptible to this treatment.

3.4.1 Lagrange

Lagrange is associated with the beginnings of the calculus of variations, where curves minimizing some quantity (such as length, for example) are studied. Certain of these problems result in a differential equation for the curve in question. One class of equations studied by Lagrange is of the form

$$y = x\, f\left(\frac{dy}{dx}\right) + g\left(\frac{dy}{dx}\right).$$

This equation appears unpromising, especially since the derivative appears as an argument in the unspecified functions. However, if one introduces $p = \frac{dy}{dx}$, then differentiating the original equation with respect to x gives an equation which is linear in the derivative of p. This can be solved to give

$$\frac{dp}{dx} = \frac{p - f(p)}{x\, f'(p) + g'(p)},$$

and if one inverts the role of the dependent and independent variables, a linear equation for x as a function of p is the result

$$\frac{dx}{dp} = -\frac{f'(p)}{f(p) - p}\, x + \frac{g'(p)}{f(p) - p}.$$

3.4.2 Riccati

The term Riccati equation is associated with quadratic differential equations arising in optimal control theory. The historical antecedent of such equations is a substitution that relates a quadratic problem to a linear one. The quadratic differential equation in question is

$$\frac{dx}{dt} + a(t)x^2 + b(t)x + c(t) = 0.$$

The substitution to make is

$$x(t) = \frac{w'(t)}{w(t)}\, \frac{1}{a(t)}.$$

Substituting in the quadratic equation gives

$$\frac{w''(t)}{a(t)w(t)} - \frac{w'(t)^2}{a(t)w(t)^2} - \frac{a'(t)w'(t)}{a(t)^2w(t)} + \frac{w'(t)^2}{a(t)w(t)^2} + \frac{b'(t)w'(t)}{a(t)w(t)} + c(t) = 0.$$

Algebra then simplifies the equation to the second order equation

$$w''(t) - \left(\frac{a'(t)}{a(t)} - b'(t)\right) w'(t) + a(t)\, c(t)\, w(t) = 0$$

for the substitution variable.

3.4.3 Mechanics

Newton's law of motion leads directly to second order differential equations, but some conservative (of energy, that is) problems in particle mechanics reduce to first order problems, and occasionally to explicit solutions.

If a particle in motion along a line is subjected to a conservative force, then the force is the derivative of a potential function, say $V(x)$. The equation of motion then will be

$$M\frac{d^2x}{dt^2} = -\frac{d}{dx}V(x).$$

If the above is multiplied through by $\frac{dx}{dt}$ and integrated over time, there results the conservation of energy relation

$$\frac{M}{2}\left(\frac{dx}{dt}\right)^2 + V(x(t)) = constant.$$

If the system starts from a state of rest, then

$$\frac{M}{2}\left(\frac{dx}{dt}\right)^2 + V(x(t)) = \frac{M}{2}0 + V(x(0)),$$

and solving this for the velocity gives the differential equation

$$\frac{dx}{dt} = \pm\frac{2}{M}\sqrt{V(x(0)) - V(x(t))}.$$

This is a separable form of first order equation.

Example

The spring and mass harmonic oscillator is an example of a mechanical system with a conservative force. The stored potential energy of a spring is

$$V(x) = \frac{1}{2}kx^2,$$

and the corresponding second order equation is

$$M \frac{d^2 x}{dt^2} = -k\,x.$$

The first order separable equation is

$$\frac{dx}{dt} = \pm \frac{2}{M} \sqrt{\frac{1}{2} k\,x_0^2 - \frac{1}{2} k\,x(t)^2}.$$

The ambiguous sign reflects the fact that the velocity is positive for part of the solution time, and negative at other times. With some care, a harmonic oscillator solution can be teased out of this approach (see the following problems), although we will see later that there are easier methods available for the harmonic oscillator problem.

3.5 Do Solutions Exist?

There is a natural tendency in the context of differential equations to expect that solutions will automatically exist, and moreover that there is a good chance at least that explicit formulas for the solution function can be obtained in short order. The expectation of existence is born of the conviction that most differential equations are produced by their author as the model for some real, physically observable process or other. Since the physical process clearly exists, and the model is *of* the physical process,...... so the argument goes.

The truth of the matter is that the "real process" and its mathematical model are logically distinct entities. The value of the model lies in its ability to reflect the observed facts of nature, and in an extreme case the model might be so far off the mark that there are no solutions at all, let alone solutions that are close to reality.

The purpose of the existence theorem industry is to provide descriptions of classes of differential equations that can be guaranteed apriori to *have* solutions, or even to have *unique* solutions. When one is formulating differential equations as candidates for physical system models, it is not a bad idea to run a theorem hypothesis check against one of the standard existence theorems. For difficult problems there often are not any closed form solutions available, so recourse must be made to numerical methods of solution. Knowing that a solution exists should be a prerequisite before investing a large amount of computer time in pursuit of something that may not be there to be found.

3.5.1 The Issue

In order to discuss existence of a solution to a differential equation such as

$$\frac{dy}{dx} = f(y, x),$$

it is necessary to be clear about what constitutes a solution.

Definition
A *solution* of the above differential equation is a differentiable function $y(\cdot)$ for which[5]

$$\frac{d}{dx}y(x) - f(y(x), x) = 0.$$

(The definition can be weakened by allowing non-differentiability on a sufficiently small set of points.)

The solution is *unique* provided that whenever $z(\cdot)$ and $w(\cdot)$ are both solutions, then $z = w$.

Example
In order to guarantee existence of a unique solution to a differential equation, some restrictions must be made on the characteristics of the slope function $f(\cdot, \cdot)$ on the right hand side of the equation. The equations need not be particularly pathological to fail in this regard; consider the following example.

The problem

$$\frac{dy}{dx} = 2y^{1/2},$$

$$y(0) = 0$$

has two solutions, namely, $z(x) = 0$, and $w(x) = x^2$. This is the case since $z(x)$ satisfies

$$\frac{dz}{dx} = 0 = 2\,(0)^{\frac{1}{2}},$$

$$z(0) = 0$$

while the other function $w(x)$ also meets the conditions

$$\frac{dw}{dx} = 2x = 2w^{1/2},$$

$$w(0) = 0.$$

This shows that it is easy to write down ordinary differential equations without unique solutions.

The source of the difficulty in that example is really due to the fact that the slope function changes too rapidly at the initial point in question. Because the derivative is

$$\frac{d}{dy}y^{1/2} = \frac{1}{2y^{1/2}},$$

---

[5]The definition of solution can be paraphrased as "it makes sense to plug it in, and it works when you do."

the slope function really has a vertical tangent at the initial point, and so changes infinitely rapidly there. It is this sort of behavior that has to be ruled out of court to get a statement of existence and uniqueness. The condition that is usually assumed is that the slope function is continuous in the x variable, and satisfies a *Lipschitz condition* with respect to the y variable. To say that $f(\cdot,\cdot)$ satisfies a Lipschitz condition means that there exists a constant K such that for all z, w we have

$$|f(z,x) - f(w,x)| \le K\,|z-w|.$$

This puts a restraint on the possible rate of change of the slope function in the y direction, and rules out the previous troubling example since the vertical tangent in that example precludes the possibility of a K satisfying the inequality above as long as the origin is kept in the problem domain.

3.5.2 Existence Theorems

The conventional existence theorem is interesting because it has a constructive proof whose mechanics are easy to understand and visualize.

As it is presented, the differential equation

$$\frac{dy}{dx} = f(y,x)$$

seems to suggest no natural means of solution. One of the problems is that the unknown appears on both sides of the equation, but is differentiated on the left. Among other effects, differentiation tends to magnify errors in approximate calculations, so the presence of the derivative is a problem. This problem is solved by converting the original differential equation to an integral equation instead, by the simple expedient of integrating both sides of the original. Assuming the initial x value is 0 for the sake of clarity, we obtain

$$\int_0^x \frac{dy}{dx}\,dx = \int_0^x f(y(x),x)\,dx,$$

$$y(x) - y(0) = \int_0^x f(y(x),x)\,dx,$$

$$y(x) = y(0) + \int_0^x f(y(x),x)\,dx.$$

The integral equation turns out to have a solution that can be found by a process of iterative substitution. The steps are

$$\begin{aligned}
y_0(x) &= y_0 \\
y_1(x) &= y_0 + \int_0^x f(y_0(x), x)\,dx \\
y_2(x) &= y_0 + \int_0^x f(y_1(x), x)\,dx \\
\vdots &= \vdots \\
y_n(x) &= y_0 + \int_0^x f(y_{n-1}(x), x)\,dx \\
\vdots &= \vdots
\end{aligned}$$

The process of constructing a solution in this fashion is called *Picard iteration*, after the originator of the method. Using this iteration, it is possible to prove an existence and uniqueness result.

Theorem: *Suppose that on a box centered on the point* (x_0, y_0) *the function* $f(x, y)$ *is continuous in* x*, and satisfies the Lipschitz condition*

$$|f(z, x) - f(w, x)| \leq K\,|z - w|$$

for all z, w *in the box. Then there exists a unique solution to*

$$\frac{dy}{dx} = f(y, x)$$

satisfying the initial condition $y(x_0) = y_0$, *and the solution can be continued up until it leaves the box.*

The iterative construction of the solution referred to in the theorem can be illustrated by using Maple to construct the sequence

$$y_0(x),\ y_1(x),\ y_2(x)\ \ldots\ y_n(x)\ \ldots$$

generated by the procedure. As long as the example is carefully enough chosen that Maple can manage the integrals in the sequence, plots can be made. A plot of the first five functions generated in an attempt to solve the simple problem

$$\frac{dy}{dx} = y$$

by this Picard iteration method is given in Figure 3.5.

Maple connection **Page: 24**

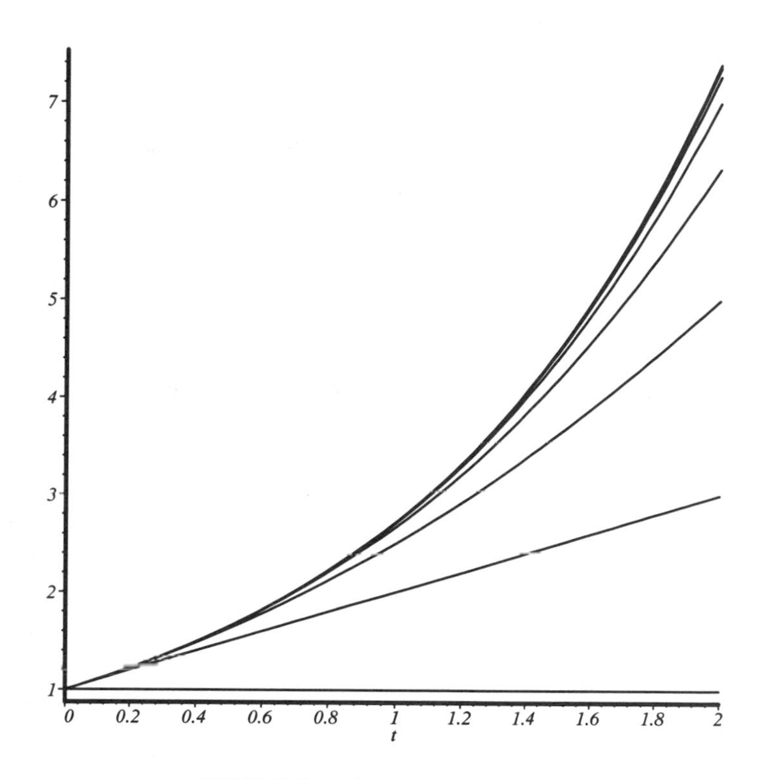

FIGURE 3.5. Picard iteration sequence

3.6 Problems

1. After doing a `with(DEtools)` make a plot of the direction field for the first order differential equation

$$\frac{dy}{dx} = y^3 x * exp(-x^2).$$

2. Plot the solution to the above which goes through the point $x = 0, y = 1$ on top of the direction field.

3. Look up `dsolve` in the Maple help, and see if it solves the problem above.

4. Find the general solution of

$$\frac{dy}{dx} = \sin(x)\sqrt{1-x^2}.$$

5. Find the general solution of

$$x\,dy - y\,dx = 0,$$

as well as the particular solution passing through the point $x = 1, y = 1$.

6. Find the general solution of

$$\frac{x}{y}\,dy + \frac{y}{x}\,dx = 0.$$

7. Solve

$$\left(\cos(xy)y - 2\sin(x^2y^2)xy^2\right)dx + \left(\cos(xy)x - 2\sin(x^2y^2)x^2y\right)dy = 0.$$

8. Write the third order differential equation

$$\frac{d^3x}{dt^3} + 2\frac{d^2x}{dt^2} + \frac{dx}{dt} = \sin(t)$$

as a vector system of three first order equations.

9. After you solve this one, look at it again.

$$\left(2x + 2x\ \ln(x^2+y^2)\right)\,dx + \left(2y + 2y\ \ln(x^2+y^2)\right)\,dy = 0.$$

10. If $(y^3 - y^2\sin(x) - x)dx + (3xy^2 + 2y\cos(x))dy = 0$ is exact, solve it. Test it first.

11. If the following are exact, solve them.

$$\left(y\ \ln(x) - e^{-xy}\right)\,dx + (1/y + x\ \ln(y))\,dy = 0$$
$$(\sin(y) - y\ \sin(x))\,dx + (\cos(x) + x\ \cos(y) - y)\,dy = 0$$
$$(4y + 2x - 5)\,dx + (6y + 4x - 1)\,dy = 0$$

12. Determine $N(x, y)$ such that $(y^{1/2}x^{-1/2} + \frac{x}{x^2+y})\,dx + N(x, y)\,dy = 0$ is exact.

13. Solve the initial value problem $x\frac{dy}{dx} + y = 2x$ with $y(1) = 0$. How about $y(0) = 1$? This last initial condition causes trouble. Why might that be?

14. Find the solution of

$$\frac{dy}{dt} - 3y = e^t,\ \ y(0) = 2.$$

15. Find the solution of

$$\frac{dy}{dt} - 3y = e^t, \ y(0) = 2,$$

by using the integrating factor method. This assumes the easy way out was taken on the previous problem. If that was not the case, do the problem the easy way this time.

16. Solve

$$\frac{dx}{dt} + e^t\, x = e^t$$

in general terms, using the weighting pattern formula for the solution.

17. Repeat the previous question, using the integrating factor method. Then find the solution which passes through the point $x = 7, t = 5$.

18. Solve the problem

$$\frac{dz}{dt} = -7z + 1, \ z(0) = 0,$$

(which falls into three different categories already) by four different methods.[6] Later chapters will provide two more methods for solving this, so it has to rank as one of the most tractable of differential equations.

19. How about $\frac{dy}{dx} + y = f(x)$, where

$$f(x) = \begin{cases} 1, & 0 \le x < 1, \\ 0, & 1 < x. \end{cases}$$

20. Although Figure 3.5 was generated using Maple to do the Picard iterations, Maple is not really required. Explicitly find a formula for the iterates obtained when solving

$$\frac{dy}{dx} = y$$

by Picard iteration. What is the error between the true solution and the iterate at stage n ?

[6]This counts integrating factors as different, even though that is debatable.

4

Introduction to Numerical Methods

In spite of a long list of analytical techniques that provide closed form solutions to certain types of differential equations, many problems of interest give rise to model equations which do not have explicit solution representations. For such problems recourse must be made to numerical methods if the desired insight is to be obtained.

Even for some problems where explicit methods exist, the models with the simple behavior may be oversimplified (e.g., linearized) versions of the "real" problem. In such cases it is useful to carry out numerical investigations of what happens when the "linear answers" are applied to a more realistic nonlinear model of the system in question, and such approaches are required when adverse consequences of over-optimistic designs have expensive side effects.

For simple problems, even simple numerical approaches appear to give satisfactory results, but on "serious" problems they are likely to give spurious and misleading output. The reference [1] has a convincing example of an earth-moon orbit calculation where Euler's method fails miserably with 24000 steps, while a high order adaptive Runge–Kutta routine gets impressive accuracy in well under 100 steps. No doubt some sort of variant of Murphy's law operates in the numerical regime, and a problem for which one *really* wants a solution probably requires some care and attention in selecting a method.

4.1 The Game and Euler's Method

One may consider the object of interest to be the solution of the first order differential equation

$$\frac{d}{dx}y(x) = f(y(x), x),$$

subject to the boundary condition $y(a) = y_a$. It is conventional to discuss numerical methods as though only scalar problems are at issue, although what is crucial is that the differential equations is of first order. The methods that are derived adapt readily to systems of first order equations (vector equations), and the issues are more easily understood in the less encumbered notation that suffices for scalar problems.

The first problem to be resolved is what one expects to obtain out of a numerical solution to the above problem. In view of the fact that any algorithm running for a finite amount of time on a computer will only produce a finite (but admittedly large) number of outputs, the relationship of these output values to the desired solution must be decided. There is a lot of scope for variation here, as one could attempt to calculate approximately the coefficients in some series expansion, for example. To be useful, a numerical method must be constructed in such a way that it is possible to somehow estimate the errors that are being made in the calculations. There is really no utility in some whiz-bang off the wall calculation if there is no way to tell how close the result of the calculation is to the desired quantity.

The quality of estimability and general utility is served by attempting to calculate the values of the actual solution $y(x)$ at a set of ordinate values $a = x_0, x_1, \ldots x_N = b$. In the case of a routine that uses a fixed step size h, what is being approximated is the sets of ordinates $y(a), y(a+h), y(a+2h), \ldots, y(a+(N-1)h), y(b)$. This is illustrated in the Figure 4.1

Once it has been decided that calculations of the sample points are desired, then the problem immediately reduces to that of computing (an approximation of) the next sample from (an approximation of) the last one. That is, determine $y(x+h)$ from the value $y(x)$. The use of that notation immediately suggests Taylor series, and in fact Taylor series play a conspicuous role in the analysis and derivation of numerical methods for solving differential equations. Recall that the basic Taylor series expansion (with remainder) takes the form

$$y(x+h) =$$
$$y(x) + \frac{dy}{dx}(x)h + \frac{d^2y}{dx^2}(x)\frac{h^2}{2!} + \ldots + \frac{d^ny}{dx^n}(x)\frac{h^n}{n!} + \frac{d^{n+1}y}{dx^{n+1}}(\zeta)\frac{h^{n+1}}{n+1!}$$

where the argument ζ of the error term lies between x and $x+h$.

The simplest (and arguably oldest) numerical method is associated with Euler, and can be viewed as arising from the above by stopping the expansion after the linear term in h and putting up with the effects of ignoring the quadratic error term.

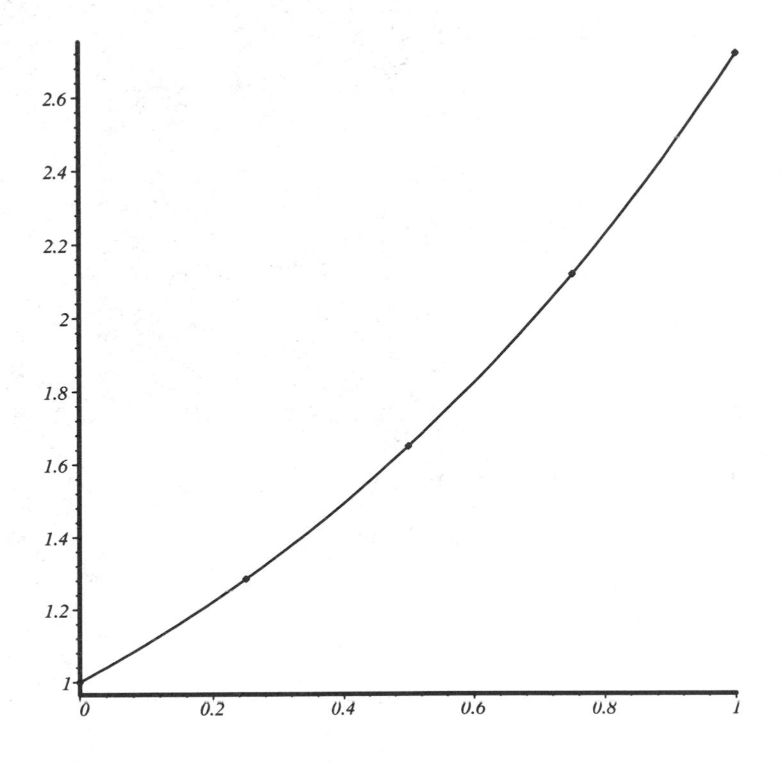

FIGURE 4.1. Solution samples

This gives the basic step as

$$\hat{y}(x+h) = \hat{y}(x) + hf(\hat{y}(x), x),$$

since the differential equation determines that $\frac{dy}{dx} = f(y(x), x)$. The Euler method algorithm amounts to following the tangent line for a distance h at a time. The iterative form for the differential equation

$$\frac{dy}{dx} = f(y(x), x)$$

is

$$y_n = y_{n-1} + h\, f(y_{n-1}, (n-1)\, h),$$

calculating the next approximate solution point in terms of the last. Given that the geometrical interpretation of this is "head in the direction of the tangent", it is clear that this algorithm is bound to lose ground on the true solution during regions of positive curvature (Figure 4.2).

By assuming a uniform upper bound of M on the size of the solution second derivative, it is easy to give a rough estimate of the total error in using Euler's

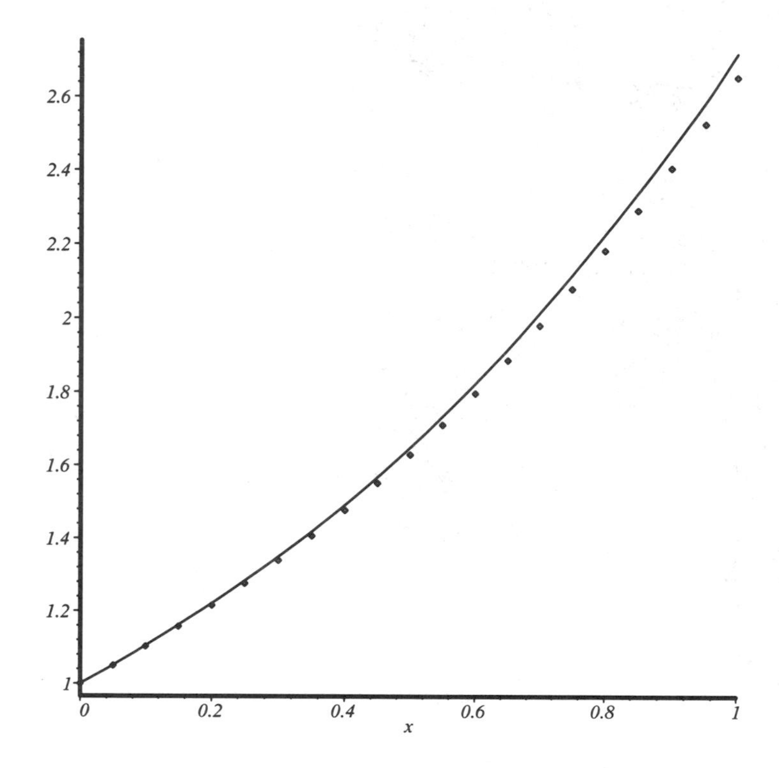

FIGURE 4.2. Euler versus the solution

method as

$$\Sigma_{i=1}^{N} \left| \frac{d^2y}{dx^2} \frac{h^2}{2} \right| \leq M \, \frac{(b-a)}{h} \frac{h^2}{2},$$

making the total global error proportional to the step size h. The above rough estimate really does not immediately stand up to scrutiny, since the numerically calculated solution does not use the values of $f(y(x), x)$ from the true solution but rather the numerically calculated values $f(\hat{y}(x), x)$. One could imagine that the divergence of the numerical solution could lead to encountering regions of the domain where $f(y(x), x)$ is very different from the values traversed by the true solution. The effect is somewhat like distortion of light rays passing through regions of variable diffractive index. Systems of differential equations displaying what is called "chaotic" character display this sort of behavior (and hence require extraordinary efforts to compute numerically.)

In spite of these reservations, it is actually possible to show that the rough estimate above (and the corresponding estimate for methods more accurate than Euler's routine) does hold, provided that the right hand side $f(y(x), x)$ of the equation is well enough behaved. Assuming something like

$$|f(y, x) - f(z, x)| \leq K|y - z|$$

(that $f(y, x)$ satisfies a uniform Lipschitz condition in y) is enough to prove that the difference between different initial conditions for the differential equation is only amplified linearly over the course of a solution on $a \leq x \leq b$. These considerations are enough to establish that the global errors for these methods do scale with h as the crude estimates indicate. The detailed proofs of these results are available in numerical analysis references, for example the book [1].

This may lead one to wonder what the news-magazine discussion of "chaos" is about, in its emphasis of the sensitivity of certain differential equation solutions to small changes in initial conditions. In terms of the above discussion, the result does not say that the amplification involves a particularly small number, and a look at the proof of the above shows that the amplification estimate is basically of the form e^K. If this sort of amplification is combined with long intervals of the independent variable ("time"), what results is chaotic behavior.

It will be seen below that it is possible to find numerical methods with much better estimates of error performance than the Euler's method above. Runge–Kutta methods are a class of widely usable methods that can be designed by looking for error estimates that scale with the step size as h^p with exponent $p > 1$.

If numerical methods are to be run against true solutions, observation of the results of a computation becomes an issue. Historically, there were long columns of printed numbers to compare; Maple can be made to generate such output, but it can also readily produce graphs of numerical results.

Maple connection **Page:** 287

The numerical solution of a simple differential equation

$$\frac{dy}{dx} = \sin(x\,y)$$

by Euler's method results in Figure 4.3.

It is instructive to try the above with $f := (y, x) \mapsto y$ and $f := (y, x) \mapsto -y$ against the true solutions. The unstable problem should display the curvature effect. Adaptive step size methods decrease their step size to maintain accuracy when the going gets steep, while Euler's method heads for the weeds along the tangent line.

4.2 Solution Taylor Series

If it is desired to get a numerical routine with more accuracy than the Euler's method, then it should be clear that more computation will be required at each step. One approach that might be tempting is to go back to the Taylor series expansion of the solution calculated above and attempt to do the calculation of $y(x+h)$ from $y(x)$ by using more terms from the Taylor polynomial than just the linear term used by Euler.

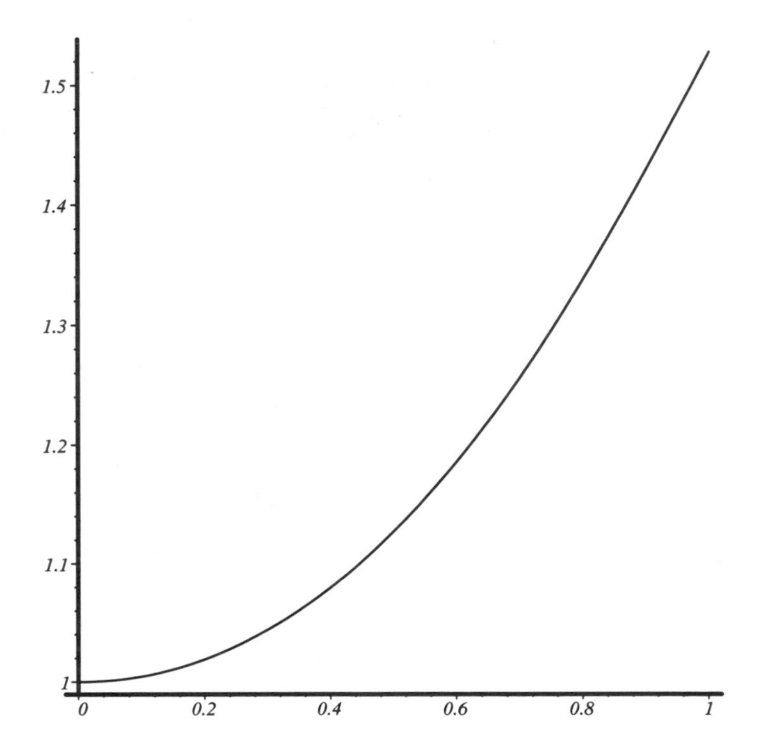

FIGURE 4.3. An Euler's method solution

In fact, given that $y(x)$ satisfies the differential equation $\frac{dy}{dx} = f(y(x), x)$ it is actually possible to compute the successive derivatives which appear in the Taylor polynomial in terms of the solution $y(x)$ and the system right hand side $f(y(x), x)$. The equations that one gets begin as

$$\begin{aligned}
(\frac{d}{dx})^0 y(x) &= y(x) \\
(\frac{d}{dx})^1 y(x) &= f(y(x), x) \\
(\frac{d}{dx})^2 y(x) &= \frac{\partial}{\partial u} f(y(x), x) \frac{dy}{dx} + \frac{\partial}{\partial v} f(y(x), x) \\
\ldots &= \ldots
\end{aligned}$$

The last displayed equation is computed by means of the chain rule in the form of

$$\frac{d}{dx} f(u(x), v(x)) = \frac{\partial f}{\partial u}\frac{\partial u}{\partial x} + \frac{\partial f}{\partial v}\frac{\partial v}{\partial x}$$

with $u(x) = y(x)$ and $v(x) = x$. The convention of making the dependent or unknown variable the first argument on the right hand side comes naturally when the independent variable is time t, but comes out non-alphabetical when the numerical analysis convention of calling the independent variable x is in play. Who wants to be a slave to notation?

It should be clear that formulas for successive derivatives can be calculated by repeatedly differentiating the above and replacing $\frac{dy}{dx}$ by $f(y(x), x)$ whenever it appears. The result will be an increasingly large sum involving $y(x)$ and various partial derivatives of $f(u, v)$. Since the required calculation amounts to a simple differentiation and substitution operation applied to the previous result, it can be viewed as a recursive calculation. To calculate using Maple just use a remember table to record the previous result:

```
> # formulas for solution derivatives: substitute f for dy/dx
>
> derivs := proc(n)
>    option remember;
>    if n = 0 then y(x)
>    elif n = 1 then f(y(x), x)
>    else
>       simplify(subs(diff(y(x),x) = f(y(x),x), diff(derivs(n-1),x)))
>    fi;
> end;
>
> # Taylor polynomial of solution:
>
> tp := proc(nterms)
>    options remember;
>    local n;
>
>    n := nterms;
>    if n = 1 then derivs(0)
>    else
>       tp(n-1)+derivs(n-1)*h^(n-1)/(n-1)!
>    fi
>   end;
>
> # Taylor series with remainder term for solution:
> # gives nterms plus remainder for cosmetic effects
>
>
> solution := proc(nterms)
>   RETURN(tp(nterms)+M*h^(nterms))
> end;
```

The annoying thing about the above code is that one is forced to decide what to use as the integer argument to the procedure. The competitors are the number of terms in the Taylor polynomial, the degree of the highest term, or the exponent in the error expression. It is easy to start using one convention for one routine and another meaning for another of the procedures. The routine produces the expected output.

```
> solution(1);
```

$$y(x) + M\,h$$

```
> solution(2);
```

$$y(x) + f(y(x),\, x)\,h + M\,h^2$$

```
> solution(3);
```

$$y(x) + f(y(x),\, x)\,h + \frac{1}{2}\left(\mathrm{D}_1(f)(y(x),\, x)\, f(y(x),\, x) + \mathrm{D}_2(f)(y(x),\, x)\right) h^2 + M\,h^3$$

```
> solution(4);
```

$$\begin{aligned} &y(x) + f(y(x),\, x)\,h \\ &\quad + \frac{1}{2}\left(\mathrm{D}_1(f)(y(x),\, x)\, f(y(x),\, x) + \mathrm{D}_2(f)(y(x),\, x)\right) h^2 \\ &+ \frac{1}{6}\Big(\mathrm{D}_{1,\,1}(f)(y(x),\, x)\, f(y(x),\, x)^2 + 2\,\mathrm{D}_{1,\,2}(f)(y(x),\, x)\, f(y(x),\, x) \\ &\quad + \mathrm{D}_1(f)(y(x),\, x)^2\, f(y(x),\, x) + \mathrm{D}_1(f)(y(x),\, x)\,\mathrm{D}_2(f)(y(x),\, x) \\ &\quad + \mathrm{D}_{2,\,2}(f)(y(x),\, x)\Big) h^3 + M\,h^4. \end{aligned}$$

The Maple output uses the differential operator notation for partial derivatives, so that $D_{2,2}(f)$ represents a second partial derivative of f, taken twice with respect to the second argument. That notation is commonly used to get out of alphabetical traps that arise from naming conventions .

The higher derivative terms in the above expressions involve computing increasingly high partial derivatives of the $f(y(x), x)$ from the differential equation. This really is a computational burden that one does not want to place on a numerical step routine. Not the least of the problems is that the right hand side may be the result of some model identification exercise, and so involve some errors. Computing high partials of the right hand side will just amplify the inaccuracies with probability close to one. Interest then lies with numerical methods that are based on only evaluating $f(y(x), x)$.

4.3 Runge–Kutta Methods

Some insight into numerical methods of the Runge–Kutta type can be obtained from an "integral trick". The differential equation we are trying to solve numerically is of the form

$$\frac{dy}{dx} = f(y(x), x),$$

and our interest is in advancing the solution from the point $(x, y(x))$ to $(x+h, y(x+h))$. An expression relating the y values can be obtained by simply integrating both sides of the differential equation between x and $x+h$. What results is

$$\int_x^{x+h} \frac{dy}{dx}\, dx = y(x+h) - y(x) = \int_x^{x+h} f(y(x), x)\, dx.$$

Multiplying and dividing the integral by h gives

$$y(x+h) = y(x) + h\left(\frac{1}{h}\int_x^{x+h} f(y(x), x)\, dx\right).$$

Now, the bracketed expression

$$\left(\frac{1}{h}\int_x^{x+h} f(y(x), x)\, dx\right)$$

actually is the average value of the slope function over the interval from x to $x+h$. If that could be calculated, then the desired value $y(x+h)$ could be calculated exactly, with no error at all, simple by multiplying the average slope by the step size h, and adding the increment to the current value $y(x)$. This indicates that what a good numerical[1] method should be doing is calculating accurate estimates of the differential equation slope function values as it proceeds across the interval of integration. The general Runge–Kutta methods discussed below explicitly construct an "average slope function" as they operate.

Judging only from geometric considerations (Figure 4.3) it would seem that more accuracy would be obtained if the right hand side of the differential equation (the "slope function") could be evaluated at a point more representative of the subinterval, say at the middle. If this were possible, then rather than the Euler step formula, one could use instead

$$y(x+h) = y(x) + h\, f\left(y(x+\frac{h}{2}),\ x+\frac{h}{2}\right)$$

If this were possible, then a mean-value theorem argument would suggest higher accuracy would be obtained. The problem of course is that computing the solution value $y(x+\frac{h}{2})$ essentially involves the same problem that is being solved, an estimation of the solution at a point at a distance form the current position. The way out of this dilemma (taken by Runge [3]) is to make do with an Euler's step

[1] At least one which operates on a step by step basis.

to compute (approximately) the required $y(x + \frac{h}{2})$. Instead of the exact value, use the Euler step value

$$y(x) + \frac{h}{2} f(y(x), x).$$

The algorithm for taking a single step of length h takes a form that evaluates the slope function twice,

$$\begin{aligned} k_1 &= f(y(x), x) \\ k_2 &= f\left(y(x) + \frac{h}{2}k_1, x + \frac{h}{2}\right) \\ y(x+h) &\approx y(x) + hk_2. \end{aligned}$$

One can do a Taylor series expansion with this set of equations to verify that it is of higher accuracy than Euler's method, at the cost of using two evaluations of $f(y(x), x)$ per step. There really is no need to do the Taylor series calculations required to verify this by hand, as Maple can readily do this. The expression for the step can be derived by substituting for k_1 and k_2 to obtain an explicit formula for the step taken by this method.

$$RK = y(x) + hf\left(y(x) + \frac{h}{2} f(y(x), x), x + \frac{h}{2}\right).$$

The Maple code to compute the difference between this and the Taylor series for the true solution is simply

```
> RK := y(x) + h* f(y(x) +( h/2)* f(y(x), x), x+(h/2)):
>
> rkseries := taylor(RK, h=0, 4);
```

$$\begin{aligned} rkseries := {} & y(x) + f(y(x), x)\,h \\ & + \left(\frac{1}{2} D_1(f)(y(x), x)\, f(y(x), x) + \frac{1}{2} D_2(f)(y(x), x)\right) h^2 \\ & + \left(\frac{1}{8} D_{1,1}(f)(y(x), x)\, f(y(x), x)^2 + \frac{1}{4} D_{1,2}(f)(y(x), x)\, f(y(x), x) \right. \\ & \left. + \frac{1}{8} D_{2,2}(f)(y(x), x)\right) h^3 + O(h^4) \end{aligned}$$

```
> solution(4);
```

The accuracy of this Runge–Kutta scheme can be evaluated by simply subtractingthe two quantities calculated above. This produces an expression for the Taylor series (as a function of the step size h) of the error between the true solution value and the numerical solution step result.

```
> taylor(rkseries - solution(4), h=0, 4);
```

$$\left(-\frac{1}{24}D_{1,1}(f)(y(x),x)\,f(y(x),x)^2-\frac{1}{12}D_{1,2}(f)(y(x),x)\,f(y(x),x)\right.$$
$$-\frac{1}{6}D_1(f)(y(x),x)^2\,f(y(x),x)-\frac{1}{6}D_1(f)(y(x),x)\,D_2(f)(y(x),x)$$
$$\left.-\frac{1}{24}D_{2,2}(f)(y(x),x)\right)h^3+O(h^4).$$

This explicit Taylor series calculation reveals that the Runge–Kutta formula (as a function of the step size h) agrees with the true solution Taylor series out through terms of order two in powers of h.

This verifies that the truncation error in Runge's original scheme actually is of order h^3, so that it is of higher accuracy than Euler's method by one order. (The `taylor` invocation in the code above serves as an easy way to force Maple to collect the terms in the h polynomial).

4.4 General Runge–Kutta Methods

The calculation of the error behavior of Runge's original scheme suggests a way to obtain even better truncation errors. The third order error was obtained by using a first order accurate Euler's method to calculate the intermediate slope, but we ended up with a second order accurate calculation for y at the end point $x+h$. If we were willing to do further calculations, we should be able to bootstrap our way into more accurate results. Rather than guessing at appropriate intermediate values to calculate (and recreating the historical evolution of Runge–Kutta routines) it turns out to be easier write down what a general such scheme should look like and see that it is possible to explicitly calculate the error performance of such routines.

The general Runge–Kutta scheme consists of a sequence of stages, each of which evaluates a trial value for the slope of the solution. The final step is to advance the solution using a weighted sum of the slopes calculated in the preceding stages.

The formulas associated with this method are

$$
\begin{aligned}
k_1 &= f(y(x), x) \\
k_2 &= f(y(x) + ha_{2,1}k_1, x + c_2h) \\
k_3 &= f(y(x) + ha_{3,1}k_1 + ha_{3,2}k_2, x + c_3h) \\
\ldots &= \ldots \\
k_s &= f(y(x) + ha_{s,1}k_1 + ha_{s,2}k_2 + \ldots + ha_{s,s-1}k_{s-1}, x + c_sh) \\
y(x+h) &= y(x) + h(b_1k_1 + b_2k_2 + \ldots + b_sk_s).
\end{aligned}
$$

where $y(x)$ (previously reserved for the exact, true solution curve) is used for the numerical solution in preference to maintaining a separate notation such as $\hat{y}(x)$ for the approximate solution produced by carrying out the numerical method algorithm. For the calculations to be made, x is conceptually fixed as the right endpoint of the current interval, and the issue is the difference between the algorithm calculation and the true solution at the subinterval endpoint $x + h$.

The parameter s in the above is the number of stages or slope evaluations. There are parameters describing the points and weightings of the method. $a_{i,j}$ gives the weighting of past slopes, and is lower triangular as an array. The c_i and b_i can be thought of as vectors determining the x points and slope weightings in the final calculation. It is conventional to describe the method by displaying a bordered array containing the parameters, and to refer to the array as a Runge–Kutta tableau. The general such array looks like

$$
\begin{array}{c|ccccc}
0 & & & & & \\
c_2 & a_{2,1} & & & & \\
c_3 & a_{3,1} & a_{3,2} & & & \\
\vdots & \vdots & \ddots & & & \\
c_s & a_{s,1} & a_{s,2} & \ldots & a_{s,s-1} & \\
\hline
 & b_1 & b_2 & \ldots & b_{s-1} & b_s\,.
\end{array}
$$

The original Runge routine from 1895 makes a quite sparse example

$$
\begin{array}{c|cc}
0 & & \\
1/2 & 1/2 & \\
\hline
 & 0 & 1\,.
\end{array}
$$

The discussion above about the accuracy of Runge's routine suggests that Maple is capable of carrying out the required Taylor series calculations. It can actually do the same sort of calculations for the general world-beating Runge–Kutta setup whose equations are above. The various parameters can be dragged through the calculations just as easily as explicit numerical examples can be evaluated, with recipes for Runge–Kutta schemes being the result.

A first pass at such a project can proceed by trying to write Maple routines that will calculate the Taylor series for the truncation error of a general form Runge–Kutta routine. Since a routine to calculate the Taylor series of the true solution is in hand from above, we need to calculate the Taylor series expansion of the Runge–Kutta result formula. This is easier than it might seem because the slope calculations take the form of an iterative procedure in which the current slope is calculated from the previous terms in the sequence.

The Maple routine to compute the slope stages is a transcription of the Runge–Kutta algorithm equations into Maple:

```
> #
> # computes the slope for stage n in a Runge--Kutta routine
> #
> slope := proc(stage)
>     local summand,i;
>     options remember;
>     if stage = 1 then RETURN(f(y(x),x))
>     else
>       summand := y(x);
>       for i from 1 to stage-1
>       do
>         summand := summand+a[stage,i]*slope(i)*h
>       od;
>       RETURN(f(summand,x+c[stage]*h))
>     fi
> end;
```

As long as the slopes (the k_i in the description) can be generated on demand, getting an expression for what the routine actually computes is a matter of forming the weighted sum for the number of stages desired.

```
> #
> # Returns the actual formula for the general Runge--Kutta step,
> # taking stages evaluations of f(y(x), x) to compute it.
> #
> GRK := proc(stages)
>     local i,summand;
>
>     summand := y(x);
>     for i from 1 to stages
>     do
>       summand := summand+b[i]*slope(i)*h
```

```
>    od;
>    RETURN(summand)
> end;
```

The results of testing the formula generating code are the following fragments.

```
> GRK(1);
```

$$y(x) + b_1 f(y(x), x)\, h$$

```
> GRK(2);
```

$$y(x) + b_1\, f(y(x), x)\, h + b_2\, f\,\big(y(x) + a_{2,1}\, f(y(x), x)h, x + c_2 h\big)\, h$$

```
> GRK(3);
```

$$y(x) + b_1 f(y(x), x)h + b_2 f\,\big(y(x) + a_{2,1} f(y(x), x)h, x + c_2 h\big)\, h$$
$$+ b_3 f\Big(y(x) + a_{3,1} f(y(x), x)h + a_{3,2} f(y(x) + a_{2,1} f(y(x), x)h, x + c_2 h)h,$$
$$x + c_3 h\Big)h$$

These expressions represent the explicit formula for a general Runge–Kutta routine. Since it is also possible to compute the Taylor series of the exact solution out to any desired order symbolically by using Maple, it is feasible to derive design rules for the coefficient sets by calculating the error Taylor polynomial and selecting the routine coefficients to make the truncation error be as high order as possible.

One might think that from this point the necessary conditions on the Runge–Kutta coefficient array could be obtained by Maple just by following the procedure used above to verify the original example. That is, form an error expression by subtracting the Runge–Kutta step formula from the true solution step. Then use the built-in Maple Taylor series routines to evaluate the order of accuracy of the general routine formula.

Although it seems plausible, this approach is a bit too simplistic to succeed. This might be suspected on the basis that the topic of Runge–Kutta order conditions has been pursued actively as a research topic in the numerical analysis literature.

The main impediments that arise are first the fact that the built-in Maple derivative and Taylor series routines are designed for dealing with scalar-valued functions, while what is desired is numerical methods that are applicable to systems of differential equations. The other issue is one that commonly occurs whenever "serious calculations" are attempted. Calculations of high order derivatives of the kind needed result in what is essentially a combinatorial explosion in the number of terms in the expressions.

The calculation contemplated actually can be carried out with Maple, but to do so requires writing derivative routines that are adapted to the problem at hand, and applying programming techniques to control the storage requirements and increase speed by avoiding repetitive calculations. These topics are pursued further in Chapter 16.

4.5 Maple Numeric Routines

Maple connection **Page:** 292

User written numerical routines can be tried out by writing simple iterative procedures (using remember tables) and plotting the numerical points directly. Maple itself has some standard routines and even a high order variable step size routine with *many* options and setup parameters. These library routines are written in a standard format, so that they can be employed by the library subroutine `DEplot` of the `DEtools` package. Routines written in a compatible form are likewise usable, so it is useful to look at the source code of an example. The library procedure `rk4` is a Maple incarnation of the RK4 discussed above.[2]

```
> proc(n,hp,pt,F)
> local h,k1,k2,k3,k4,t,x,i,k,y,z;
> options 'Copyright 1993 by Waterloo Maple Software';
>   h := hp;
>   t := pt[1];
>   x := array(1 .. n);
>   for i to n do  x[i] := pt[i+1] od;
>
>      k1 := array(1 .. n);
>      k2 := array(1 .. n);
>      y := array(1 .. n);
>      k3 := array(1 .. n);
>      k4 := array(1 .. n);
>      z := array(1 .. n);
>      F(k1,x,t);
>      for k to n do  y[k] := x[k]+1/2*h*k1[k] od;
>      F(k2,y,t+1/2*h);
>      for k to n do  z[k] := x[k]+1/2*h*k2[k] od;
>      F(k3,z,t+1/2*h);
>      for k to n do  y[k] := x[k]+h*k3[k] od;
>      F(k4,y,t+h);
>      for i to n do  x[i] := x[i]+1/6*h*(k1[i]+2*k2[i]+2*k3[i]+k4[i]) od;
>      t := t+h
>
>   pt[1] := t;
>   for i to n do  pt[i+1] := x[i] od
> end
>
```

[2]The ease with which the Maple code for built-in numerical methods can be found seems to vary substantially from release to release. Consider locating the subroutine rk4 as a rainy day adventure.

To discuss the above subroutine, it is useful to establish the notational conventions upon which the code is based. The independent variable is "time" t, and the solution sought is for a system of vector equations

$$\frac{d\mathbf{x}}{dt} = \mathbf{F}(\mathbf{x}(t), t).$$

In terms of individual components the equations would be written as

$$\frac{dx_i}{dt} = F_i(x_1(t), \ldots, x_n(t), t),$$

for $i = 1, \ldots, n$.

The routine is designed to take a single RK4 step with step size `hp` passed as one of the arguments. The code works with vector systems of first order differential equations, but the only difference from the scalar case is that the calculations are carried out separately for each component of the vector solution. The argument `n` is the dimension of the vector system of equations, and is the limit for all the loops over vector quantities.

The solution vector itself is carried in the array `pt[ ]`, in positions 2 through $n + 1$. This array is workspace handed down from the driver routine calling `rk4`, and the values of the $\mathbf{x}$ (dependent variables) passed in as arguments are overwritten with the updated values when the subroutine returns. The same comment applies to the current time value, passed as the first entry of the `pt[ ]` array.

The arguments above are associated with the overhead and mechanical operation of the subroutine, and are independent of the particular system of differential equations being numerically solved. The definition of the problem is carried in the argument `F`, which is the name of a subroutine that the user of the routine must supply. What `F` computes is exactly the usual right hand side of the system of differential equations, with the wrinkle that the slope values are returned in the entries of the array supplied as the first argument to `F`. It is more efficient to copy the values into storage used by the calling `rk4` than to return a copy of an array.

As an example of a routine that could be used in the same way as as `rk4` (and Euler's method, for that matter), one could build an adaptive step size method using a three step – four step Runge–Kutta combination. Since `rk4` is already available as a library routine, all that is required is to get an implementation of Heun's `rk3`, and then follow this with a Maple version of the step size change logic. This ought to adhere to the `DEplot` specification for numerical method add-ons. In short, it should use the same arguments as the `rk4` example.

The three stage `rk3` is a simple matter of editing the `rk4` code to remove a stage and adjust the method parameters. The result is

```
> rk3 := proc(n,hp,pt,F)
> local h,k1,k2,k3,t,x,i,k,y,z;
> options 'Copyright K. Heun, 1900, better late than never';
>     h := hp;
>     t := pt[1];
```

```
>    x := array(1 .. n);
>
>       k1 := array(1 .. n);
>       k2 := array(1 .. n);
>       k3 := array(1 .. n);
>       y := array(1 .. n);
>       z := array(1 .. n);
>
>       F(k1,x,t);
>       for k to n do  y[k] := x[k]+1/3*h*k1[k] od;
>       F(k2,y,t+1/3*h);
>       for k to n do  z[k] := x[k]+2/3*h*k2[k] od;
>       F(k3,z,t+2/3*h);
>
>       for i to n do  x[i] := x[i]+1/4*h*(k1[i]+3*k3[i]) od;
>       t := t+h
>
>    pt[1] := t;
>    for i to n do  pt[i+1] := x[i] od
> end;
```

4.6 Calling all RK4's (and relatives)

Using a subroutine of the standard Maple format of the RK4 and RK3 examples of the previous section, a numerical integration of a vector differential equation can be undertaken. The method of choice is Heun's third order Runge–Kutta method, coded in the Maple required format for numerical method step routines.

```
> interface(verboseproc=2);
> read(‘rk3.txt‘);


rk3 := proc(n,hp,pt,F)
     local h,k1,k2,k3,t,x,i,k,y,z;
     options ‘Copyright K. Heun, 1900, better late than never‘;
        h := hp;
        t := pt[1];
        x := array(1 .. n);
        for i to n do  x[i] := pt[i+1] od;
           k1 := array(1 .. n);
           k2 := array(1 .. n);
           k3 := array(1 .. n);
           y := array(1 .. n);
           z := array(1 .. n);
           F(k1,x,t);
           for k to n do  y[k] := x[k]+1/3*h*k1[k] od;
           F(k2,y,t+1/3*h);
```

```
        for k to n do  z[k] := x[k]+2/3*h*k2[k] od;
        F(k3,z,t+2/3*h);
        for i to n do  x[i] := x[i]+1/4*h*(k1[i]+3*k3[i]) od;
        t := t+h
    pt[1] := t;
    for i to n do  pt[i+1] := x[i] od
end
```

The use of the numerical step procedure above requires a routine (F internally in `rk3`) that evaluates the right hand side of the system of differential equations ("the slopes") and returns the values in the first argument passed. Note that the variables involved are arrays, so that they are passed by *name* in Maple, and the operations in the F routine modify the "k" values back in $rk3$. The right hand side routine defined below corresponds to the harmonic oscillator discussed as an example above. For a numerical solution of the problem, the harmonic oscillator must be written in the form of a first order system. Using x for the position, and v as the velocity, the form of the system of equations is

$$\frac{d}{dt}\begin{bmatrix} x \\ v \end{bmatrix} = \begin{bmatrix} v \\ -\frac{k}{M}x \end{bmatrix}.$$

The subroutine needed for the numerical method must evaluate the slope function of the differential equation, which in this case amounts to filling in the vector components on the right hand side of the differential equation. From the use of the slope function in the `rk3` code, the first subroutine argument carries the slope values, the second is the current state variable, and the last is the problem independent variable.

```
> springmass := proc(k, x, t)
>  global K, M;
>
>   k[1] := x[2];
>   k[2] := -K/M * x[1];
>
> end;
```

The format of the Maple numerical stepping routines enforces the convention that the first component of the stored vector is the "time" value, followed by the state components in sequence. The iteration procedure is the enforcer of this, and subsequent plotting must respect it as well. The iteration routine uses a remember table for storage, and fills in the current time on every call. It returns the initial condition when $n = 0$, and otherwise calls RK3 to make the step from the previous to current values. Storage space for the solution points is allocated by the `array` call at each step. The 2 in the `rk3` call is the dimension of the harmonic oscillator state vector. The use of the `eval` in the `RETURN` statement is caused by the fact that the variable is an array, which will be returned as a name unless the evaluation is forced.

```
> iteration := proc(n)
>  option remember;
>  local pt;
>
>  pt := array(1..3);
>  pt[1] := n*h;
>  if n = 0 then
>   do
>     pt[1] := 0;
>     pt[2] := 0;
>     pt[3] := 1;
>     RETURN(eval(pt));
>   od;
>  else
>   do
>     pt[2] := iteration(n-1)[2];
>     pt[3] := iteration(n-1)[3];
>     rk3(2, h, pt, springmass);
>     RETURN(eval(pt));
>   od;
>  fi;
> end;
```

Set the global variables to try this out:

```
> h:= .01;
>
> K := 1;
>
> M := 1;
```

The harmonic oscillator position variable is the first state component, but is the second component in the Maple iteration arrays (the first is "time".) To get something plottable by Maple, generate a list of solution time-position pairs (each pair a list of length 2.)

```
> pospts := [seq([iteration(n)[1], iteration(n)[2]], n=0..200)];
```

pospts := [[0, 0], [.02, .009999833333], [.03, .01999866668],
[.04, .02999550017], [.05, .03998933413], [.06, .04997916918],
[.07, .05996400635], [.08, .06994284716], [.09, .07991469373],
....
[1.93, .9396453983], [1.94, .9361769768], [1.95, .9326149384],
[1.96, .9289596393], [1.97, .9252114450], [1.98, .9213707303],
[1.99, .9174378793], [2.00, .9134132853], [2.00, .9134132853]]

To plot the numerical result along with the true solution, save the corresponding PLOT structure in a Maple variable.

```
> rk3plot := plot(pospts);
```

rk3plot := PLOT(CURVES([[0, 0], [.02, .009999833333],
[.03, .01999866668], [.04, .02999550017], [.05, .03998933413],
[.06, .04997916918], [.07, .05996400635], [.08, .06994284716],
....
[1.92, .9430198560], [1.93, .9396453983], [1.94, .9361769768],
[1.95, .9326149384], [1.96, .9289596393], [1.97, .9252114450],
[1.98, .9213707303], [1.99, .9174378793], [2.00, .9134132853],
[2.00, .9134132853]], COLOUR(*RGB*, 0, 0, 0)), AXESLABELS(,),
TITLE( ), AXESTICKS(*DEFAULT*, *DEFAULT*),
VIEW(*DEFAULT*, *DEFAULT*))

For the problem parameters and initial conditions chosen, the "real solution" is just a sine curve.

```
> sinplot := plot(sin(x), x=0..2);
```

sinplot := PLOT(CURVES([[0, 0], [.03985881624, .03984826296050123],
[.08251638915, .08242277930203681],
[.1254567241, .1251278809526790],
[.1666666666, .1658961326276722],
....
[1.833333333, .9657346538440102],
[1.874439869, .9542534099326293],
[1.914877537, .9413857870025076],
[1.959778370, .9252955959589811], [2., .9092974268256817]],
COLOUR(*RGB*, 0, 0, 0)), AXESLABELS(*x*,), TITLE( ),
AXESTICKS(*DEFAULT*, *DEFAULT*), VIEW(0..2., *DEFAULT*))

```
> with(plots);
```

Display will superimpose the two plots.

```
> display({rk3plot, sinplot});
```

The plot of the numerical and true solutions is quite uninformative. One obtains the default (large, on a numerical scale) step size for the sine plot, and finer resolution for the numerical solution. Really, the error should be plotted at the numerical resolution to see the performance of the method.

4.7 Variable Step Size Methods

Maple connection **Page:** 293

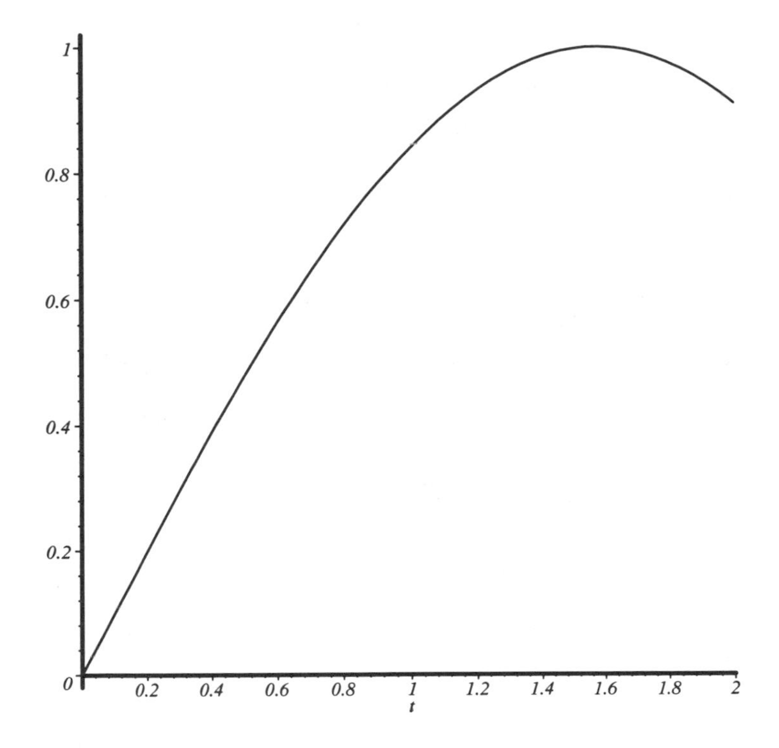

FIGURE 4.4. RK3 solution against a true one

The numerical methods discussed in the previous sections are known as fixed step size methods. Although the truncation error calculations make clear that (within limits imposed by roundoff) making step sizes small tends to make overall errors smaller, it also will increase the running time of the routine directly as the number of required steps increases. Typical solution profiles for problems reveal profiles that may change slowly over large regions of the domain, with localized more rapid variations. This is easily imagined if one considers solving the problem of calculating a satellite orbit with a close approach to a large planet. The velocity of the satellite can be expected to show rapid variation during the close encounter, perhaps resulting in a graph as in Figure 4.5.

It will be necessary to use a small step size on this problem in order to cope with the large slopes and rapid changes during the near collision, but the small steps are not required for the great part of the orbit where the variations are slow.

In order to avoid the inefficiencies of using small step sizes throughout, the step size h must be changed during the course of the calculation. Of course it would be possible to write codes that changed the step size more or less appropriately for a particular problem on the basis of laborious numerical experimentation, but

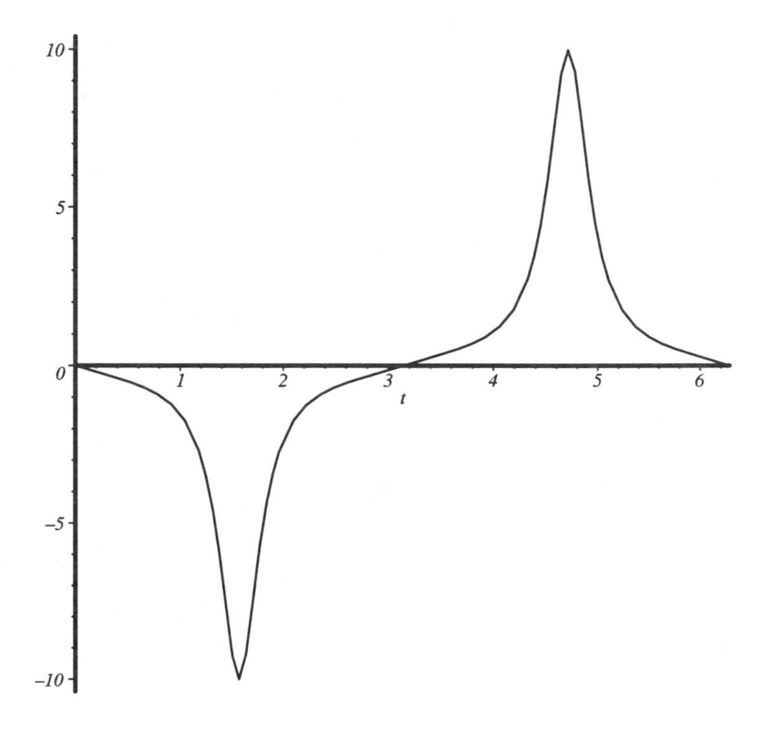

FIGURE 4.5. Horizontal satellite velocity

it is more useful to have a routine that automatically adjusts its step size as the calculation proceeds. It is possible to do such a thing, and the basis of such schemes relies on the ability to compute an estimation of the truncation error at each stage of the calculation. One way to do this is to do the calculation by running two different methods in parallel. The numerical solution is propagated with one of the methods, while the second is calculated solely to estimate and hence control the truncation error. To see how this is possible, consider carrying this out with two Runge–Kutta routines of different orders.

4.7.1 *Running RK4 With RK3*

A popular four stage Runge–Kutta scheme called RK4 is accurate to fourth order, and is described by the tableau (this is due to Kutta, 1901):

$$\begin{array}{c|cccc} 0 & & & & \\ 1/2 & 1/2 & & & \\ 1/2 & 0 & 1/2 & & \\ 1 & 0 & 0 & 1 & \\ \hline & 1/6 & 2/6 & 2/6 & 1/6. \end{array}$$

What the accuracy statement means of course is that the relationship between the true solution step and the result of a single step of RK4 is given by

$$y(x+h) = RK4(x,h) + Mh^5 + O(h^6),$$

since being accurate to order four[3] requires that the Taylor coefficients match to that exponent order. If we symbolically refer to the third order method described by

$$\begin{array}{c|ccc} 0 & & & \\ 1/3 & 1/3 & & \\ 2/3 & 0 & 2/3 & \\ \hline & 1/4 & 0 & 3/4 \end{array}$$

as RK3, then the corresponding relation takes the form

$$y(x+h) = RK3(x,h) + Kh^4 + O(h^5).$$

If these two equations are subtracted, the result is

$$0 = RK3(x,h) - RK4(x,h) + Kh^4 + O(h^5).$$

Since Kh^4 is essentially the truncation error for RK3 at that step, the truncation error can be estimated (not exactly calculated: if it were exactly calculated this way we could make the numerical calculation have no error at all) by

$$Kh^4 = RH4 - RK3 \ (+O(h^5)).$$

This can be interpreted as saying that if you drag along an RK4 calculation in addition to propagating RK3, you can tell basically how well RK3 is doing on the truncation front. More than that, the information can be used to modify the step size as you proceed.

Maple connection **Page:** 293

4.7.2 *Adjusting h*

Suppose that you have decided that the quantity TOL is a level of truncation error that you are willing to tolerate at each step. (This is a bit hard to estimate apriori, since the total number of steps taken by a variable step size routine is unknown at the start.) If we have used the current h to take what should be thought of as a trial step, the truncation error is estimated by

$$|K|\,h^4 = \Delta = |RK4 - RK3|.$$

---

[3]The truncation order is 5, one greater than the accuracy.

If we had taken a step with step size h_1, then the resulting truncation error should have been

$$|K|\, h_1^4$$

on the grounds that the variation of the derivative value in K between using h or h_1 only affects the higher order terms, which we are systematically ignoring. If our estimated truncation error is larger than TOL, then we calculate what h_1 should be in order that we meet the TOL criterion. We want

$$|K| h_1^4 = TOL,$$

while our calculation of the trial step with h says $|K| = \frac{\Delta}{h^4}$. This allows us to eliminate the mysterious K in the above formula in favor of the calculated quantity $\frac{\Delta}{h^4}$. Solving for the h_1 that would meet the tolerance TOL, we have

$$h_1 = \left(\frac{TOL}{\Delta}\right)^{\frac{1}{4}} h.$$

It is silly to attempt to hit the tolerance exactly, especially in view of the fact that higher order terms were neglected in arriving at the above results. In practice, the recipe is adjusted by a "safety factor" $S = 0.9$, so that if $\Delta > TOL$, the calculation

$$h_1 = S \left(\frac{TOL}{\Delta}\right)^{\frac{1}{4}} h.$$

is used to adjust the step size downward. If the estimated error beats the tolerance, then one could use the raw calculation above to expand the h, or more often something like

$$h_1 = 1.5h,$$

representing a somewhat aggressive reach for longer steps. An overreach will be reigned in on the next cycle, so the overall scheme may be hoped to track along the tolerance limit. A pseudo code description of the operation of the routine is

start procedure

trycount = 0
while trycount < MAXTRIES
begin loop
    compute RK4 step
    compute RK3 step
    $\Delta = |RK4 - RK3|$
    if $\frac{\Delta}{TOL} < 1$ **then**
        $h = 1.5 * h$
        return RK3
    else
        $h = 0.9 * (TOL/\Delta)^{(1/4)}$
        trycount = trycount + 1
    end if
    if trycount > 20 **then**
        return RK3 and complain loudly
    end if
end loop

end procedure

To implement an adaptive step size method, one must arrange that the step size be carried as a variable (effectively as an additional function of time beyond the variables being solved for.) Probably the easiest way to do this is to increase the dimension of the solution vector by one component, and carry the step size as this last component. An adaptive routine also requires a certain number of control parameters. The most obvious is the truncation tolerance which is the basis of the routine rationale, but some provisions to keep the method sane are also advisable. The step size must be kept comfortably above the roundoff level, so a lower limit on the step size is required. For the sake of plotting the final result in an intelligible fashion, there is no point in letting the step size increase without limit, even if the solution curve is "flat" in the region. The parameter set includes an upper limit on the step size to account for this.

The adaptation process attempts to adjust the step size downward to meet the truncation tolerance. With a lower step size limit in place, it is possible that this process would loop forever and never reach a point where the estimated truncation was smaller than the tolerance. By placing a limit on the number of reduction loops attempting to meet the tolerance, we prevent the method from getting stuck in a loop over a problem so badly behaved that the tolerance will not be met.

If an adaptive routine is to be written, it must be decided how these control parameters are to be made available to the numerical stepping routine. The style of solution adopted for many of the Maple library routines is to append a series of `this=that` parameters to the procedure argument list, but then the called routine has to "pull apart" the arguments looking for equations containing keywords. The solution adopted below is to put these control values in a table, and arrange that it be passed as the last argument to the code. If `adrk4` is used with a hand written driver, the required table can just be passed directly to the routine.

```
> # adaptive rk4/rk3 step
>
>  adrk4:=proc(n,hp,pt,F,mparm)
>
>  local i, x1sav, x2sav, h, retries, tsav,
>       argseq, EE, HMAX, HMIN, TOL, RETRIES,temp;
>
>  x1sav := array(1..n+1);
>  x2sav := array(1..n+1);
>  tsav := pt[1];
>  x1sav[1] := tsav;
>  x2sav[1] := tsav;
>
>  for i to n
>  do
>    x1sav[i+1] := pt[i+1];
>    x2sav[i+1] := pt[i+1];
>  od;
>
> # for adaptive scheme, keep the variable step size in extra variable for
> # a record of what it did. Treated like a state variable. You'd put the
> # initial h in with the initial conditions.
>
>  h := pt[n+2];
>
>  HMAX := mparm[hmax];
>  HMIN:= mparm[hmin];
>  TOL := mparm[tol];
>  RETRIES := mparm[tries];
>
>  retries := 0;
>  while retries < RETRIES
>  do
> # take one rk4 and one rk3 step
>
>    rk4(n, h, x1sav, F);
>    rk3(n, h, x2sav, F);
>
> # take worst component as error estimate and scale it by the tolerance
> # use the built in Maple max to avoid problems
>
>    argseq := NULL;
```

```
>     for i from 1 to n
>     do
>       temp := eval(abs(x1sav[i+1] - x2sav[i+1]));
>       argseq := argseq,temp;
>     od;
> #
>     EE := max(argseq);
>     EE := EE/TOL;
> #
> # if we beat the error limit
> #
>     if EE < 1.0 then
>       pt[1] := tsav+h;
>       for i from 1 to n
>       do
>         pt[i+1] := x2sav[i+1];
>       od;
>
> # this code really is only for h>0 in spite of Maple being willing to
> # integrate backwards using negative h.  Fix it if you want to.
> # ad hoc expansion of h
> #
>       h := h * 1.5;
>       if h > HMAX then h:= HMAX; fi;
>
>       pt[n+2] := h;
> #
> # break out of the while loop
>       break;
>     fi;
>
> # otherwise shrink the h and try again
> #
>     retries := retries +1;
>     if retries < RETRIES then
>       h := .7 * h * exp(- .25 * log(EE));  # magic safety factor
>       if h < HMIN then h := HMIN; fi;
>     else
>       pt[1] := tsav+h;
>       for i from 1 to n
>       do
>         pt[i+1] := x2sav[i+1];
>       od;
>       pt[n+2] := h;
>       print(`WARNING: Exceeded retry limit in adrk4 routine`);
> #
> # better luck next iteration
>       break;
>     fi;
```

```
> od;
> end;
```

4.8 Serious Methods

For serious numerical computation, numerical methods that are both accurate and computationally efficient are needed. The computational burden of a numerical method is dominated by the cost of repeated slope evaluations, which typically may involve complicated evaluations of transcendental functions. It is of interest then to find adaptive step size methods of high order (to minimize the number of steps taken), and which are efficient from the point of view of floating point evaluations.

Each tableau row in a Runge–Kutta scheme corresponds to a slope evaluation. When an adaptive scheme is in use, two sets of tableau rows are involved. Fehlberg is credited with discovering that there are a number of pairs of Runge–Kutta schemes of differing truncation order, which have the property that the *leading rows* of the tableaus are identical.

The implication of this is enormous. Some of these *Fehlberg pairs* have only one extra slope evaluation in the more accurate method of the pair. This means that an adaptive version of a Runge–Kutta method can be run with only a single extra slope evaluation per step.

The tableau for a certain Runge–Kutta numerical method is given by

$$
\begin{array}{c|ccccc}
0 & & & & & \\
2/9 & 2/9 & & & & \\
1/3 & 1/12 & 1/4 & & & \\
3/4 & 69/128 & -243/128 & 135/64 & & \\
1 & -17/12 & 27/4 & -27/5 & 16/5 & \\
\hline
 & 1/9 & 0 & 9/20 & 16/45 & 1/12
\end{array} .
$$

This method turns out to have a truncation order of 5. The Runge–Kutta method described by

$$
\begin{array}{c|cccccc}
0 & & & & & & \\
2/9 & 2/9 & & & & & \\
1/3 & 1/12 & 1/4 & & & & \\
3/4 & 69/128 & -243/128 & 135/64 & & & \\
1 & -17/12 & 27/4 & -27/5 & 16/5 & & \\
5/6 & 65/432 & -5/16 & 13/16 & 4/27 & 5/144 & \\
\hline
 & 47/450 & 0 & 12/25 & 32/225 & 1/30 & 6/25
\end{array}
$$

has a truncation order of 6. If they are combined in the natural way to form an adaptive scheme, the result is known as Fehlberg's rk4(5).[4] There are a number of pairs of this type, discovered by Fehlberg and other numerical investigators.

4.9 Problems

1. Get Maple to make plots of the difference between the true solution to $\frac{dx}{dt} = x(t)$, $x(0) = 1$, and the numerical approximation calculated using Euler's method over the interval $0 \leq t \leq 1$. Try various h. What difference does the choice seem to make?

2. Repeat the experiment using Runge's version of a Runge–Kutta scheme. What are the differences between the two numerical methods as well?

3. Code up a variable step size routine in Maple, and figure out how to end up being able to plot the step sizes it uses along with the solution it calculates. Try it out. Can you find an example problem that shows the routine "shortening up" on the steep portions of the solution curve?

4. Adaptive step size routines arise whenever two calculations produce truncation errors that can be combined to estimate the truncation error of the less accurate one. Convince your self that running one Runge–Kutta routine, and in parallel taking two steps of length $h/2$ with the same Runge–Kutta step algorithm also leads to an adaptive routine.

5. Derive the step size adjustment rule for Fehlberg's rk4(5).

6. Using the rk4/rk3 adaptive method of the text as a model, code a Maple version of Fehlberg's rk4(5).

---

[4]The notation reflects the order to which the sub-methods are accurate, rather than the (one greater) truncation order.

5

Higher Order Differential Equations

5.1 Equations of Order N

Some ordinary differential equations "arrive" in a form involving derivatives of higher than first order. A conspicuous source of second order problems is mechanics, where the form of Newton's second law leads directly to a second order equation. An example of this is the harmonic oscillator equation coming from a spring mass system.

$$\frac{d^2}{dt^2}y(t) + \frac{K}{M}\,y(t) = 0,$$

with initial conditions

$$y(0) = y_0, \quad \frac{dy}{dt}(0) = v_0.$$

The general linear differential equation of order n has the form

$$a_n(t)\,\frac{d^n y}{dt^n} + a_{n-1}(t)\,\frac{d^{n-1}y}{dt^{n-1}} + a_{n-2}(t)\,\frac{d^{n-2}y}{dt^{n-2}} + \ldots + a_0(t)\,y(t) = f(t),$$

where $f(t)$ is referred to as the *forcing function* of the equation. The expression above is referred to as the *inhomogeneous form* of the differential equation. The corresponding equation

$$a_n(t)\,\frac{d^n y}{dt^n} + a_{n-1}(t)\,\frac{d^{n-1}y}{dt^{n-1}} + a_{n-2}(t)\,\frac{d^{n-2}y}{dt^{n-2}} + \ldots + a_0(t)\,y(t) = 0$$

obtained by setting the forcing function to zero is called the *homogeneous form* of the equation. There is also a fundamental difference between the cases where the coefficients in the equation actually vary with the independent variable t, or in fact are constants. In the case of *constant coefficients* much more explicit calculations of solutions can be made, and Laplace transforms may be used to advantage.

5.1.1 Vector Equations

As suggested in earlier discussions, high order differential equations can be made to appear as first order equations, provided that one is willing to use vector variables instead of always restricting attention to scalar quantities. This point of view is profitable, as first order systems are of greater generality than high order scalar equations, as well as a naturally occurring form.

Starting from the linear equation of order n above, it is easy to rewrite it in the form of a first order vector system. The method is to create a vector whose components are the successive derivatives of the original solution. That is, define $\mathbf{x}(t)$ by

$$\mathbf{x}(t) = \begin{bmatrix} y(t) \\ \frac{dy}{dt} \\ \frac{d^2y}{dt^2} \\ \vdots \\ \frac{d^{n-1}y}{dt^{n-1}} \end{bmatrix}.$$

Then using the original equation in the form

$$\frac{d^n y}{dt^n} = \frac{-1}{a_n(t)} \left(a_{n-1}(t) \frac{d^{n-1}y}{dt^{n-1}} + a_{n-2}(t) \frac{d^{n-2}y}{dt^{n-2}} + \ldots + a_0(t)\, y(t) \right)$$

the time derivative of the vector $\mathbf{x}$ is expressed as

$$\frac{d}{dt}\mathbf{x}(t) = \begin{bmatrix} \frac{dy}{dt} \\ \frac{d^2y}{dt^2} \\ \vdots \\ \frac{d^n y}{dt^n} \end{bmatrix}$$

$$= \begin{bmatrix} \frac{dy}{dt} \\ \frac{d^2y}{dt^2} \\ \vdots \\ -\frac{a_0(t)}{a_n(t)} y(t) - \frac{a_1(t)}{a_n(t)} \frac{dy(t)}{dt} \cdots - \frac{a_{n-1}(t)}{a_n(t)} \frac{d^{n-1}y(t)}{dt^{n-1}} \end{bmatrix}$$

$$= \begin{bmatrix} 0 & 1 & 0 & \dots & 0 \\ 0 & 0 & 1 & \dots & 0 \\ \vdots & & & & \vdots \\ -\frac{a_0(t)}{a_n(t)} & -\frac{a_1(t)}{a_n(t)} & \dots & \dots & -\frac{a_{n-1}(t)}{a_n(t)} \end{bmatrix} \begin{bmatrix} y(t) \\ \frac{dy}{dt} \\ \frac{d^2y}{dt^2} \\ \vdots \\ \frac{d^{n-1}y}{dt^{n-1}} \end{bmatrix}$$

If $\mathbf{A}(t)$ is used as notation for the (time varying, in general) coefficient matrix in the above equation, the vector equation can be compactly written in what is referred to as *state vector* form

$$\frac{d}{dt}\mathbf{x}(t) = \mathbf{A}(t)\,\mathbf{x}(t).$$

The representation of the n-th order original equation in this vector form might be termed the *standard representation*, characterized by the fact that the components of the vector are the successive derivatives of the original solution. The form of the coefficient matrix, with the value 1 on the super diagonal, and the equation coefficients arranged in the bottom row, is referred to as a *companion matrix*. What the above shows is that given the solution of the n-th order equation, we can produce a solution of the standard form vector equation. One might also wonder whether the converse relation holds. Does solving the vector equation (in standard form) also provide a solution to the single equation of order n that we started with?

5.1.2 *Equivalent Formulations*

The conversion between scalar n-th order equations and the standard form vector equation actually results in equivalent formulations of the problem. To see this, we have to verify that if we have a solution of the vector equation with the standard representation coefficient matrix, then this provides a solution of the original scalar

n-th order problem. Hence we assume that we have a vector function

$$\mathbf{x}(t) = \begin{bmatrix} x_1(t) \\ x_2(t) \\ x_3(t) \\ \vdots \\ x_n(t) \end{bmatrix}$$

that satisfies

$$\frac{d}{dt} \begin{bmatrix} x_1(t) \\ x_2(t) \\ x_3(t) \\ \vdots \\ x_n(t) \end{bmatrix} = \begin{bmatrix} 0 & 1 & 0 & \ldots & 0 \\ 0 & 0 & 1 & \ldots & 0 \\ \vdots & & & & \vdots \\ -\frac{a_0(t)}{a_n(t)} & -\frac{a_1(t)}{a_n(t)} & \ldots & \ldots & -\frac{a_{n-1}(t)}{a_n(t)} \end{bmatrix} \begin{bmatrix} x_1(t) \\ x_2(t) \\ x_3(t) \\ \vdots \\ x_n(t) \end{bmatrix}.$$

In view of the construction of the standard form, we expect that the solution of the n-th order problem is just the first component of the vector solution. That is, we simply define the function $y(t)$ by declaring it as the first component of the vector solution

$$y(t) = x_1(t).$$

It then remains to verify that this $y(t)$ actually satisfies the original scalar equation. The first row of the vector equation actually reduces to the equation

$$\frac{d}{dt} x_1(t) = x_2(t),$$

making

$$x_2(t) = \frac{dy}{dt}.$$

The second row of the vector equation (assuming $n > 2$) says

$$\frac{d}{dt} x_2(t) = x_3(t),$$

so that

$$x_3(t) = \frac{d^2y}{dt^2},$$

and like considerations lead to

$$x_n(t) = \frac{d^{n-1}y}{dt^{n-1}}.$$

The final row of the vector equation written out in isolation is

$$\frac{d}{dt}x_n(t) = \frac{-1}{a_n(t)}\left(a_{n-1}(t)\frac{d^{n-1}y}{dt^{n-1}} + a_{n-2}(t)\frac{d^{n-2}y}{dt^{n-2}} + \ldots + a_0(t)\,y(t)\right)$$

and since

$$\frac{d}{dt}x_n(t) = \frac{d^n y}{dt^n}$$

we see that

$$\frac{d^n y}{dt^n} = \frac{-1}{a_n(t)}\left(a_{n-1}(t)\frac{d^{n-1}y}{dt^{n-1}} + a_{n-2}(t)\frac{d^{n-2}y}{dt^{n-2}} + \ldots + a_0(t)\,y(t)\right).$$

This amounts to saying $y(t)$ as defined by the first component of the vector equation solves the original n-th order equation. The two points of view are therefore equivalent in the sense that a solution to one problem provides a solution to the other.

5.2 Linear Independence and Wronskians

The general n-th order equation can be written as

$$\frac{d^n}{dt^n}y(t) = \frac{-1}{a_n(t)}\left(a_{n-1}(t)\frac{d^{n-1}y}{dt^{n-1}} + a_{n-2}(t)\frac{d^{n-2}y}{dt^{n-2}} + \ldots + a_0(t)\,y(t)\right),$$

and read as giving $\frac{d^n}{dt^n}y(t)$ as a function of the lower derivatives. This point of view suggests that in order to "start up" the solution of the n-th order equation, initial values for all of

$$y(0),\ \frac{dy}{dt}(0),\ \ldots,\ \frac{d^{n-1}y}{dt^{n-1}}(0)$$

will have to be specified. This point raises the vexing question of how we propose to meet n initial value constraints when psychologically we are contemplating only a single solution of the homogeneous n-th order equation

$$a_n(t)\frac{d^n y}{dt^n} + a_{n-1}(t)\frac{d^{n-1}y}{dt^{n-1}} + a_{n-2}(t)\frac{d^{n-2}y}{dt^{n-2}} + \ldots + a_0(t)\,y(t) = 0.$$

Clearly, we need "n" solutions, and "different" ones at that. The ideas and definitions that make these fuzzy notions precise are really in the province of linear algebra.

5.2.1 Linear Independence

Definition
Given a set of functions $\{f_1(t),\ f_2(t),\ \ldots,\ f_n(t)\}$, the set is *linearly dependent* if there exist constants $c_1,\ c_2,\ \ldots,c_n$, not all zero, such that

$$c_1\, f_1(t) + c_2\, f_2(t) + \ldots + c_n\, f_n(t) = 0.$$

If no such set of constants exist, then the set of functions is *linearly independent*.

One way to test for linear independence is by constructing the *Grammian* matrix whose entries are the pairwise inner products of the functions, but for this approach to be taken, the machinery of inner product spaces has to be invoked. When the functions under discussion are solutions of an n-th order differential equation, they are actually n-times differentiable (since they *are* solutions). This lets manipulations be made that are not available for less well behaved functions. A determinant condition called the Wronskian condition can then be found for testing linear dependence.

5.2.2 Wronskians

The idea behind Wronskians is to find a system of linear equations satisfied by the dependence coefficients $c_1,\ c_2,\ \ldots,c_n$ when the set of $n-1$ times differentiable functions $\{f_1(t),\ f_2(t),\ \ldots,\ f_n(t)\}$ is linearly dependent. Starting from the hypothesis that

$$c_1\, f_1(t) + c_2\, f_2(t) + \ldots + c_n\, f_n(t) = 0$$

differentiation gives

$$c_1 \frac{d}{dt}\, f_1(t) + c_2 \frac{d}{dt}\, f_2(t) + \ldots + c_n \frac{d}{dt}\, f_n(t) = 0,$$

and then

$$c_1 \frac{d^2}{dt^2}\, f_1(t) + c_2 \frac{d^2}{dt^2}\, f_2(t) + \ldots + c_n \frac{d^2}{dt^2}\, f_n(t) = 0,$$

with the sequence ending with

$$c_1 \frac{d^{n-1}}{dt^{n-1}}\, f_1(t) + c_2 \frac{d^{n-1}}{dt^{n-1}}\, f_2(t) + \ldots + c_n \frac{d^{n-1}}{dt^{n-1}}\, f_n(t) = 0.$$

Writing this set of equations in matrix form amounts to

$$\begin{bmatrix} f_1 & f_2 & \cdots & f_n \\ \frac{d}{dt} f_1 & \frac{d}{dt} f_2 & \cdots & \frac{d}{dt} f_n \\ \vdots & \vdots & \vdots & \vdots \\ \frac{d^{n-1}}{dt^{n-1}} f_1 & \frac{d^{n-1}}{dt^{n-1}} f_2 & \cdots & \frac{d^{n-1}}{dt^{n-1}} f_n \end{bmatrix} \begin{bmatrix} c_1 \\ c_2 \\ \vdots \\ c_n \end{bmatrix} = \begin{bmatrix} 0 \\ 0 \\ \vdots \\ 0 \end{bmatrix}.$$

If the set of functions is linearly dependent, then the determinant of the coefficient matrix must vanish.

Definition
The *Wronskian* of the set of ($n-1$ times differentiable) functions is defined as the determinant of the system

$$W(f_1, f_2, \ldots, f_n) = \det \begin{bmatrix} f_1 & f_2 & \cdots & f_n \\ \frac{d}{dt} f_1 & \frac{d}{dt} f_2 & \cdots & \frac{d}{dt} f_n \\ \vdots & \vdots & \vdots & \vdots \\ \frac{d^{n-1}}{dt^{n-1}} f_1 & \frac{d^{n-1}}{dt^{n-1}} f_2 & \cdots & \frac{d^{n-1}}{dt^{n-1}} f_n \end{bmatrix}.$$

If the Wronskian vanishes identically, then the functions are dependent, otherwise they are independent.

Example
The set of functions $\sin(t)$, $\cos(t)$ would appear to be linearly independent, and a calculation of the Wronskian shows that to be the case:

$$\begin{aligned} W(\sin(t), \cos(t)) &= \det \begin{bmatrix} \sin(t) & \cos(t) \\ \cos(t) & -\sin(t) \end{bmatrix} \\ &= -sin^2(t) - cos^2(t) \\ &= -1. \end{aligned}$$

Example
A similar calculation shows that

$$W(e^t, t\,e^t, t^2\,e^t) = 2\,e^{3t},$$

so that $e^t, t\,e^t, t^2\,e^t$ is a linearly independent triple of functions.

Calculating a Wronskian for anything but a small set of functions is tedious. Maple (surprise) has a `wronskian` procedure, which can be used if such a calculation is getting out of hand.

5.3 Fundamental Solutions

If $y_1(t)$ and $y_2(t)$ are both solutions of the homogeneous equation

$$a_n(t)\,\frac{d^n y}{dt^n} + a_{n-1}(t)\,\frac{d^{n-1} y}{dt^{n-1}} + a_{n-2}(t)\,\frac{d^{n-2} y}{dt^{n-2}} + \ldots + a_0(t)\,y(t) = 0,$$

then due to the linearity of the governing equation it is easy to verify that the linear combination

$$y(t) = c_1\, y_1(t) + c_2\, y_2(t)$$

also satisfies the same equation. By similar reasoning (the linearity of the equation) arbitrary linear combinations of any set $\{y_1(t), y_2(t), \ldots, y_n(t)\}$ of solutions also satisfy the homogeneous equation. The solutions count the zero function among their number, so these observations are enough to conclude that the solutions form a vector space. It is this which leads to the linear algebraic aspects of the problem.

Definition
A *fundamental set of solutions* of the homogeneous equation

$$a_n(t)\,\frac{d^n y}{dt^n} + a_{n-1}(t)\,\frac{d^{n-1} y}{dt^{n-1}} + a_{n-2}(t)\,\frac{d^{n-2} y}{dt^{n-2}} + \ldots + a_0(t)\, y(t) = 0$$

is a linearly independent set which spans the set of solutions.

If $y_1(t), y_2(t), \ldots, y_n(t)$ is a fundamental set then every solution of the homogeneous equation is of the form

$$y(t) = c_1\, y_1(t) + c_2\, y_2(t) + \ldots + c_n\, y_n(t)$$

for some set of coefficients $c_1,\ c_2\ \ldots, c_n$, while the Wronskian

$$W(y_1(t), y_2(t), \ldots, y_n(t)) \neq 0,$$

since the functions are linearly independent.

In the terminology of linear algebra, a fundamental set of solutions is simply a basis for the solution space. Of course, this may be a somewhat vacuous definition as it stands, as nothing that has been mentioned so far gives any clue that the solution space is finite dimensional (it is), or hints at what the fundamental set might look like. These questions are considered in the following section.

5.3.1 Existence of Fundamental Solutions

There are two aspects to the question of existence of a set of fundamental solutions. One is the question of logical or theoretical existence of such a set, and the other is the possibility of making explicit calculations of the functions in question.

A widely studied special case of the n-th order homogeneous problem is the constant coefficient version

$$a_n\,\frac{d^n y}{dt^n} + a_{n-1}\,\frac{d^{n-1} y}{dt^{n-1}} + a_{n-2}\,\frac{d^{n-2} y}{dt^{n-2}} + \ldots + a_0\, y(t) = 0.$$

This is the case where both aspects can be handled expeditiously, because Laplace transforms easily (naturally, in fact) compute a fundamental solution set. See the following Chapter.

In the case where the equation coefficients are actually time-varying, explicit simple calculations of formulas for the solution functions are not available except in special circumstances. (This comment arbitrarily classifies power series representations of solutions as being beyond the simple range.)

This means that the existence of fundamental solutions in the general time varying case must rest on a constructive existence theorem. Actually, it is not a "new" existence theorem, but an adaptation of the previously described iterative solution method to the present circumstances. The secret is to establish existence for a particularly chosen set of solutions. It actually is the "same" solution set in the general case as the Laplace transform method naturally produces in the special situation of constant coefficients.

We seek a set of functions $\{y_1(t), y_2(t), \ldots, y_n(t)\}$ satisfying the following conditions. First, each of the solutions *is* a solution, and hence is required to satisfy the homogeneous equation

$$a_n(t)\,\frac{d^n y}{dt^n} + a_{n-1}(t)\,\frac{d^{n-1} y}{dt^{n-1}} + a_{n-2}(t)\,\frac{d^{n-2} y}{dt^{n-2}} + \ldots + a_0(t)\,y(t) = 0.$$

The n solutions differ from each other in that they satisfy different initial conditions at the initial time $t = 0$. The conditions for the different functions are the following.

y_1 : At $t = 0$, we have $y_1(0) = 1$, while $\frac{dy_1}{dt}(0)$, $\frac{d^2 y_1}{dt^2}(0)$, $\ldots$, $\frac{d^{n-1} y_1}{dt^{n-1}}(0)$ all vanish.

y_2 : At $t = 0$ we have $y_2(0) = 0$, with $\frac{dy_2}{dt}(0) = 1$, but $\frac{d^2 y_2}{dt^2}(0)$, $\ldots$, $\frac{d^{n-1} y_2}{dt^{n-1}}(0)$ all vanishing.

y_3 : At $t = 0$ we have $y_3(0) = \frac{dy_3}{dt}(0) = 0$, $\frac{d^2 y_3}{dt^2}(0) = 1$, and the rest of the initial derivatives are all vanishing.

y_k : $\ldots$

y_n : At $t = 0$, we arrange that $y_n(0)$, $\frac{dy_n}{dt}(0)$, $\frac{d^2 y_n}{dt^2}(0)$, $\ldots$, $\frac{d^{n-2} y_1}{dt^{n-2}}(0)$ all vanish, but $\frac{d^{n-1} y_1}{dt^{n-1}}(0) = 1$.

Leaving aside for a moment the existence argument for these solutions, it is easy to verify that a set of solutions with those properties will automatically be linearly independent.

What has to be shown that if it is possible to find coefficients c_1, c_2, $\ldots$, c_n such that

$$0 = c_1\,y_1(t) + c_2\,y_2(t) + \ldots + c_n\,y_n(t),$$

then it must follow that $c_1 = c_2 = \ldots = c_n = 0$, so all the coefficients vanish. Assuming that such constants exist, substitute $t = 0$ in the equation. This produces

$$\begin{aligned} 0 &= c_1\, y_1(0) + c_2\, y_2(0) + \ldots + c_n\, y_n(0) \\ 0 &= c_1\, 1 + c_2\, 0 + \ldots + c_n\, 0 \\ 0 &= c_1. \end{aligned}$$

Differentiate the equation and then substitute $t = 0$. There results

$$\begin{aligned} 0 &= c_1 \frac{dy_1}{dt}(0) + c_2 \frac{dy_2}{dt}(0) + \ldots + c_n \frac{dy_n}{dt}(0) \\ 0 &= c_2, \end{aligned}$$

because the second solution has 1 as the first derivative value. Successively differentiating the original equation and evaluating the resulting expression at $t = 0$ in turn gives

$$c_3 = c_4 = \ldots = c_n = 0,$$

proving the coefficients all vanish as required. If the solutions exist, they are therefore linearly independent.

The set of solutions meeting the successive derivative conditions as described above are exactly the appropriate vehicles to provide a solution to the general *initial value problem* for the homogeneous equation

$$a_n(t) \frac{d^n y}{dt^n} + a_{n-1}(t) \frac{d^{n-1} y}{dt^{n-1}} + a_{n-2}(t) \frac{d^{n-2} y}{dt^{n-2}} + \ldots + a_0(t)\, y(t) = 0,$$

with the given initial condition constraints

$$\begin{aligned} y(0) &= B_1 \\ \frac{dy}{dt}(0) &= B_2 \\ \vdots &= \vdots \\ \frac{d^{n-1} y}{dt^{n-1}}(0) &= B_n. \end{aligned}$$

The linear combination meeting those constraints is simply the solution

$$y(t) = B_1\, y_1(t) + B_2\, y_2(t) + \ldots + B_n\, y_n(t).$$

This works because the arrangement of initial properties of $\{y_1(t),\ y_2(t),\ \ldots,\ y_n(t)\}$ effectively decouples the initial condition equations, as one can easily verify by applying the initial conditions to the expression.

For any other fundamental set of solutions, satisfying a given set of initial conditions ends up requiring the solution of a set of linear equations. To see that this is the case, consider the following.

If another basis for the solution set $\{z_1(t),\ z_2(t),\ \ldots,z_n(t)\}$ is used, then meeting the initial condition constraint involves writing a general solution in the form

$$y(t) = c_1\,z_1(t) + c_2\,z_2(t) + \ldots + c_n\,z_n(t),$$

and calculating the initial values in terms of the unknowns $c_1,\ \ldots,\ c_n$. This gives the equations for the solution derivatives as

$$\begin{aligned}
y(0) &= c_1\,z_1(0) + c_2\,z_2(0) + \ldots + c_n\,z_n(0),\\
\frac{dy}{dt}(0) &= c_1\,\frac{dz_1}{dt}(0) + c_2\,\frac{dz_2}{dt}(0) + \ldots + c_n\,\frac{dz_n}{dt}(0),\\
\vdots &= \vdots,\\
\frac{d^{n-1}y}{dt^{n-1}}(0) &= c_1\,\frac{d^{n-1}z_1}{dt^{n-1}}(0) + c_2\,\frac{d^{n-1}z_2}{dt^{n-1}}(0) + \ldots + c_n\,\frac{d^{n-1}z_n}{dt^{n-1}}(0).
\end{aligned}$$

If these values are equated with the given initial values

$$\begin{aligned}
y(0) &= B_1\\
\frac{dy}{dt}(0) &= B_2\\
\vdots &= \vdots\\
\frac{d^{n-1}y}{dt^{n-1}}(0) &= B_n,
\end{aligned}$$

the result is an $n \times n$ system of equations for the unknown constants $c_1,\ \ldots,\ c_n$.

$$\begin{bmatrix}
z_1(0) & z_2(0) & \ldots & z_n(0)\\
\frac{d}{dt}\,z_1(0) & \frac{d}{dt}\,z_2(0) & \ldots & \frac{d}{dt}\,z_n(0)\\
\vdots & \vdots & \vdots & \vdots\\
\frac{d^{n-1}}{dt^{n-1}}\,z_1(0) & \frac{d^{n-1}}{dt^{n-1}}\,z_2(0) & \ldots & \frac{d^{n-1}}{dt^{n-1}}\,z_n(0)
\end{bmatrix}
\begin{bmatrix} c_1\\ c_2\\ \vdots\\ c_n \end{bmatrix}
=
\begin{bmatrix} B_1\\ B_2\\ \vdots\\ B_n \end{bmatrix}.$$

Note that the coefficient matrix in this system is the one used to define the Wronskian determinant for the solution basis, and hence it is nonsingular if $\{z_1(t),\ z_2(t),\ldots,z_n(t)\}$ is a fundamental set of solutions for the equation. In the case in which the special basis $\{y_1(t), y_2(t), \ldots, y_n(t)\}$ is used, the system of equations is diagonal (the coefficient matrix is an identity matrix) and the solution of the corresponding set of equations is immediate.

Example
One may verify that $\cos(2t)$ and $\frac{1}{2}\sin(2t)$ are both solutions of

$$\frac{d^2y}{dt^2} + 4y = 0,$$

and also satisfy the initial conditions that make satisfaction of a set of initial conditions easy. The solution that satisfies

$$\begin{aligned} y(0) &= 2, \\ \frac{dy}{dt}(0) &= 3 \end{aligned}$$

is

$$y(t) = 2\cos(2t) + \frac{3}{2}\sin(2t).$$

Example
The differential equation

$$\frac{d^2y}{dt^2} - 4y = 0$$

has the linearly independent set $\{e^{2t}, e^{-2t}\}$ as solutions. The general solution of the homogeneous equation is thus

$$y(t) = c_1 e^{2t} + c_2 e^{-2t}.$$

In order to satisfy the initial conditions

$$\begin{aligned} y(0) &= 2, \\ \frac{dy}{dt}(0) &= 3 \end{aligned}$$

the constants c_1, c_2 must meet the conditions

$$\begin{aligned} c_1 + c_2 &= 2 \\ 2c_2 - 2c_2 &= 3. \end{aligned}$$

The equations in this case must be solved to provide $c_1 = \frac{7}{4}$, $c_2 = \frac{1}{4}$, and the solution in the form

$$y(t) = \frac{7}{4}e^{2t} + \frac{1}{4}e^{-2t}.$$

5.3.2 Constructing Fundamental Solutions

The existence of the desired set of solutions can be established in a constructive fashion by adapting the Picard iteration method to the present circumstances. The starting point is to realize that the Picard iteration is based on converting a first order differential equation to an integral equation as a first step. In order to apply this to the problem at hand, the problem has to be approached as a first order system. From Section 5.1.1, the n-th order single equation can be written in the form of an equivalent vector system as

$$\frac{d}{dt}\mathbf{x} = \mathbf{A}(t)\,\mathbf{x}(t).$$

In this representation, earlier called the standard representation, the natural choice of components of the state vector $\mathbf{x}(t)$ makes these the successive derivatives of the original equation solutions. The first component of $\mathbf{x}(t)$ identifies with $y(t)$ in the original equation. Because of these identifications it is possible to devise a Picard iteration (more precisely, n separate Picard iterations, one for each initial condition) converging to the solution basis described above.

A vector system form of Picard iteration follows naturally by simply integrating both sides of the vector system of equation. This gives the corresponding integral equation system

$$\mathbf{x}(t) = \mathbf{x}(0) + \int_0^t \mathbf{A}(t)\,\mathbf{x}(t)\,dt.$$

Conversion of this to a convergent iteration follows the same successive substitution pattern as the original Picard process. That is, substitute the previous approximation into the integral term of the equation, giving the *Picard iteration*

$$\mathbf{x}_n(t) = \mathbf{x}(0) + \int_0^t \mathbf{A}(t)\,\mathbf{x}_{n-1}(t)\,dt.$$

Since the equation is linear, as long as the coefficient matrix is bounded, the iteration will converge to a solution of the vector equation, no matter what initial value is used for $\mathbf{x}(0)$.

To generate the solution basis, it only remains to describe the appropriate initial vectors $\mathbf{x}(0)$ which produce the solutions with the required initial conditions. But these are easily described, since the scalar problem initial conditions are the successive components of the $\mathbf{x}(0)$ vector.

The solution $y_1(t)$ is obtained by iterating the above Picard scheme with the initial condition

$$\mathbf{x}(0) = \begin{bmatrix} 1 \\ 0 \\ \vdots \\ 0 \end{bmatrix}.$$

The second basis solution $y_2(t)$ comes from

$$\mathbf{x}(0) = \begin{bmatrix} 0 \\ 1 \\ \vdots \\ 0 \end{bmatrix},$$

and so on. Finally $y_n(t)$ is obtained as the (first component of) the vector solution resulting from

$$\mathbf{x}(0) = \begin{bmatrix} 0 \\ 0 \\ \vdots \\ 1 \end{bmatrix}.$$

With some care in the choice of problem, these iterations can actually be carried out with Maple for illustrative purposes.

Maple connection **Page:** 293

5.4 General and Particular Solutions

The terminology of *general* and *particular* solutionsis commonly used in connection with linear differential equations of order n.

A particular solution of an inhomogeneous differential equation

$$a_n(t)\,\frac{d^n y}{dt^n} + a_{n-1}(t)\,\frac{d^{n-1} y}{dt^{n-1}} + a_{n-2}(t)\,\frac{d^{n-2} y}{dt^{n-2}} + \ldots + a_0(t)\,y(t) = f(t),$$

is *any* function $y_p(t)$ satisfying

$$a_n(t)\,\frac{d^n y_p}{dt^n} + a_{n-1}(t)\,\frac{d^{n-1} y_p}{dt^{n-1}} + a_{n-2}(t)\,\frac{d^{n-2} y_p}{dt^{n-2}} + \ldots + a_0(t)\,y_p(t) = f(t)$$

for the given forcing function $f(t)$.

Particular solutions are far from being unique. If we had two particular solutions $y_{p1}(t)$ and $y_{p2}(t)$, then

$$a_n(t)\,\frac{d^n y_{p1}}{dt^n} + a_{n-1}(t)\,\frac{d^{n-1} y_{p1}}{dt^{n-1}} + a_{n-2}(t)\,\frac{d^{n-2} y_{p1}}{dt^{n-2}} + \ldots + a_0(t)\,y_{p1}(t) = f(t),$$

$$a_n(t)\,\frac{d^n y_{p2}}{dt^n} + a_{n-1}(t)\,\frac{d^{n-1} y_{p2}}{dt^{n-1}} + a_{n-2}(t)\,\frac{d^{n-2} y_{p2}}{dt^{n-2}} + \ldots + a_0(t)\,y_{p2}(t) = f(t),$$

and subtracting the two equations,

$$a_n(t)\,\frac{d^n(y_{p1}-y_{p2})}{dt^n}+a_{n-1}(t)\,\frac{d^{n-1}(y_{p1}-y_{p2})}{dt^{n-1}}$$

$$+\,a_{n-2}(t)\,\frac{d^{n-2}(y_{p1}-y_{p2})}{dt^{n-2}}+\ldots+a_0(t)\,(y_{p1}-y_{p2})=0.$$

This means that the difference between any two particular solutions is a solution of the corresponding homogeneous equation. A particular solution is negotiable up to an arbitrary addition of a solution of the homogeneous equation. Since we discovered in the previous section that solutions of the homogeneous equation form an n dimensional vector space with a fundamental solution set as the basis, an arbitrary solution of

$$a_n(t)\,\frac{d^n y}{dt^n}+a_{n-1}(t)\,\frac{d^{n-1}y}{dt^{n-1}}+a_{n-2}(t)\,\frac{d^{n-2}y}{dt^{n-2}}+\ldots+a_0(t)\,y(t)=f(t),$$

will take the form

$$y(t)=y_p(t)+B_1\,y_1(t)+B_2\,y_2(t)+\ldots+B_n\,y_n(t),$$

where $y_p(t)$ is *any* particular solution of the inhomogeneous equation, and the collection $\{y_1(t),\ y_2(t),\ \ldots,\ y_n(t)\}$ is a fundamental set of solutions of the homogeneous equation.

This expression represents the *general solution* of the inhomogeneous equation.

5.5 Constant Coefficient Problems

While the general properties of n-th order linear equations are independent of the nature of the equation coefficients, a complete description of the solutions becomes possible if one assumes that the coefficients are constant, independent of the variables in the equation. In this section we outline the "classical" treatment of such problems, which basically describes the fastest route to an answer for "small" versions of such problems. The style of discussion runs a bit toward recipes as to what to try in which situation; the whys of the recipes become clear when Laplace transform methods for solving such problems are considered.

5.5.1 Homogeneous Case

A homogeneous n-th order constant coefficient differential equation is of the form

$$a_n\,\frac{d^n y}{dt^n}+a_{n-1}\,\frac{d^{n-1}y}{dt^{n-1}}+a_{n-2}\,\frac{d^{n-2}y}{dt^{n-2}}+\ldots+a_0\,y(t)=0,$$

where each coefficient a_i is constant, independent of t as well as y. Solutions of the equation may be discovered by guessing an appropriate form (with some free parameters in the assumed form) and determining conditions required for the "trial solution" to actually work.

In the case of the equation above, the guess is motivated by the observation that differentiating an exponential function

$$y(t) = C\, e^{\lambda t}$$

an arbitrary number of times produces a multiple of the original function. Substituting this form into the differential equation gives

$$\left(a_n \lambda^n + a_{n-1} \lambda^{n-1} + \ldots + a_0\right) C\, e^{\lambda t} = 0$$

as the requirement for the assumed form to work. This means that the constant λ in the exponential function (referred to as the *time constant*) must satisfy the *characteristic equation*

$$a_n \lambda^n + a_{n-1} \lambda^{n-1} + \ldots + a_0 = 0.$$

For each root of the characteristic equation, we obtain a solution of the homogeneous equation. Since linear combinations of homogeneous solutions are again solutions, a number of solutions are obtained in this fashion.

Maple connection **Page:** 295

Example
The simplest example is of order 1,

$$a_1 \frac{dy}{dt} + a_0\, y(t) = 0,$$

with characteristic equation linear in the unknown

$$a_1\, \lambda + a_0 = 0.$$

The solution is

$$y(t) = C\, e^{-\frac{a_0}{a_1} t},$$

which can be also obtained by separation of variables.

Example
A spring mass system has the equation of motion

$$\frac{d^2 y}{dt^2} + \frac{K}{M} y = 0,$$

for which the characteristic equation is

$$\lambda^2 + \frac{K}{M} = 0.$$

If we declare $\omega_0 = \sqrt{\frac{K}{M}}$, the roots of the characteristic equation are $\pm i\,\omega_0$. Solutions are then obtained in the form of

$$y(t) = C_1 e^{i\omega_0 t} + C_2 e^{-i\omega_0 t}.$$

In this form, the coefficients C_1 and C_2 will in most cases end up as complex numbers, complex conjugates in problems where a real valued $y(t)$ is sought. Solutions involving real valued functions can be obtained by making use of Euler's equations :

$$\begin{aligned} e^{it} &= \cos(t) + i\,\sin(t) \\ e^{-it} &= \cos(t) - i\,\sin(t) \\ \cos(t) &= \frac{e^{it} + e^{-it}}{2} \\ \sin(t) &= \frac{e^{it} - e^{-it}}{2i}. \end{aligned}$$

(These can be obtained from Taylor's series). Using Euler's relations, solutions of the homogeneous equation can also be written in the form

$$y(t) = A\,\cos(\omega_0 t) + B\,sin(\omega_0 t).$$

After solutions have been obtained through the auspices of the characteristic equation, it is natural to wonder whether the solutions obtained are actually linearly independent, so that a basis for the solution space might be available. The easy case to consider is when the characteristic equation

$$a_n\lambda^n + a_{n-1}\lambda^{n-1} + \ldots + a_0 = 0$$

has no repeated roots, so that n distinct roots $\{\lambda_1, \lambda_2, \ldots, \lambda_n\}$ are obtained. We then obtain n solutions

$$\begin{aligned} y_1(t) &= e^{\lambda_1 t}, \\ y_2(t) &= e^{\lambda_2 t}, \\ y_3(t) &= e^{\lambda_3 t}, \\ \vdots &= \vdots \\ y_n(t) &= e^{\lambda_n t}. \end{aligned}$$

To check that these are linearly independent, compute the Wronskian. The matrix with columns of solution derivatives is

$$= \begin{bmatrix} e^{\lambda_1 t} & e^{\lambda_2 t} & \dots & e^{\lambda_n t} \\ \lambda_1 e^{\lambda_1 t} & \lambda_2 e^{\lambda_2 t} & \dots & \lambda_n e^{\lambda_n t} \\ \vdots & \vdots & \vdots & \vdots \\ {\lambda_1}^{n-1} e^{\lambda_1 t} & {\lambda_2}^{n-1} e^{\lambda_2 t} & \dots & {\lambda_n}^{n-1} e^{\lambda_n t} \end{bmatrix},$$

and since each entry in a column contains the same exponential factor, the Wronskian determinant is

$$W(y_1, y_2, \dots, y_n) = e^{\lambda_1 t} \dots e^{\lambda_n t} \det \left(\begin{bmatrix} 1 & 1 & \dots & 1 \\ \lambda_1 & \lambda_2 & \dots & \lambda_n \\ \vdots & \vdots & \vdots & \vdots \\ {\lambda_1}^{n-1} & {\lambda_2}^{n-1} & \dots & {\lambda_n}^{n-1} \end{bmatrix} \right).$$

The determinant in the above is also encountered in polynomial interpolation problems, and is referred to as a Vandermonde matrix determinant. In case any two of the λ_i are the same, the determinant will have two identical columns, and hence will vanish. Elaborating that observation can be shown to give the conclusion that the determinant is nonzero as long as the λ_i are all distinct.

The final conclusion is that the exponential solutions are a fundamental set as long as the characteristic equation has distinct roots. This raises the question of what the situation is when there are repeated roots in the characteristic equation.

The rule is that repeated roots introduce factors of "t" into the original solutions. That is, if the characteristic equation contains a factor

$$p(\lambda) = (\lambda - \lambda_1)^m \dots (\lambda - \lambda_k),$$

then the solutions corresponding to λ_1 must be taken as

$$\begin{aligned} y_1(t) &= e^{\lambda_1 t}, \\ y_2(t) &= t\, e^{\lambda_1 t}, \\ y_3(t) &= t^2 e^{\lambda_1 t}, \\ y_m(t) &= t^{m-1} e^{\lambda_1 t} \\ \vdots &= \vdots \end{aligned}$$

This can be justified directly from the "guess and substitute" formulas (see the problems below), or viewed as an outcome of solving the problem with Laplace transforms (see the following Chapter).

Example
A damped harmonic oscillator is governed by the equation

$$\frac{d^2}{dt^2}y(t) + 2\,\zeta\,\omega_n \frac{d}{dt}y(t) + {\omega_n}^2 y(t) = 0.$$

The constant ζ is referred to as the *damping ratio*, and the parameter ω_n is called the *fundamental frequency*, although it will be seen from the calculations below that solutions do not oscillate at that frequency unless the damping coefficient happens to vanish. The effect of increasing damping is to decrease the actual frequency, or even to halt oscillations altogether. All of these effects become evident from examining the characteristic equation

$$\lambda^2 + 2\,\zeta\,\omega_n \lambda + {\omega_n}^2 = 0.$$

The roots of the characteristic equation follow from the quadratic formula and take the form

$$\lambda_{1,2} = -\zeta\,\omega_n \pm i\,\omega_n\,\sqrt{1-\zeta^2}.$$

Some comments are in order. The first is that the stilted form of the original equation is a conventional form, adopted specifically so that the roots of the characteristic equation take the simple forms encountered above. Secondly, the formula above is written with the expectation that $-1 < \zeta < 1$. It is not that the formulas are wrong in other cases, but just inconvenient and misleading. The form above is appropriate for the oscillatory solutions which are obtained for $-1 < \zeta < 1$. For values outside that range, solutions involve real time constant exponentials.

With the above roots, solutions of the homogeneous equation are of the form

$$y(t) = A\,e^{-\zeta\,\omega_n\,t}\cos(\omega_n\,\sqrt{1-\zeta^2}\,t) + B\,e^{-\zeta\,\omega_n\,t}\sin(\omega_n\,\sqrt{1-\zeta^2}\,t).$$

This expression represents a damped sinusoid. The oscillation frequency of the solution is given by the expression $\omega_n\,\sqrt{1-\zeta^2}$, while the amplitude decreases exponentially with the time constant $-\zeta\,\omega_n$. The presence of the damping makes the oscillation slower than it would be in the undamped case. See Figure 5.1 for an illustration of the damped oscillation.

For values of the damping coefficient ζ outside of the range covered above, the solutions take a different form. A boundary case is $\zeta = 1$.

In that case, the characteristic equation has a double root since

$$\lambda_{1,2} = -\zeta\,\omega_n \pm 0.$$

Accordingly, solutions take the form

$$y(t) = A\,e^{-\zeta\,\omega_n\,t} + B\,t\,e^{-\zeta\,\omega_n\,t}.$$

This is referred to as the *critically damped case*. The terminology actually arises from what happens when the equation is subjected to a step function input, rather

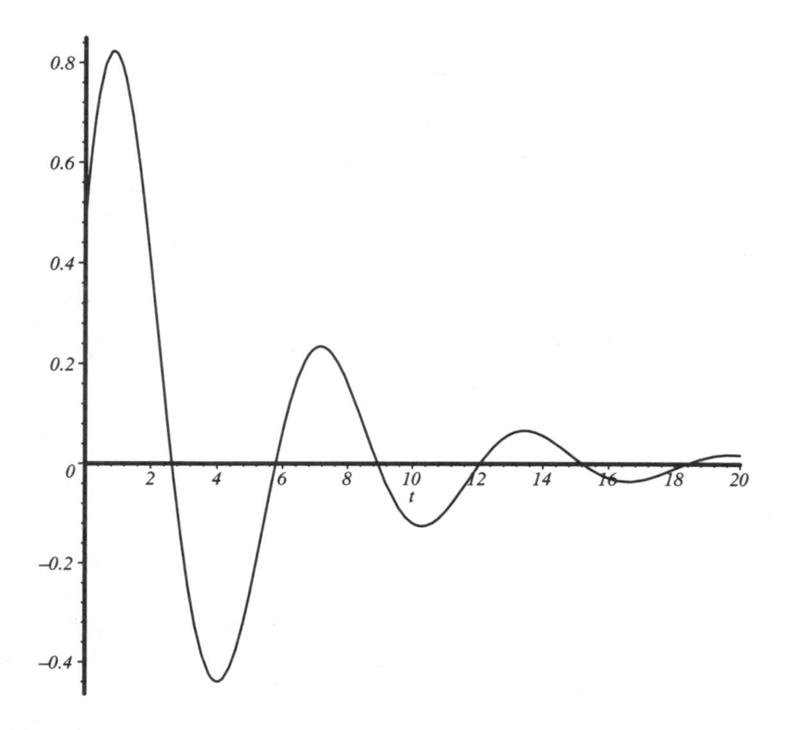

FIGURE 5.1. Damped sinusoid

than from properties of the homogeneous solutions. The solutions above are on the dividing line between oscillatory and exponential solutions. The above and the exponential solutions have the property that they can change sign only once, while the oscillatory solutions of more lightly damped systems in principle cross zero an infinite number of times.

An exceptional case of the damped oscillatory regime occurs when $\zeta = 0$. For $\zeta = 0$, the solutions are pure sinusoids at the frequency ω_n. For $\zeta < 0$, but $\zeta > -1$, the solutions are oscillatory, but the amplitude of the oscillations grows exponentially with time. (This is really the same case as above, except that the sign of the time constant is reversed.) Finally, at $\zeta = -1$ the solution crosses into the region where the solutions are a sum of exponentials with real, positive time constants.

5.5.2 *Undetermined Coefficients*

Undetermined coefficients is a term for guessing the form of a particular solution to a forced constant coefficient ordinary differential equation. Effectively, the process is limited to the case in which the forcing function consists of sums and products of sinusoids, exponentials, and polynomials.

Coincidentally, this is the case for which Laplace transform methods provide "back of an envelope" length solutions, so that major investments in undetermined coefficients are probably not required.

The generic problem under consideration is the forced constant coefficient equation

$$a_n \frac{d^n y}{dt^n} + a_{n-1} \frac{d^{n-1} y}{dt^{n-1}} + a_{n-2} \frac{d^{n-2} y}{dt^{n-2}} + \ldots + a_0 \, y(t) = f(t),$$

and what is sought is *any* particular solution. (Recall that particular solutions are by no means unique: they may differ by any solution of the homogeneous equation.)

The motivation for undetermined coefficients is closely related to that of trial solutions of the homogeneous equation (i.e., it looks like it ought to work if you plug it in.) The easiest approach is to consider the case where the forcing function is an exponential, say

$$a_n \frac{d^n y}{dt^n} + a_{n-1} \frac{d^{n-1} y}{dt^{n-1}} + a_{n-2} \frac{d^{n-2} y}{dt^{n-2}} + \ldots + a_0 \, y(t) = e^{\alpha t}.$$

By the same reasoning used in the homogeneous case, one is lead to try

$$y(t) = C \, e^{\alpha t}.$$

If this is substituted in as a trial solution, there results

$$\left(a_n \, \alpha^n + a_{n-1} \, \alpha^{n-1} + \ldots + a_0\right) C \, e^{\alpha t} = e^{\alpha t},$$

$$C = \frac{1}{a_n \, \alpha^n + a_{n-1} \, \alpha^{n-1} + \ldots + a_0},$$

at least as long as α is not a root of the characteristic equation. (If α is a root, then the "solution" above involves a division by 0, so the assumed solution form is invalid.)

If the forcing function is of the form

$$f(t) = \sin(\beta t),$$

then the appropriate first guess is

$$y(t) = A \, \cos(\beta t) + B \, \sin(\beta t).$$

Both terms are required since differentiation will produce both sines and cosines. Again, substituting this form in the equation leads to problems if the time constants $i \, \beta, -i \, \beta$ happen to be roots of the characteristic equation. Barring this, solutions emerge as expected.

By continuing in this fashion (guess and try) a table of particular solution forms can be generated.

forcing function	**particular solution**
e^{at}	$C\, e^{at}$
t^n	$a_n t^n + a_{n-1} t^{n-1} + \ldots + a_0$
$\sin(\beta t + \phi)$	$A\,\cos(\beta t) + B\,\sin(\beta t)$
$t^n e^{at}$	$(a_n t^n + a_{n-1} t^{n-1} + \ldots + a_0)\, e^{at}$
$t^n \sin(\beta t + \phi)$	$(a_n t^n + a_{n-1} t^{n-1} + \ldots + a_0)\cos(\beta t)$ $+(b_n t^n + b_{n-1} t^{n-1} + \ldots + b_0)\,\sin(\beta t)$

The particular solution forms above are based on the assumption that the exponent of the forcing function is not a root of the characteristic equation. If this is the situation, then the form of the guess should be multiplied by t raised to a power. The power required is the order of the root in the characteristic equation. Each entry in the particular solution column of the table really should be hedged by that caveat.

Rather than attempting to memorize the chart and the adjustment rules, it is simpler to treat these problems by Laplace transforms. See the following Chapter for this approach, and the explanation of the reason for the "throw in a power of t" adjustment rule.

Example
For the differential equation

$$\frac{dy}{dx} + y = e^{3x},$$

the form of the trial particular solution is

$$y(x) = A\, e^{3x}.$$

Substitution of this leads to

$$\begin{aligned} 3\,Ae^{3x} + A\,e^{3x} &= e^{3x}, \\ A &= \frac{1}{4}, \\ y(x) &= \frac{1}{4}\, e^{3x}. \end{aligned}$$

Example
Undetermined coefficients is a sufficient tool to compute the response of a simple linear system to a sinusoidal forcing function. The first order version of this is

$$\frac{dy}{dt} + ay = \cos(\omega\, t).$$

The assumed form for this problem is [1]

$$y(t) = A\,\cos(\omega t) + B\,\sin(\omega t).$$

Substituting, this is a solution provided that

$$\begin{aligned} -A\omega\sin(\omega t) + B\omega\cos(\omega t) & \\ +aA\cos(\omega t) + aB\sin(\omega t) &= \cos(\omega t). \end{aligned}$$

Matching coefficients of the linearly independent functions produces

$$\begin{aligned} aA + B\omega &= 1, \\ -A\omega + aB &= 0, \end{aligned}$$

with solutions

$$\begin{aligned} A &= \frac{a}{\omega^2 + a^2}, \\ B &= \frac{\omega}{\omega^2 + a^2}. \end{aligned}$$

The particular solutions then takes the form

$$y(t) = \frac{a}{\omega^2 + a^2}\cos(\omega t) + \frac{\omega}{\omega^2 + a^2}\sin(\omega t).$$

Particular solutions of constant coefficient problems with sinusoidal forcing functions generically (barring resonance or instability) are sinusoids of the same frequency as the forcing function, but different amplitude and phase. It is of interest to characterize the phase and amplitude responses because of its connection with stability of the systems when feedback is applied to achieve control of the response.

For the problem above (and other examples as well) the calculations already made will yield that information with a small amount of trigonometric manipulation. The basis is the trigonometric identity

$$\cos(A - B) = \cos(A)\,\cos(B) + \sin(A)\,\sin(B).$$

Rewrite the solution in the form

$$y(t) = \frac{1}{\sqrt{\omega^2 + a^2}}\left(\frac{a}{\sqrt{\omega^2 + a^2}}\cos(\omega t) + \frac{\omega}{\sqrt{\omega^2 + a^2}}\sin(\omega t)\right).$$

Then identifying

$$\begin{aligned} A &= \omega t, \\ \tan(B) &= \frac{\omega}{a}, \\ B &= \tan^{-1}(\frac{\omega}{a}) \end{aligned}$$

---

[1] In this and the following examples, what is being computed is a *particular solution*, so perhaps $y_p(t)$ throughout would be a better notation.

allows the answer to be rewritten in the form

$$y(t) = \frac{1}{\sqrt{\omega^2 + a^2}} \cos(\omega t - \phi),$$

where the phase lag ϕ is

$$\phi = \tan^{-1}(\frac{\omega}{a}).$$

Example
The phenomenon of resonance refers to the occurrence of a large amplitude response in a physical system subjected to a periodic forcing input. In mechanical systems such situations are usually undesirable, since large amplitude vibrations stress the materials.

Generally resonance is a result of driving a system at close to its natural frequency of vibration. The starting point for study of such effects is the forced harmonic oscillator model

$$\frac{d^2}{dt^2}y(t) + 2\zeta\omega_n \frac{d}{dt}y(t) + {\omega_n}^2 y(t) = \cos(\omega t).$$

What is of interest here is the relation of the amplitude of the response to the other parameters in the problem, in particular the damping ratio and natural frequency.

A particular solution to this problem will take the form of a sinusoidal oscillation at the driving frequency, so

$$y(t) = A\ \cos(\omega t) + B\ \sin(\omega t).$$

From this assumed form we can compute the first and second derivatives

$$y'(t) = -\omega A\ \sin(\omega t) + \omega B\ \cos(\omega t),$$

$$y''(t) = -\omega^2 A\ \cos(\omega t) - \omega^2 B\ \sin(\omega t).$$

Substitute these expressions into the equation and then match coefficients of the sines and cosines. This produces the expected simultaneous equations for the unknown coefficients. These are

$$\begin{aligned} A\ (\omega_n^2 - \omega^2) + 2\zeta\omega\omega_n B &= 1, \\ B\ (\omega_n^2 - \omega^2) - 2\zeta\omega\omega_n A &= 0. \end{aligned}$$

Solving these by matrix methods gives the solution in the form

$$\begin{bmatrix} A \\ B \end{bmatrix} = \begin{bmatrix} \frac{\omega_n^2 - \omega^2}{(\omega_n^2 - \omega^2)^2 + 4\zeta^2\omega^2\omega_n^2} \\ \frac{2\zeta\omega\omega_n}{(\omega_n^2 - \omega^2)^2 + 4\zeta^2\omega^2\omega_n^2} \end{bmatrix}$$

Finally, the amplitude of the response is given by the relatively simple expression

$$\sqrt{A^2 + B^2} = \frac{1}{\sqrt{(\omega_n^2 - \omega^2)^2 + 4\zeta^2\omega^2\omega_n^2}}.$$

This expression is (nearly) a function of ζ and the normalized frequency $\frac{\omega}{\omega_n}$. By taking a factor of $\frac{1}{\omega_n^2}$ out of the denominator it becomes so, and takes a form which is easy to plot. The resonance phenomenon can be seen by plotting

$$\frac{1}{\sqrt{(1 - \frac{\omega^2}{\omega_n^2})^2 + 4\zeta^2 \frac{\omega^2}{\omega_n^2}}}.$$

as a function of the driving frequency ratio $\frac{\omega}{\omega_n}$ for various values of the damping coefficient. Such a plot is in Figure 5.2.

Maple connection Page: 286

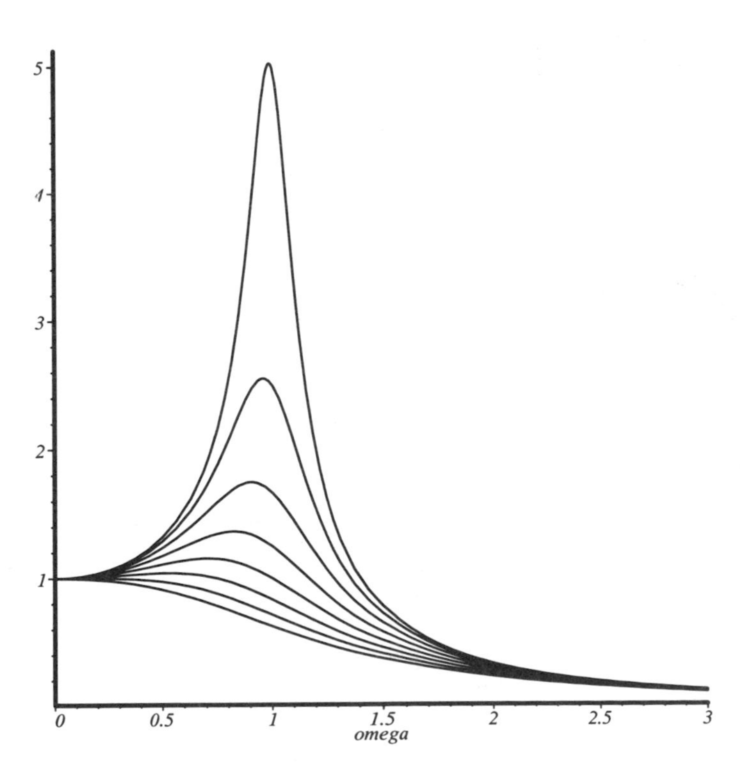

FIGURE 5.2. Resonance curves

5.6 Problems

1. Write

$$\frac{d^2y}{dt^2} + \sin(t)\,\frac{dy}{dt} + 49\,y = 0$$

in the form of an equivalent first order vector differential equation.

2. Find the general solutions of

(a)

$$\frac{dx}{dt} + 11\,x = 0.$$

(b)

$$\frac{d^2y}{dt^2} + 3\,\frac{dy}{dt} + 2\,y = 0.$$

(c)

$$\frac{d^2z}{dt^2} + 22\frac{dz}{dt} + 121\,z = 0.$$

3. Find a particular solution of

(a)

$$\frac{dx}{dt} - 11\,x = e^{12\,t}.$$

(b)

$$\frac{dx}{dt} = 2\,x + \sin(5\,t).$$

(c)

$$\frac{dx}{dt} = 2\,x + t + e^t + \sin(t).$$

(d)

$$\frac{dx}{dt} = 2\,x + e^{2t}.$$

4. Give the general solution for each of the parts of Problem 3.

5. Find the general solution of

$$\frac{d^2y}{dt^2} + 3\frac{dy}{dt} + 2y = t.$$

6. Find the general solution of

$$\frac{d^2y}{dt^2} + 3\frac{dy}{dt} + 2y = A\cos(\omega t + \phi).$$

7. Compute the Wronskians of

   (a) $\{e^t\}$.

   (b) $\{e^t, e^{2t}\}$.

   (c) $\{e^t, e^{2t}, e^{3t}\}$.

   (d) $\{e^t, \cos(t)\}$.

8. Find the pair of solutions $\{y_1(t), y_2(t)\}$ of

$$\frac{d^2y}{dt^2} + 7\frac{dy}{dt} + 12y\ 0$$

satisfying the initial conditions

$$\begin{aligned} y_1(0) &= 1, \\ \frac{dy_1}{dt}(0) &= 0, \\ y_2(0) &= 0, \\ \frac{dy_2}{dt}(0) &= 1. \end{aligned}$$

9. Find the (unique) solution of

$$\frac{d^2y}{dt^2} + 7\frac{dy}{dt} + 12y\ 0$$

meeting the initial conditions

$$\begin{aligned} y(0) &= 2, \\ \frac{dy}{dt}(0) &= -2. \end{aligned}$$

10. Find the general solution of

$$\frac{d^2x}{dt^2} + \omega_n^2 x = \sin(2t), \ \omega_n \neq \pm 2.$$

11. Find the general solution of

$$\frac{d^2x}{dt^2} + \omega_n^2 x = \sin(\omega t), \ \ \omega_n \neq \pm\omega.$$

12. Find the general solution of

$$\frac{d^2x}{dt^2} + \omega_n^2 x = \sin(\omega_n t).$$

13. Can you derive the answer to Problem 12 by computing a $\lim_{\omega \to \omega_n}$ of the answer you obtained in Problem 11? Pick a convenient numerical value for ω_n, and use Maple to plot a sequence of solutions as $\omega \to \omega_n$.

14. The solution to

$$\frac{dx}{dt} = -x(t) + \sin(\omega t), \ \ x(0) = 1$$

consists of a transient term which tends to zero as $t \to \infty$ plus a sinusoidal oscillation whose amplitude and phase shift differ from that of the forcing term. Calculate the amplitude and phase of the response as a function of the forcing frequency ω.

15. An n-th order linear constant coefficient homogeneous differential equation looks like

$$a_n \frac{d^n x}{dt^n} + a_{n-1} \frac{d^{n-1}x}{dt^{n-1}} + \ldots + a_0 x(t) = 0.$$

Under what circumstances is $x(t) = Cexp(\lambda t)$ a solution of that equation?

16. An n-th order linear constant coefficient homogeneous difference equation is something that you presumably have never seen before. The unknowns form a sequence of values $\{x_k\}$ that satisfy the linear equation

$$a_n\, x_{k+n} + a_{n-1}\, x_{k+n-1} + \ldots + a_0\, x_k = 0,$$

where k runs through the integers ≥ 0. Under what circumstances is $x_k = C\rho^k$ a solution of that equation? The $a_0, a_1, \ldots, a_n$ are constants that define the problem.

17. Find the general solution of

$$\frac{d^2y}{dt^2} + 4\frac{dy}{dt} + y = 0.$$

18. Find the general solution of

$$\frac{dy}{dt} + 5\, y = 0,$$

treating it as a constant coefficient problem.

19. The following is a constant coefficient vector differential equation:

$$\frac{d}{dt}\begin{bmatrix} x_1(t) \\ x_2(t) \end{bmatrix} = \begin{bmatrix} 0 & 1 \\ -4 & -3 \end{bmatrix}\begin{bmatrix} x_1(t) \\ x_2(t) \end{bmatrix}.$$

Under what circumstances does the above have solutions of the form

$$\begin{bmatrix} x_1(t) \\ x_2(t) \end{bmatrix} = \begin{bmatrix} \alpha \\ \beta \end{bmatrix} \cdot e^{\lambda t}$$

for some Greek constants as above?

20. Find the particular and general solutions of

$$\frac{d^2y}{dt^2} + 4\frac{dy}{dt} + y = \sin(\omega t) + e^t.$$

21. Find the particular and general solutions of

$$\frac{dy}{dt} + 5\,y = t^2\,\sin(12t).$$

6

Laplace Transform Methods

Laplace transforms are one of a class of "operational methods" that are useful for solving differential equations, primarily equations with constant coefficients. The hallmark of operational methods is supposed to be that a "hard" problem is replaced by an "easier" equivalent calculation. In the case of differential equations the operational effect is to replace a differential equation problem with an algebraic one.

Laplace transforms are a pervasive tool on a conceptual as well as a computational basis. They provide a shorthand method for describing the behavior of "linear time invariant systems" with the Laplace transform notation. The most pervasive use of this is in the analysis of linear electrical circuits, where transform methods (referred to as *impedance analysis*) almost completely replace time domain descriptions of the problems.

6.1 Basic Definition

Definition

Laplace transforms replace one function with another, in a reversible fashion. The "original functions" are defined on the positive half-line $[0, \infty)$, with the independent variable usually thought of as time, denoted by t. For such a function $f(t)$, the corresponding Laplace transform function $F(s)$ is defined through the formula

$$\mathcal{L}\{f(t)\} = F(s) = \int_0^\infty f(t)\, e^{-st}\, dt.$$

In order for this formula to actually define a function of the variable s, the transformed function $f(t)$ must be such that the integral indicated can actually be evaluated (that is, the integral converges for some values of s). Typical assumptions that serve to guarantee this are to the effect that $f(t)$ is ultimately exponentially bounded, so that

$$|f(t)| \leq M\, e^{K\,t}$$

for some sufficiently generously chosen constants $M,\ K$. The classical first example is the transform of an exponential function, say

$$f(t) = e^{5\,t}.$$

Then

$$\begin{aligned} F(s) &= \int_0^\infty e^{5t}\, e^{-st}\, dt \\ &= \left. \frac{e^{-(s-5)\,t}}{-(s-5)} \right|_0^\infty \\ &= \frac{1}{s-5}, \end{aligned}$$

provided at least that $s > 5$. (Actually, s must be allowed to take complex values, so the condition really is $Re(s) > 5$.)

Not all functions *have* Laplace transforms. For the function

$$f(t) = e^{t^2},$$

the integral

$$\int_0^\infty e^{t^2}\, e^{-st}\, dt$$

diverges no matter what the value of s is, so there is no Laplace transform function to define.

The general scenario for the use of Laplace transform is that the problem under pursuit is converted to a simpler (e.g., algebraic) one in terms of the transformed variables, where it is solved in those terms. What is desired of course is the solution to the problem in terms of the original "time" domain variable, not the " s-domain" version. This means that the transforming process has to be reversed (inverted) to get back to the time domain.

In terms of the operational notation for the operation of taking the Laplace transform, what is required is to compute

$$\mathcal{L}^{-1}\{F(s)\} = f(t).$$

In order for this symbolism to make sense, it has to be known that there is an essentially unique $f(t)$ corresponding to a given $F(s)$. This is actually true, but proper demonstrations of the fact are part of the theory of Fourier transforms and integrals.

There actually are inversion methods and theorems based on complex variable integration results, but they require a knowledge of complex analysis to be used in more than a superficial way. The alternative way to invert the transforms is to rely on the known uniqueness of the process, and build up a repertoire of examples and manipulation rules to allow inversion by brute force recognition (or tables and algebraic manipulation). The examples that can be easily handled this way are coincidentally co-extensive with the forcing functions encountered in computing particular solutions by means of undetermined coefficients. Knowing the corresponding Laplace transforms is actually easier than memorizing the undetermined coefficients rules, and is sufficiently useful to allow one to recreate civilization while stranded on a desert island.

6.1.1 Examples

Example
The constant function 1 has a simple transform.

$$\begin{aligned} F(s) &= \int_0^\infty e^{-st}\,dt \\ &= \left.\frac{e^{-st}}{-s}\right|_0^\infty \\ &= \frac{1}{s}. \end{aligned}$$

The corresponding inversion formula takes the form

$$\mathcal{L}^{-1}\left\{\frac{1}{s}\right\} = 1.$$

Example
The general exponential case can be calculated as well as the numerical example above:

$$\begin{aligned} F(s) &= \int_0^\infty e^{at}\,e^{-st}\,dt \\ &= \left.\frac{e^{-(s-a)t}}{-(s-a)}\right|_0^\infty \\ &= \frac{1}{s-a}. \end{aligned}$$

Inverting the transform

$$\mathcal{L}^{-1}\left\{\frac{1}{s-a}\right\} = e^{at}.$$

Example
Euler's formula allows sines and cosines to be handled from the above result. Since

$$\cos(\omega t) = \frac{e^{i\omega t} + e^{-i\omega t}}{2}.$$

we get the transform of the cosine as

$$\begin{aligned}
\mathcal{L}\{\cos(t)\} &= \int_0^\infty \left(\frac{e^{i\omega t} + e^{-i\omega t}}{2}\right) dt \\
&= \frac{1}{2}\left(\frac{1}{s - i\omega} + \frac{1}{s + i\omega}\right) \\
&= \frac{s}{s^2 + \omega^2}.
\end{aligned}$$

Inverting,

$$\mathcal{L}^{-1}\left\{\frac{s}{s^2 + \omega^2}\right\} = cos(\omega t)$$

Example
The sine transform is computed in the same way, with the result that

$$\begin{aligned}
\mathcal{L}\{\sin(t)\} &= \int_0^\infty \left(\frac{e^{i\omega t} - e^{-i\omega t}}{2i}\right) dt \\
&= \frac{1}{2i}\left(\frac{1}{s - i\omega} - \frac{1}{s + i\omega}\right) \\
&= \frac{\omega}{s^2 + \omega^2},
\end{aligned}$$

while

$$\mathcal{L}^{-1}\left\{\frac{\omega}{s^2 + \omega^2}\right\} = \sin(\omega t).$$

Example
Until some of the rules that allow computation of new transforms from old are established, the only way to transform the "ramp" function $f(t) = t$ is to use integration by parts

$$\begin{aligned}
\mathcal{L}\{t\} &= \int_0^\infty t\, e^{-st}\, dt \\
&= \left.\frac{(-s\,t - 1)\, e^{-st}}{(-s)^2}\right|_0^\infty \\
&= \frac{1}{s^2} \\
\mathcal{L}^{-1}\left\{\frac{1}{s^2}\right\} &= t.
\end{aligned}$$

Example
With perseverance, the function $f(t) = t^2$ can be transformed to produce

$$\mathcal{L}\{t^2\} = \frac{2}{s^3}.$$

Before getting carried away with these calculations, it is better to know a few of the manipulation rules for Laplace transforms. It is arguable that the transform of $f(t) = 1$ is the only "hand calculation" that needs to be done for the above examples. All of the other formulas follow from applying "rules" to the transform $\mathcal{L}\{1\}$.

6.2 New Wine From Old

The fact that the Laplace transform involves an exponential function leads to a number of manipulation rules. One is the effect of multiplying a given function by an exponential one. By direct calculation

$$\begin{aligned} \mathcal{L}\{f(t)\, e^{at}\} &= \int_0^\infty f(t)\, e^{at}\, e^{-st}\, dt \\ &= \int_0^\infty f(t)\, e^{(-s+a)\,t}\, dt \\ &= F(s-a), \end{aligned}$$

leading to the exponential shift rule written as

$$\mathcal{L}\{f(t)\, e^{at}\}(s) = \mathcal{L}\{f(t)\}(s-a).$$

Note that the parentheses denote function evaluation at the indicated argument, not a multiplicative factor. Perhaps the result might be written as

$$\mathcal{L}\{f(t)\, e^{at}\}(s) = \mathcal{L}\{f(t)\}|_{s=s-a}$$

to indicate that the original transform argument is replaced by $s - a$.

Example

$$\mathcal{L}\{t\, e^{at}\} = \frac{1}{(s-a)^2}.$$

For that matter,

Example

$$\mathcal{L}\{t^2\, e^{at}\} = \frac{2}{(s-a)^3}.$$

Generally speaking, differentiation of an integral under under the integral sign is an operation that is subject to restricted validity, but the restriction generally amounts to verifying that the integral obtained converges. If the integral

$$\mathcal{L}\{f(t)\}(s) = \int_0^\infty f(t)\, e^{-st}\, dt$$

converges, then an extra t in the integrand can be compensated with a larger value of the real part of s. This line of reasoning leads to the rule for t multiplication

$$\begin{aligned} \mathcal{L}\{t\, f(t)\}(s) &= \int_0^\infty t\, f(t)\, e^{-st}\, dt \\ &= \int_0^\infty -\frac{d}{ds}\left(f(t)\, e^{-st}\right) dt \\ &= -\frac{d}{ds}\mathcal{L}\{f(t)\}(s). \end{aligned}$$

Example

This rule allows the transforms of the powers of t to all be derived from the transform of 1:

$$\begin{aligned} \mathcal{L}\{t\} &= -\frac{d}{ds}\frac{1}{s} \\ &= \frac{1}{s^2}, \\ \mathcal{L}\{t^2\} &= -\frac{d}{ds}\frac{1}{s^2} \\ &= \frac{2}{s^3}, \\ \mathcal{L}\{t^3\} &= -\frac{d}{ds}\frac{2}{s^3} \\ &= \frac{3 \cdot 2 \cdot 1}{s^4}, \\ \mathcal{L}\{t^n\} &= \frac{n!}{s^{n+1}}. \end{aligned}$$

Finally, computing

$$\mathcal{L}\{t\, \sin(\omega\, t)\} = -\frac{d}{ds}\frac{\omega}{s^2 + \omega^2} = \frac{2\, \omega\, s}{(s^2 + \omega^2)^2}$$

is a much more pleasant task than explicitly evaluating the integral by parts several times.

Many problems of interest involve forcing functions that are defined in piecemeal fashion, with one recipe for one part of the time domain and another on the remainder. An example occurs in calculating the response of a physical system to a pulse input. If the system is a simple RC circuit, then the model is

$$RC\,\frac{d}{dt}v(t) + v(t) = f(t),$$

where

$$f(t) = \begin{cases} 1 & \text{if } 0 < t < T \\ 0 & \text{if } t > T. \end{cases}$$

A function referred to either as a unit step or Heaviside function makes it easy to represent this and similar situations with a formula. The definition of this function follows.

Definition
The *unit step function* (or *Heaviside function*) is defined by

$$U(t) = \begin{cases} 0 & \text{if } t < 0 \\ 1 & \text{if } t \geq 0. \end{cases}$$

Many problems involving piecewise defined functions are conveniently represented by using combinations of delayed unit step functions. Since

$$U(t-T) = \begin{cases} 0 & \text{if } t < T \\ 1 & \text{if } t \geq T, \end{cases}$$

the pulse response model can be written as

$$RC\,\frac{d}{dt}v(t) + v(t) = U(t) - U(t-T).$$

The first step defines the leading edge of the pulse at $t = 0$, and the step at $t = T$ cancels the input for $t > T$.

The Laplace transform of the delayed unit step $U(t-T)$ can be calculated as

$$\begin{aligned} \mathcal{L}\{U(t-T)\} &= \int_0^\infty U(t-T)\,e^{-st}\,dt \\ &= \int_T^\infty U(t-T)\,e^{-st}\,dt \\ &= \frac{e^{-sT}}{s}. \end{aligned}$$

This indicates the delay by T introducing a factor of e^{-sT} in the original transform. It is described in this fashion because the same result holds with functions other than just the unit step. Consider

$$U(t-T)\,f(t-T) = \begin{cases} 0 & \text{if } t < T \\ f(t-T) & \text{if } t \geq T, \end{cases}$$

which should be thought of as a delayed version of $f(t)$. More precisely, it is the function obtained by delaying $f(t)$ by T, and then filling in the initial segment between 0 and T with 0. Note that this region corresponds to "negative f arguments", and since Laplace transforms really are concerned with functions defined on the positive time axis, these values would be "outside of the domain" of f.

The Laplace transform of such a delayed function can be calculated according to the delay law

$$\begin{aligned}
\mathcal{L}\{U(t-T)\, f(t-T)\} &= \int_0^{\infty} U(t-T)\, f(t-T)\, e^{-st}\, dt \\
&= \int_T^{\infty} U(t-T)\, f(t-T)\, e^{-st}\, dt \\
&= \int_T^{\infty} f(t-T)\, e^{-st}\, dt \\
&= e^{-sT} \int_0^{\infty} f(\tau)\, e^{-s\tau}\, d\tau \\
&= e^{-sT}\, \mathcal{L}\{f(t)\}.
\end{aligned}$$

This calculation involves the substitution $\tau = t - T$, with $d\tau = dt$, and the corresponding change of limits.

Example

The Laplace transform of a sinusoid starting "one period late" can be calculated as

$$\mathcal{L}\left\{U\left(t-\frac{2\pi}{\omega}\right)\sin\left(\omega(t-\frac{2\pi}{\omega})\right)\right\} = e^{-s\,\frac{2\pi}{\omega}}\,\frac{\omega}{s^2+\omega^2}.$$

A picture of a delayed damped sinusoid is in Figure 6.1

6.2.1 *Algebra and Tables*

To invert Laplace transforms without recourse to complex variable integrals requires essentially rewriting the transform algebraically until a familiar transform is recognized. To invert transforms using inversion integrals actually amounts to virtually identical calculations. The techniques available for doing this in the case of rational functions of s amount to a combination of completing the square (tuning up small problems) and partial fractions (breaking apart large problems).

Partial fractions can be regarded as a normal form for rewriting rational functions. As such, it finds a use in many contexts beyond Laplace transform inversion. The fundamental facts of partial fractions are a consequence of the theory of Laurent series (basically Taylor series with negative power exponents allowed) for functions of a complex variable. The form of the partial fraction expansion for a proper (higher degree denominator than numerator) rational function depends

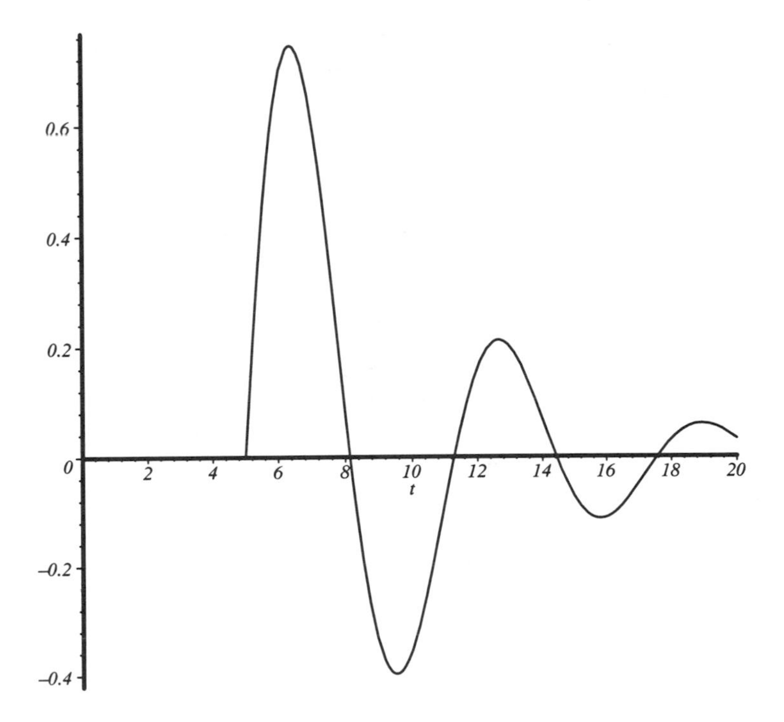

FIGURE 6.1. Delayed damped sinusoid

entirely on the zeroes of the denominator. If

$$R(s) = \frac{q(s)}{p(s)}$$

and

$$p(s) = (s - p_1)^{m_1} (s - p_2)^{m_2} \dots (s - p_k)^{m_k}$$

is the factorization of the denominator into distinct (complex, if need be) factors, then the partial fractions expansion of $R(s)$ will consist of a separate sum of terms, each of these determined by the zeroes of the denominator. Corresponding to the factor

$$(s - p_1)^{m_1}$$

in the denominator, the partial fractions expansion will contain terms of the form

$$\frac{A_{m_1}}{(s - p_1)^{m_1}} + \frac{A_{m_1 - 1}}{(s - p_1)^{m_1 - 1}} + \dots + \frac{A_2}{(s - p_1)^2} + \frac{A_1}{(s - p_1)}.$$

Similar terms must be added for each of the zeroes of the denominator.

The simple case of course is when all of the roots of the denominator polynomial are distinct, so that all of the denominators in the expansion occur to the first power only. In this case the expansion looks like

$$\frac{B_1}{s-p_1}+\frac{B_2}{s-p_2}+\frac{B_3}{s-p_3}+\ldots+\frac{B_n}{s-p_n},$$

appropriate for a rational function of the form

$$R(s)=\frac{q(s)}{(s-p_1)(s-p_2)\ldots(s-p_n)}.$$

It should be mentioned that the above expressions are based on the use of complex numbers in the expansion. In the case where the original problem involves only real coefficients, the complex valued denominator roots will occur in complex conjugate pairs. The expansion coefficients in turn will be complex conjugate in pairs, and the conjugate pairs might be combined to obtain a real quadratic in s in the denominator.

For the purposes of determining the expansion coefficients, it is advantageous to leave the expansion in complex form. This advice is based on use of the Heaviside Cover up Method for computing the coefficients.

Heaviside Cover up Method

Amateur partial fraction operators often attempt to determine the unknown coefficients of a partial fractions expansion by a technique of cross multiplying and attempting to match coefficients in the resulting polynomial equations. Needless to say, this is full of hazard, and leads only to a system of linear equations that must be row reduced to obtain the solution. The system to be solved becomes more difficult with increase in the degrees of the polynomials involved.

The cover up method is most easily understood on the basis of the distinct root case, where the expansion coefficients can be written down directly. If the expansion is

$$\frac{q(s)}{(s-p_1)(s-p_2)\ldots(s-p_n)}=\frac{B_1}{s-p_1}+\frac{B_2}{s-p_2}+\frac{B_3}{s-p_3}+\ldots+\frac{B_n}{s-p_n},$$

then

$$B_1=\frac{q(p_1)}{(p_1-p_2)(p_1-p_3)\ldots(p_1-p_n)}.$$

This is obtained rigorously by multiplying the expansion equation through by $(s-p_1)$, and then substituting $s=p_1$ in what results. The $(s-p_1)$ term cancels from the rational function, while all the terms except the first are left with a factor of $(s-p_1)$, so they drop out when $s=p_1$ is substituted. Clearly this technique will evaluate all of the constants in the expansion with the same minimal effort.

The cover up method gets the name from the fact that calculations are made as though one were "covering up" terms in the expansion involving $(s-p_1)$.

The calculation in that mode is

$$B_1 = \frac{q(s)}{\boxed{(s-p_1)}(s-p_2)\dots(s-p_n)}\Bigg|_{s=p_1} = \frac{q(p_1)}{(p_1-p_2)(p_1-p_3)\dots(p_1-p_n)},$$

$$B_2 = \frac{q(s)}{(s-p_1)\boxed{(s-p_2)}\dots(s-p_n)}\Bigg|_{s=p_2} = \frac{q(p_2)}{(p_2-p_1)(p_2-p_3)\dots(p_2-p_n)},$$

$$\vdots$$

Example

$$\frac{1}{(s+1)(s+2)(s+3)} = \frac{B_1}{s+1} + \frac{B_2}{s+2} + \frac{B_3}{s+3},$$

and

$$\begin{aligned} B_1 &= \frac{1}{(-1+2)(-1+3)} = \tfrac{1}{2} \\ B_2 &= \frac{1}{(-2+1)(-2+3)} = -1 \\ B_3 &- \frac{1}{(-3+1)(-3+2)} = \tfrac{1}{2}. \end{aligned}$$

Actually, in problems like this, the B_j might as well be written down directly along with the right hand side of the expansion.

Example
The forms of a partial fractions expansion are dependent on the use of complex roots for the polynomials involved, which means that the expansion coefficients will generally be complex numbers. For

$$R(s) = \frac{1}{(s+1)(s^2+1)}$$

the denominator roots are $s = 1,\ s = \pm i$, so that the form of the expansion must be

$$\frac{1}{(s+1)(s+i)(s-1)} = \frac{B_1}{s+1} + \frac{B_2}{s+i} + \frac{B_3}{s-i}.$$

By coverup,

$$\begin{aligned} B_1 &= \frac{1}{2}, \\ B_2 &= \frac{1}{-2i(-i+1)}, \\ B_3 &= \frac{1}{2i(i+1)}. \end{aligned}$$

The method is not quite as direct when the zeroes of the denominator are of higher than first order, although it is still far more efficient than multiplying things out. The method is to observe that in

$$\frac{A_{m_1}}{(s-p_1)^{m_1}} + \frac{A_{m_1-1}}{(s-p_1)^{m_1-1}} + \ldots + \frac{A_2}{(s-p_1)^2} + \frac{A_1}{(s-p_1)}.$$

the coefficient A_{m_1} (the coefficient of the highest term) can be directly computed as in the case of the first order factor. *After* this is done, then A_{m_1} is a known quantity, and so it can be moved to the left side of the equation. But then

$$R(s) - \frac{A_{m_1}}{(s-p_1)^{m_1}} = \frac{A_{m_1-1}}{(s-p_1)^{m_1-1}} + \ldots + \frac{A_2}{(s-p_1)^2} + \frac{A_1}{(s-p_1)}$$

presents the same opportunity. A_{m_1-1} can be evaluated by cover up, and the process will continue until all of the expansion coefficients are known.

In truth, the situation is even better than one might think, in that the process is actually self-checking as it proceeds. After the first step

$$R(s) - \frac{A_{m_1}}{(s-p_1)^{m_1}},$$

in spite of appearing to have a factor of $(s-p_1)^{m_1}$ in the denominator, must end up with a clear factor of $(s-p_1)$ in the numerator in order that one factor of $(s-p_1)$ cancels to make the resulting degree in the denominator correct.

Example

$$\frac{1}{(s+1)^2(s+2)(s+3)} = \frac{A}{(s+1)^2} + \frac{B}{s+1} + \frac{C}{s+2} + \frac{D}{s+3}$$

leads to

$$A = \frac{1}{(-1+2)(-1+3)} = \frac{1}{2}.$$

Then

$$\begin{aligned} R(s) - \frac{1}{2(s+1)^2} &= \frac{2}{2(s+1)^2(s+2)(s+3)} - \frac{(s+2)(s+3)}{2(s+1)^2(s+2)(s+3)} \\ &= \frac{2-(s^2+5s+6)}{2(s+1)^2(s+2)(s+3)} \\ &= \frac{-(s+1)(s+4)}{2(s+1)^2(s+2)(s+3)} \\ &= \frac{-(s+4)}{2(s+1)(s+2)(s+3)}. \end{aligned}$$

The remaining coefficients then follow as in a first order case.

6.3 Maple Facilities

Maple contains routines for the manipulation of Laplace transforms. In addition to routines for computing Laplace and inverse Laplace transforms, the general manipulation facilities include a partial fraction calculator. The partial fraction calculations are part of a general Maple facility for conversions. Try

```
> ?convert
```

to see what conversions are available.

See Chapter 14 for further examples of Maple Laplace transform calculations.

Maple reference **Page:** 297

6.4 Derivatives and Laplace

The utility of Laplace transforms in differential equations comes about because of the simple relationships between derivatives and transforms. To obtain the relationship, simply attempt to calculate the transform of a derivative.

$$
\begin{aligned}
\mathcal{L}\{\frac{df}{dt}\} &= \int_0^\infty \frac{df}{dt}\, e^{-st}\, dt \\
&= \lim_{T\to\infty} \int_0^T \frac{df}{dt}\, e^{-st}\, dt \\
&= \lim_{T\to\infty} \left\{ f(t)\, e^{-st}\Big|_0^T - \int_0^T f(t)\,(-s)\, e^{-st}\, dt \right\} \\
&= s \int_0^\infty f(t)\, e^{-st}\, dt - f(0) \\
&= s\, \mathcal{L}\{f(t)\} - f(0).
\end{aligned}
$$

These calculations go through, for instance, assuming that $f(t)$ is exponentially bounded, so that the "boundary terms" from the integration by parts vanish as $T \to \infty$.

Example

Knowing only the rule for Laplace transforming a first derivative is sufficient to solve linear (constant coefficient) first order equations. Since first order equations are de rigueur, this really ought to be all one needs to know. A simple example is the equation

$$\frac{dy}{dt} + 5\, y = 0,$$

together with the initial condition $y(0) = y_0$. Use the notation

$$\mathcal{L}\{y(t)\} = Y(s),$$

and apply the transform to both sides of the above equation. In view of the effect of transforming a derivative, the result is (the s domain equation)

$$s\,Y(s) - y(0) + 5\,Y(s) = 0.$$

Solving for the transform of the solution

$$\begin{aligned} (s+5)\,Y(s) &= y_0, \\ Y(s) &= \frac{y_0}{s+5}, \end{aligned}$$

and inverting the transform gives

$$y(t) = \mathcal{L}^{-1}\left\{\frac{y_0}{s+5}\right\} = y_0\,e^{-5t}.$$

Example
The results we have so far are sufficient to relatively easily solve what is otherwise a somewhat messy differential equation problem. The problem is to calculate the response of a resistor-capacitor-voltage source circuit, given that there is an initial voltage on the capacitor, and that the source driving voltage is a single pulse arriving at some later time. We take the pulse to be of unit ampltude, to begin at time t_0 and end at time $t_1 > t_0$.

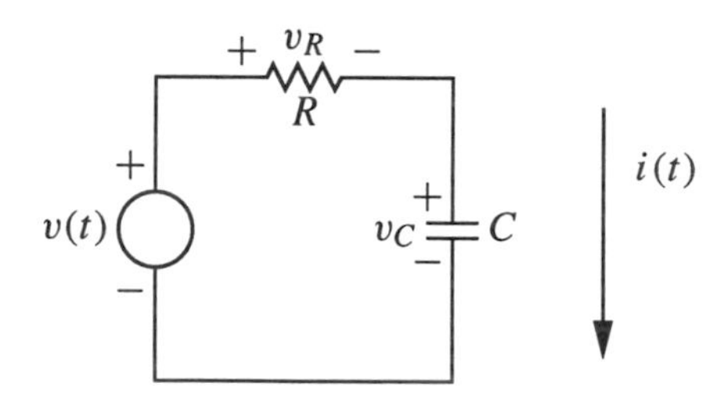

FIGURE 6.2. RC circuit - pulse input

The system is diagrammed in Figure 6.2, and the governing equation takes the form

$$RC\frac{dv_C}{dt} + v_C(t) = v(t) = U(t-t_0) - U(t-t_1),$$

together with the initial equation $v_C(0) = v_0$.

Transforming both sides of the equation gives

$$\begin{aligned}
RC(s\,V_C(s) - v_0) + V_C(s) &= \mathcal{L}\{v(t)\}, \\
&= \frac{e^{-st_0} - e^{-st_1}}{s}, \\
(RC\,s + 1)\,V_C(s) &= RC\,v_0 + \frac{e^{-st_0} - e^{-st_1}}{s}, \\
V_C(s) &= \frac{RC\,v_0}{RCs + 1} + \frac{e^{-st_0} - e^{-st_1}}{s\,(RC\,s + 1)}.
\end{aligned}$$

The transform domain solution can be identified in familiar terms. The first term is due to the initial voltage (initial condition), and corresponds to a solution to the homogeneous version of the equation. The second term has arisen from the transform of the forcing function, and so represents a particular solution to the equation for that pulse shaped input. The terms can be inverted separately, and the results added.

The initial condition response takes the form

$$\begin{aligned}
\mathcal{L}^{-1}\left\{\frac{RC\,v_0}{RCs + 1}\right\} &= \mathcal{L}^{-1}\left\{\frac{v_0}{s + \frac{1}{RC}}\right\} \\
&= v_0\,e^{-\frac{t}{RC}}.
\end{aligned}$$

The forcing term is more complicated on two counts. It involves a partial fractions problem, and there are also exponential delay factors. Partial fractions is a technique for rewriting expressions that are *rational functions* of a variable. The exponential delay factors in the forced part of the answer *do not* participate in the partial fraction rewriting process. The rational function part of the expression is handled separately, and the presence of the exponential delay factors is accommodated after the expansion.

The partial fraction problem underlying the forced response is

$$\begin{aligned}
\frac{1}{s\,(RCs + 1)} &= \frac{1}{RC}\frac{1}{s(s + \frac{1}{RC})} \\
&= \frac{A}{s} + \frac{B}{s + \frac{1}{RC}} \\
&= \frac{1}{s} + \frac{-1}{s + \frac{1}{RC}}.
\end{aligned}$$

The inverse transform of this expression is

$$\mathcal{L}^{-1}\left\{\frac{1}{s\,(RCs + 1)}\right\} = U(t) - e^{-\frac{t}{RC}}\,U(t).$$

(This expression represents an exponential rise toward 1, with $\frac{1}{RC}$ as the time constant.) The forced part of the response has the transform

$$e^{-st_0}\left(\frac{1}{s}+\frac{-1}{s+\frac{1}{RC}}\right)-e^{-st_1}\left(\frac{1}{s}+\frac{-1}{s+\frac{1}{RC}}\right),$$

and so consists of the sum of two delayed versions of this function. That is,

$$U(t-t_0)-e^{-\frac{t-t_0}{RC}}\,U(t-t_0)-\left(U(t-t_1)-e^{-\frac{t-t_1}{RC}}\,U(t-t_1)\right).$$

The complete solution to the problem is the sum of the initial condition response and the particular solution due to the pulse input. That is,

$$v_C(t)=v_0\,e^{-\frac{t}{RC}}+U(t-t_0)-e^{-\frac{t-t_0}{RC}}\,U(t-t_0)-\left(U(t-t_1)-e^{-\frac{t-t_1}{RC}}\,U(t-t_1)\right).$$

6.5 High Order Problems by Laplace

The rules for computing the Laplace transform of derivatives of order higher than one can be obtained in a sequential fashion from the first order result. For the second derivative case,

$$\begin{aligned}\mathcal{L}\left\{\frac{d^2f}{dt^2}\right\} &= s\,\mathcal{L}\left\{\frac{df}{dt}\right\}-\frac{df}{dt}(0)\\ &= s\,(s\mathcal{L}\{f(t)\}-f(0))-\frac{df}{dt}(0)\\ &= s^2\,F(s)-s\,f(0)-\frac{df}{dt}(0).\end{aligned}$$

The transform of a third derivative is similarly

$$\mathcal{L}\left\{\frac{d^3f}{dt^3}\right\}=s^3\,F(s)-s^2\,f(0)-s\,\frac{df}{dt}(0)-\frac{d^2f}{dt^2}(0).$$

From this pattern, the general result

$$\mathcal{L}\left\{\frac{d^nf}{dt^n}\right\}=s^n\,F(s)-s^{n-1}\,f(0)-s^{n-2}\,\frac{df}{dt}(0)-\cdots-\frac{d^{n-1}f}{dt^{n-1}}(0),$$

follows as a conjecture, verifiable through an induction argument if the pattern is not sufficiently convincing.

Example
Transforming the equation

$$\frac{d^2y}{dt^2}+3\,\frac{dy}{dt}+2\,y=0$$

gives

$$
\begin{aligned}
(s^2 + 3s + 2)Y(s) &= (s+3)y(0) + \frac{dy}{dt}(0) \\
Y(s) &= \frac{s+3}{s^2+3s+2}y(0) + \frac{1}{s^2+3s+2}\frac{dy}{dt}(0).
\end{aligned}
$$

Using the partial fraction expansions

$$
\begin{aligned}
\frac{s+3}{s^2+3s+2} &= \frac{2}{s+1} - \frac{1}{s+2}, \\
\frac{1}{s^2+3s+2} &= \frac{1}{s+1} - \frac{1}{s+2},
\end{aligned}
$$

we can write the general solution of the homogeneous equation as

$$
y(t) = \left(2\,e^{-t} - e^{-2t}\right)\, y(0) + \left(e^{-t} - e^{-2t}\right)\frac{dy}{dt}(0).
$$

Example

The damped harmonic oscillator problem serves as a prototype for studies of vibrations of mechanical systems, and as a model of the "ideal system response" in classical control theory. A representation of the general form of the solution to the homogeneous equation

$$
\frac{d^2}{dt^2}y(t) + 2\,\zeta\,\omega_n\frac{d}{dt}y(t) + {\omega_n}^2 y(t) = 0
$$

can be obtained with Laplace transforms and a little algebraic effort.

The transformed form of the equation (using the derivative rules) is

$$
s\,(s\,Y(s) - y(0)) - \frac{dy}{dt}(0) + 2\,\zeta\,\omega_n(s\,Y(s) - y(0)) + {\omega_n}^2\,Y(s) = 0.
$$

Solving this expression for the transform of the solution function gives

$$
Y(s) = -\frac{(-2\,\zeta\,\omega_n - s)y(0)}{s^2 + 2\,\zeta\,\omega_n s + {\omega_n}^2} + \frac{\frac{dy}{dt}(0)}{s^2 + 2\,\zeta\,\omega_n s + {\omega_n}^2}
$$

Inverting this transform by tabular methods requires "completing the square" to put the algebraic expressions in a standard form. Something like

$$
\frac{s}{s^2 + 2\,\zeta\,\omega_n s + {\omega_n}^2}
$$

cannot be inverted by sight. Rather, it must be rewritten in the form

$$
\frac{s}{(s + \zeta\,\omega_n)^2 - \zeta^2\,\omega_n^2 + {\omega_n}^2}
$$

for the denominator terms to be compatible with the exponential shift form. In order to apply this to the inversion, the variable s must be "shifted" in all occurrences in the expression; this requires adding and subtracting $\zeta\,\omega_n$ in the numerator. At this stage the expression is

$$\frac{s + \zeta\,\omega_n - \zeta\,\omega_n}{(s + \zeta\,\omega_n)^2 - \zeta^2\,\omega_n^2 + {\omega_n}^2}.$$

The term

$$\frac{s + \zeta\,\omega_n}{(s + \zeta\,\omega_n)^2 - \zeta^2\,\omega_n^2 + {\omega_n}^2}$$

will invert directly to give

$$e^{-\zeta\,\omega_n\,t}\,\cos(\sqrt{1 - \zeta^2}\,\omega_n\,t).$$

The other term nearly inverts to the corresponding sine, but must be adjusted to account for the fact that the "natural frequency" associated with the expression is $\sqrt{1 - \zeta^2}\,\omega_n$, and this must appear in the numerator to have the transform of the sine. Chasing through all these algebraic annoyances, the final solution inversion produces

$$\frac{\zeta\,y(0)e^{(-\zeta\,\omega_n t)}\sin\left(\sqrt{1 - \zeta^2}\omega_n t\right)}{\sqrt{1 - \zeta^2}} + y(0)e^{(-\zeta\,\omega_n t)}\cos\left(\sqrt{1 - \zeta^2}\omega_n t\right)$$
$$+ \frac{\frac{dy}{dt}(0)e^{(-\zeta\,\omega_n t)}\sin\left(\sqrt{1 - \zeta^2}\omega_n t\right)}{\sqrt{1 - \zeta^2}\omega_n}.$$

6.5.1 *Fundamental Solutions*

The form of answer that emerges from a Laplace transform solution of a constant coefficient differential equation problem allows easy identification of a fundamental set of solutions for the homogeneous equation.

The general form of a constant coefficient homogeneous equation is

$$a_n\,\frac{d^n y}{dt^n} + a_{n-1}\,\frac{d^{n-1} y}{dt^{n-1}} + a_{n-2}\,\frac{d^{n-2} y}{dt^{n-2}} + \ldots + a_0\,y(t) = 0.$$

If this is solved by Laplace transforms, the derivative rule ensures that what results will be of the form

$$(a_n\,s^n + a_{n-1}\,s^{n-1} + \ldots + a_1\,s + a_0)\,Y(s) = MESS(s),$$

where the term on the right hand side of the equation is a polynomial (of degree at most $n - 1$) in the variable s. This term arises from the (sum of) initial condition terms which "drop out" when the derivatives are transformed. The form of the these contributions is such that all expressions present are *linear* in the equation initial conditions. The term on the right hand side of the above equation really takes the form

$$MESS(s) = y(0)\, q_0(s) + \frac{dy}{dt}(0)\, q_1(s) + \ldots + \frac{d^{n-1}}{dt^{n-1}}(0)\, q_{n-1}(s).$$

The transform of the general homogeneous solution is hence

$$Y(s) = y(0)\, \frac{q_0(s)}{p(s)} + \frac{dy}{dt}(0)\, \frac{q_1(s)}{p(s)} + \ldots + \frac{d^{n-1}}{dt^{n-1}}(0)\, \frac{q_{n-1}(s)}{p(s)},$$

where the denominator polynomial is the characteristic polynomial of the equation.

$$p(s) = a_n\, s^n + a_{n-1}\, s^{n-1} + \ldots + a_1\, s + a_0.$$

Recall that a basis for the solution space of the homogeneous equation can be obtained by selecting a set of solutions that have one initial derivative equal to 1, and the rest zero. From the form of the solution above, the Laplace transform representation of these special solutions can be read directly from the expansion above. The basis consists of each of the terms in the solution equation; the solution formula above amounts to the solution expanded in the set of basis functions we are looking for, expressed in the Laplace transform domain. To get the "time domain" version of the solutions, only a transform inversion is required. If the fundamental set is denoted by $\{y_1(t), y_2(t), \ldots, y_n(t)\}$, then

$$\begin{aligned}
y_1(t) &= \mathcal{L}^{-1}\{\frac{q_0(s)}{p(s)}\}, \\
y_2(t) &= \mathcal{L}^{-1}\{\frac{q_1(s)}{p(s)}\}, \\
\vdots &= \vdots \\
y_n(t) &= \mathcal{L}^{-1}\{\frac{q_{n-1}(s)}{p(s)}\}.
\end{aligned}$$

Example
If the differential equation in question is

$$\frac{d^2 y}{dt^2} + 3\, \frac{dy}{dt} + 2\, y = 0,$$

then (as above) solving for the solution transform in terms of the initial condition parameters gives

$$Y(s) = \frac{s+3}{s^2+3s+2}\, y(0) + \frac{1}{s^2+3s+2}\, \frac{dy}{dt}(0).$$

Using the partial fraction expansions

$$\frac{s+3}{s^2+3s+2} = \frac{2}{s+1} - \frac{1}{s+2},$$
$$\frac{1}{s^2+3s+2} = \frac{1}{s+1} - \frac{1}{s+2},$$

identifies the fundamental set components as

$$y_1(t) = \mathcal{L}^{-1}\left\{\frac{s+3}{s^2+3s+2}\right\} = \left(2e^{-t} - e^{-2t}\right),$$
$$y_2(t) = \mathcal{L}^{-1}\left\{\frac{1}{s^2+3s+2}\right\} = \left(e^{-t} - e^{-2t}\right).$$

Example
If the order of the equation is high, the process of rearranging terms as suggested above is laborious, and better left to `Maple`. Examples of these problems done in that form appear in Chapter 14.

Maple reference **Page: 299**

The calculations done previously also allow one to read off the basis for the case of the general harmonic oscillator. Since the transformed form of the solution in that case was

$$Y(s) = -\frac{(-2\zeta\omega_n - s)y(0)}{s^2 + 2\zeta\omega_n s + \omega_n{}^2} + \frac{\frac{d}{dt}(y)(0)}{s^2 + 2\zeta\omega_n s + \omega_n{}^2},$$

the fundamental set takes the form

$$y_1(t) = \frac{\zeta e^{(-\zeta\omega_n t)} sin\left(\sqrt{1-\zeta^2}\omega_n t\right)}{\sqrt{1-\zeta^2}} + e^{(-\zeta\omega_n t)} cos\left(\sqrt{1-\zeta^2}\omega_n t\right),$$
$$y_2(t) = \frac{e^{(-\zeta\omega_n t)} sin\left(\sqrt{1-\zeta^2}\omega_n t\right)}{\sqrt{1-\zeta^2}\omega_n}.$$

6.5.2 *Undetermined Coefficients Redux*

The mysteries of undetermined coefficient rules can be easily understood on the basis of partial fractions. Recall that undetermined coefficients amounts to a method for guessing the form of a particular solution to an equation of the form

$$a_n \frac{d^n y}{dt^n} + a_{n-1}\frac{d^{n-1}y}{dt^{n-1}} + a_{n-2}\frac{d^{n-2}y}{dt^{n-2}} + \ldots + a_0\, y(t) = f(t).$$

Undetermined coefficient rules do not apply to arbitrary forcing functions, but rather to sums and products of polynomials and exponentials. That is, the forcing functions considered are those whose Laplace transform is a proper rational function of the transform variable s:

$$\mathcal{L}\{f(t)\} = R_f(s) = \frac{q_f(s)}{p_f(s)}.$$

Particular solutions of linear equations are not unique, since they are ambiguous to the extent of an arbitrary added solution of the homogeneous equation. If we aim to compute a particular solution, the calculations might as well be as simple as possible. If we look for a solution of

$$a_n \frac{d^n y}{dt^n} + a_{n-1} \frac{d^{n-1} y}{dt^{n-1}} + a_{n-2} \frac{d^{n-2} y}{dt^{n-2}} + \ldots + a_0\, y(t) = f(t)$$

satisfying the conditions that

$$y(0) = \frac{dy}{dt}(0) = \ldots = \frac{d^{n-1} y}{dt^{n-1}}(0) = 0,$$

then initial condition terms will be absent from the Laplace transformed version of the problem, so it will have the form

$$p(s)\, Y(s) = \frac{q_f(s)}{p_f(s)},$$

where $p(s)$ is the characteristic polynomial,

$$p(s) = a_n\, s^n + a_{n-1}\, s^{n-1} + \ldots + a_1\, s + a_0.$$

The transform of the particular solution then is

$$Y(s) = \frac{1}{p(s)} \frac{q_f(s)}{p_f(s)}.$$

This is the key to the undetermined coefficient rules. Since

$$Y(s) = \frac{q_f(s)}{p(s)\, p_f(s)},$$

the form of the solution is determined jointly by the zeroes of the characteristic polynomial $p(s)$ and the zeroes of the forcing transform denominator $p_f(s)$. The zeroes of $p_f(s)$ are the time constants (exponents of the exponential factors) from the forcing function, while the zeroes of $p(s)$ are the time constants of the solutions of the homogeneous equation.

The simple situation is the case in which $p(s)$ and $p_f(s)$ have no common roots. Then a partial fractions expansion of $Y(s)$ will simply contain a sum of the terms corresponding to factors in $p(s)$ and $p_f(s)$. In particular, the highest order denominator in the terms due to zeroes of $p_f(s)$ will be the order of the factor in $p_f(s)$ itself.

Example
If the equation is

$$\frac{d^2y}{dt^2} + 2\frac{dy}{dt} + y = t\,e^{-2t},$$

then the characteristic equation is

$$p(s) = s^2 + 2s + 1 = (s+1)^2.$$

Since

$$\mathcal{L}\{t\,e^{-2t}\} = \frac{1}{(s+2)^2},$$

the denominator of the forcing function transform is

$$p_f(s) = (s+2)^2.$$

The particular solution described above is of the form

$$Y(s) = \frac{A}{(s+1)^2} + \frac{B}{s+1} + \frac{C}{(s+2)^2} + \frac{D}{s+2}.$$

The first two terms of this are transforms of solutions of the homogeneous equation, and may as well be ignored for the purposes of computing a particular solution. The last two terms arise from the forcing function. Two terms arise because partial fractions requires all descending powers to be present. This effect will occur in any similar problem, and accounts for the "polynomial" entries in an undetermined coefficient chart. The transform of the particular solution thus can be taken to have the form

$$\frac{C}{(s+2)^2} + \frac{D}{s+2}.$$

The more complicated cases arise when the polynomials $p(s)$ and $p_f(s)$ have common factors. Then the solution rational function

$$Y(s) = \frac{q_f(s)}{p(s)\,p_f(s)}$$

will have a higher order zero in the denominator than would be expected from looking at the forcing function transform alone. As a consequence, the solution partial fraction expansion will contain terms starting down in order from the degree of the factor in the product $p(s)\,p_f(s)$. This is the "rule" that applies to undetermined coefficients when the forcing function contains terms which are solutions of the homogeneous equation.

Example
If the equation is

$$\frac{d^2y}{dt^2} + 2\frac{dy}{dt} + y = t\,e^{-t} + e^{-2t},$$

then

$$p(s) = s^2 + 2s + 1 = (s+1)^2,$$

and

$$p_f(s) = (s+1)^2\,(s+2)$$

once the forcing function transform is put over a common denominator. The solution denominator will therefore contain a factor of $(s+1)^4$, and the partial fractions expansion form will be

$$Y(s) = \frac{A}{(s+1)^4} + \frac{B}{(s+1)^3} + \frac{C}{(s+1)^2} + \frac{D}{(s+1)} + \frac{E}{(s+2)}.$$

The form of this in the time domain is

$$y(t) = (a_3t^3 + a_2t^2 + a_1t + a_0)\,e^{-t} + b\,e^{-2t},$$

illustrating that the order of the polynomial associated with the e^{-t} terms is 2 higher than the degree associated with the forcing function. The cause of this is the fact that the characteristic polynomial has a double root with the same time constant as that of the forcing function.

Convince yourself that if you are only interested in a particular solution of the problem, then you might as well drop the C and D terms (equivalently the a_0 and a_1 in the time domain form). The reasoning is that the form above corresponds to a particular solution "cooked" to have zero initial conditions, while the terms that are proposed for dropping are solutions of the homogeneous equation.

6.6 Convolutions

Convolutions are in essence the mathematical representation of the response of a linear time invariant system to an input. It is possible to "invent" the representation from the above description provided that one is prepared to contemplate more or less abstract linear input-output systems. As an introduction, it is probably better to start from simple examples.

We have a representation in hand for the general solution of a first order (scalar) linear differential equation. The linear equation has the form

$$\frac{dy}{dt} + a(t)\,y(t) = f(t).$$

To make contact with time invariant systems and Laplace transforms, this has to be specialized to remove the time variation in the coefficient, leaving the model as

$$\frac{dy}{dt} + a\,y(t) = f(t).$$

This simple problem can be solved by two distinct methods now at our disposal. In the first case, since it is a special case of a linear first order problem, that solution method can be applied. What results is the formula

$$y(t) = y(0)\,e^{-at} + \int_0^t e^{-a\,(t-\tau)}\,f(\tau)\,d\tau,$$

since the integrating factor for the problem is e^{at}. This represents the general solution as the sum of an initial condition response, and a *forced response* due to the *forcing function* $f(t)$. The integral represents the forced response as a "weighted combination" of the "past" values of the forcing function. The fact that the original problem model is *time invariant* is reflected in the fact that the weighting function $e^{-a(t-\tau)}$ is a function only of the difference between the current time (t) and the time of occurrence of the input (τ). The limits of integration run $0 \to t$ to ensure that the *current* response depends only on the *past* input values, and reflect what is called the *causality* of the system being modeled: the system does not jump before it is kicked.

The integral in the convolution above involves the product of the forcing function, and a reflection of an exponential about the axis $\tau = t$. The time variation in the "answer" arises from the sliding of the exponential weighting function past the graph of the forcing function. This geometrical interpretation of the convolution is illustrated in Figure 6.3. The figure corresponds to a damped cosine forcing function with an exponential function as weighting pattern.

While the above may seem a lot to read into such a simple system, the characteristics are shared by more complicated linear time invariant systems where the forced part of the response takes the form

$$\int_0^t g(t-\tau)\,f(\tau)\,d\tau.$$

The second approach that might be taken to the solution of the first order time invariant problem is to use Laplace transforms. If the original equation is transformed, what results is

$$(s+a)\,Y(s) \quad = \quad y(0) + F(s),\, Y(s) = \frac{y(0)}{s+a} + \frac{1}{s+a}\,F(s).$$

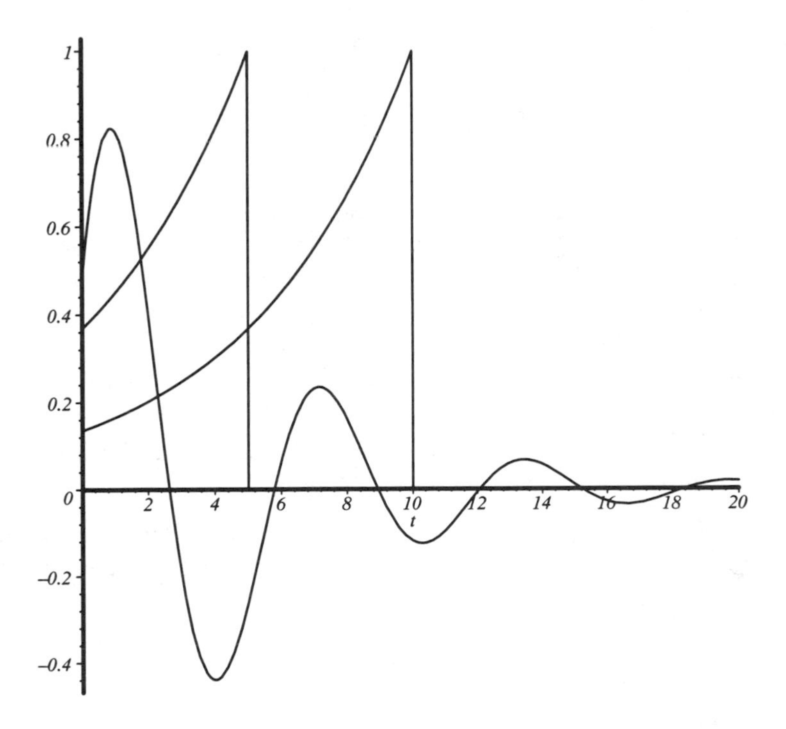

FIGURE 6.3. Convolution integral integrands

If the solution is completed by inverse transforming, then we obtain

$$\begin{aligned} y(t) &= \mathcal{L}^{-1}\{Y(s)\}, \\ &= \mathcal{L}^{-1}\left\{\frac{y(0)}{s+a}\right\} + \mathcal{L}^{-1}\left\{\frac{1}{s+a}\,F(s)\right\} \\ &= y(0)\,e^{-at} + \mathcal{L}^{-1}\left\{\frac{1}{s+a}\,F(s)\right\}. \end{aligned}$$

Matching the two solution representations, we conclude that

$$\mathcal{L}^{-1}\left\{\frac{1}{s+a}\,F(s)\right\} = \int_0^t e^{-a(t-\tau)}\,f(\tau)\,d\tau.$$

What is striking about the above formula is that it somehow manages to invert a Laplace transform without ever explicitly knowing what the forcing function $f(t)$ (or its transform $F(s)$) really might be.

Definition
The expression

$$\int_0^t g(t-\tau)\, f(\tau)\, d\tau$$

is referred to as the *convolution* of the functions $g(t)$ and $f(t)$. It ought to be thought of as a sort of complicated way of multiplying two functions to produce a third. This idea leads to the use of a multiplication-like notation for the operation of convolving two functions. The notation is

$$(g * f)(t) = \int_0^t g(t-\tau)\, f(\tau)\, d\tau,$$

which can be read as " the value at time t of the convolution between g and f is given by the integral ... ".

It turns out that convolutions have a much simpler representation in the Laplace domain than they do in the time domain. To see this, attempt to compute the transform of a general convolution. This takes the form of

$$\mathcal{L}\{(g * f)(t)\} = \int_0^\infty \left(\int_0^t g(t-\tau)\, f(\tau)\, d\tau \right) e^{-st}\, dt.$$

This can be evaluated by viewing it as an iterated integral, and exchanging the order of integration (see Figure 6.4).

If the t integration is performed first, the integral is

$$\int_0^\infty \left(\int_\tau^\infty g(t-\tau)\, f(\tau)\, e^{-st}\, dt \right) d\tau.$$

This simplifies if a change of variable of integration is made in the inside integral. The substitution is

$$u = t - \tau,$$

which causes the u limits to become $0 \to \infty$. In these terms

$$\begin{aligned}
\mathcal{L}\{(g * f)(t)\} &= \int_0^\infty \left(\int_0^\infty g(u)\, f(\tau)\, e^{-s(u+\tau)}\, du \right) d\tau \\
&= \left(\int_0^\infty g(u)\, e^{-su}\, du \right) \left(\int_0^\infty f(\tau)\, e^{-s\tau}\, d\tau \right) \\
&= \mathcal{L}\{g(t)\} \cdot \mathcal{L}\{f(t)\}.
\end{aligned}$$

This result is referred to as the *Convolution Theorem*. It can be interpreted as saying that convolution really does amount to a multiplication operation, with the multiplication taking place in terms of the s domain quantities.

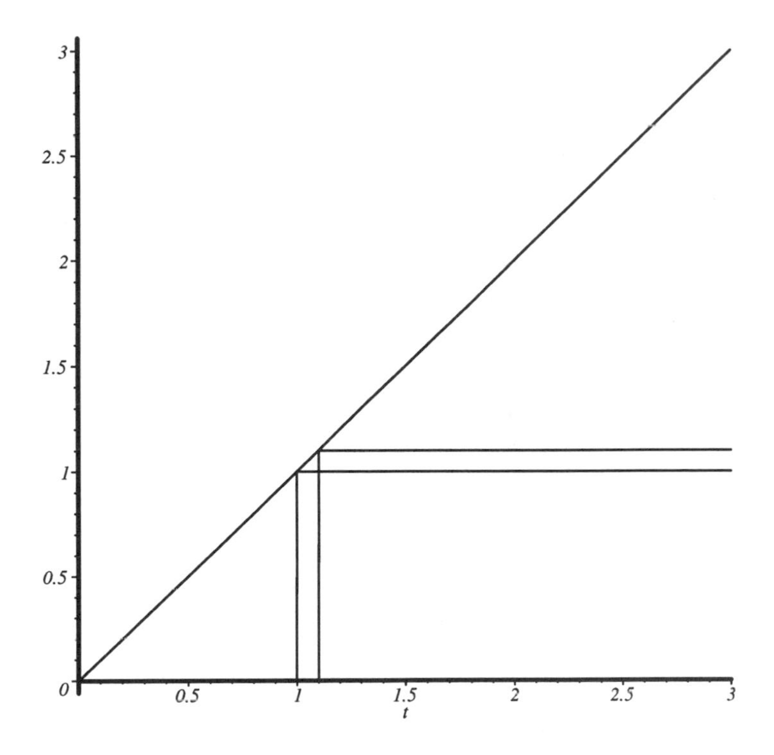

FIGURE 6.4. Exchanging integral order

Example
One could easily argue that the primary use of the convolution theorem is as a conceptual device devoted to the understanding of linear time invariant systems. These are comparatively simple when viewed in terms of transforms instead of in the time domain.

If one wants a simple example of convolution, a pair of functions that results in closed form anti-derivatives is required; the simplest of these is the convolution of a pair of exponentials.

If $g(t) = e^{-at}$, and $f(t) = e^{-bt}$, then the convolution computed directly is

$$
\begin{aligned}
(g * f)(t) &= \int_0^t e^{-a(t-\tau)} e^{-b\tau}\, d\tau, \\
&= e^{-at} \int_0^t e^{(a-b)\tau}\, d\tau, \\
&= \frac{e^{-bt} - e^{-at}}{a-b}.
\end{aligned}
$$

The product of the transforms can be expanded in partial fractions as

$$\begin{aligned}\frac{1}{(s+a)(s+b)} &= \frac{A}{s+a} + \frac{B}{s+b} \\ &= \frac{1}{b-a}\frac{1}{s+a} + \frac{1}{a-b}\frac{1}{s+b} \\ &= \frac{1}{a-b}\left(\frac{1}{s+b} - \frac{1}{s+a}\right).\end{aligned}$$

The inverse transform of this last expression is easily seen to be the same result as obtained by direct calculation of the convolution, so that a contradiction of the Convolution Theorem is averted.

6.6.1 *Inhomogeneous Problems*

In a certain sense there is an appalling lack of variety among the particular solutions of a forced, linear, constant coefficient, ordinary differential equation. Because such an equation is of the form

$$a_n \frac{d^n y}{dt^n} + a_{n-1} \frac{d^{n-1} y}{dt^{n-1}} + a_{n-2} \frac{d^{n-2} y}{dt^{n-2}} + \ldots + a_0\, y(t) = f(t),$$

it is in principle easily solvable by use of Laplace transforms. Simply transforming both sides will produce

$$p(s)\, Y(s) = MESS(s) + F(s),$$

where $p(s)$ is the characteristic polynomial,

$$p(s) = a_n\, s^n + a_{n-1}\, s^{n-1} + \ldots + a_1\, s + a_0,$$

$Y(s)$ represents the transform of the solution, while

$$F(s) = \mathcal{L}\{f(t)\}$$

is the transform of the forcing function. As above, $MESS(s)$ represents an initial condition polynomial, and when divided by the characteristic polynomial is the Laplace transform of a general solution of the associated homogeneous equation.

The forced part of the solution is obtained by setting $MESS(s) = 0$ and solving for $Y(s)$. This gives (the transform of) a particular solution corresponding to the solution of the inhomogeneous equation with all of the initial values $y(0), \frac{dy}{dt}(0), \ldots, \frac{d^{n-1}y}{st^{n-1}}$ set to 0. This is

$$Y(s) = \frac{1}{p(s)}\, F(s).$$

Since this is a product in the transform domain, the solution back in the time domain is a convolution. Hence

$$y(t) = \int_0^t g(t-\tau)\, f(\tau)\, d\tau,$$

where the "weighting function" g is determined from

$$g(t) = \mathcal{L}^{-1}\{\frac{1}{p(s)}\}.$$

Convolutions are therefore a universal recipe for the forced response of linear time-invariant differential equations, and hence of the physical systems (linear circuits, linear mechanical problems, ...) represented by such models.

Example
The forced version of the harmonic oscillator problem is

$$\frac{d^2}{dt^2}y(t) + 2\,\zeta\,\omega_n \frac{d}{dt}y(t) + {\omega_n}^2 y(t) = f(t).$$

This system represents the model for the RLC circuit of Figure 2.18. The variable $y(t)$ in this case is the capacitor voltage.

From the discussion immediately above, the forced part of the solution to the above is a convolution between the forcing function and the weighting function of the system, obtained as the inverse Laplace transform of the inverse of the characteristic polynomial. The weighting function for this problem becomes

$$g(t) = \mathcal{L}^{-1}\{\frac{1}{s^2 + 2\,\zeta\,\omega_n s + {\omega_n}^2}\},$$

which is a scaling and exponential shift away from a sine function. In short,

$$g(t) = \frac{e^{(-\zeta\omega_n t)} \sin\left(\sqrt{1-\zeta^2}\omega_n t\right)}{\sqrt{1-\zeta^2}\omega_n}.$$

The explicit form of the forced response of a generic harmonic oscillator is then

$$y(t) = \int_0^t \frac{e^{(-\zeta\omega_n(t-\tau))} \sin\left(\sqrt{1-\zeta^2}\omega_n (t-\tau)\right)}{\sqrt{1-\zeta^2}\omega_n} f(\tau)\, d\tau,$$

no matter what the forcing function may be.

6.7 Problems

1. Compute Laplace transforms of the following functions:

(a) $\sinh(at) = \frac{e^{at} - e^{-at}}{2}$

(b) $t \cos(\omega t)$

(c) $t^2 e^{bt} \sin(\omega t)$

2. Compute Laplace transforms of the following:

(a) $t^2 \sin(\omega t)$

(b) $t\, e^{-2t} \cos(\omega t)$

(c) $1 + t + \frac{t^2}{2!} + \ldots + \frac{t^n}{n!}$

(d) $t^1 2\, e^{-4t}$

3. Solve

$$\frac{d^2}{dt^2} + 2\frac{dy}{dt} + y = e^{-t}.$$

4. Expand in partial fractions

$$R(s) = \frac{(s+1)(s+2)}{(s+12)(s+13)(s+14)}.$$

5. Expand in partial fractions

$$R(s) = \frac{1}{(s+1)^2\,(s+2)^2\,(s+3)}.$$

6. Expand in partial fractions

$$R(s) = \frac{1}{s^2 + 2\,\zeta\,\omega_n\, s + {\omega_n}^2}.$$

7. The following problem can be solved in a multitude of ways. This time find the general solution of

$$\frac{dy}{dt} = 7y + 1$$

by using Laplace transforms.

8. Solve by Laplace transforms

$$\frac{dy}{dt} + y = \sin(\omega t).$$

9. Use Laplace transforms to solve

$$\frac{d^2y}{dt^2} + 3\frac{dy}{dt} + 2y = e^{-3t}.$$

Is this faster than undetermined coefficients?

10. Use Laplace transforms to solve

$$\frac{d^2y}{dt^2} + 3\frac{dy}{dt} + 2y = t\,e^{-3t}.$$

11. Use the method credited to Laplace to find the general solution of

$$\frac{d^2y}{dt^2} + 6\frac{dy}{dt} + 9y = t\,e^{-3t}.$$

12. For each of the following inhomogeneous problems, determine the form of the particular solution without evaluating any constants involved.

    (a) $\frac{d^2y}{dt^2} + 4\,y = e^{-t}$

    (b) $\frac{d^2y}{dt^2} + 4\,y = t\,e^{-t}$

    (c) $\frac{d^2y}{dt^2} + 4\,y = \sin(t)$

    (d) $\frac{d^2y}{dt^2} + 4\,y = \sin(2t)$

    (e) $\frac{d^2y}{dt^2} + 4\,y = t\,\cos 2t$

13. Find the particular solutions for the above problems by partial fractions methods.

14. Solve by Laplace transforms, assuming that $\omega \neq \pm\omega_n$,

$$\frac{d^2y}{dt^2} + \omega_n^2\,y = \sin(\omega t).$$

15. How much harder is it to solve

$$\frac{d^2y}{dt^2} + \omega_n^2\,y = \sin(\omega_n t)$$

by Laplace transforms? What is the general solution?

16. Can you solve

$$\frac{d^2y}{dt^2} + \omega_n^2\,y = e^{(i\omega_n t)}$$

by Laplace transforms, and then use the phasor insight to find the solution to Problem 15?

17. Solve by Laplace transforms:

$$\frac{d^2y}{dt^2} + 2\,\zeta\,\omega_n\,\frac{dy}{dt} + \omega_n^2\,y = \sin(\omega t).$$

18. Compute the convolution

$$e^{-3t} * e^{-4t}$$

by evaluating the defining integral.

19. Solve Problem 18 by evaluating

$$\mathcal{L}^{-1}\left\{\frac{1}{s+3}\,\frac{1}{s+4}\right\}.$$

20. Compute the convolution

$$e^{-t} * \sin(\omega t)$$

by evaluating the integral. This is an occasion in which Euler's formula comes in really handy.

21. Evaluate the convolution

$$e^{-t} * \sin(\omega t)$$

by using Laplace transforms and partial fractions.

22. Define $pulse(t)$ by

$$pulse(t) = \begin{cases} 0, & \text{if } 0 \le t < t_0 \\ 1, & \text{for } t_0 \le t < t_1 \\ 0, & \text{for } t > t_1. \end{cases}$$

Compute the following two ways: by explicitly evaluating the integrals and by using the Laplace transform convolution theorem.

(a) The convolution $e^{-t} * pulse(t)$.

(b) The convolution $pulse(t) * pulse(t)$.

23. Solve the differential equation problem

$$\frac{d^2y}{dt^2} + 2\,\zeta\,\omega_n\,\frac{dy}{dt} + {\omega_n}^2\,y = pulse(t),$$
$$\frac{dy}{dt}(0) = 0,\ \ y(0) = 0$$

by two methods. Evaluate the solution by Laplace transforms and partial fraction methods, and then use the convolution theorem.

24. Can you obtain the answer to Problem 23 by using undetermined coefficients? This is actually hard to do honestly, and requires thinking about time invariance, as well as the conditions satisfied by solutions calculated by Laplace transform methods.

7
Systems of Equations

The topic of systems of linear time invariant differential equations is entirely intertwined with topics of linear algebra. The issues that come up include change of basis, canonical (standard) forms, eigenvalues and eigenvectors, and functions of matrices. The availability of this array of algebraic machinery allows constant coefficient systems to be "completely analyzed", with their solutions and properties categorized and described.

7.1 Linear Systems

Previous chapters have made a point of including vector formulations of the differential equations encountered. Some problems are naturally formulated as vector equations. Problems in coupled vibrations of structures take the form

$$\frac{d^2}{dt^2}\mathbf{x} = \mathbf{K}\mathbf{x},$$

where $\mathbf{x}$ represents a displacement vector, and $\mathbf{K}$ is referred to as the "stiffness matrix" of the problem. Another class of problems that arrives on the doorstep in vector format arises from models of linear electric circuits. The energy storage variables in such a circuit consist of the inductor currents and capacitor voltages. The equations of motion for such a system take the form

$$\begin{aligned} C_j \frac{d}{dt} v_j &= I_j(t) \\ L_k \frac{d}{dt} i_k &= E_k(t) \end{aligned}$$

where j ranges over the number of capacitors in the circuit, and k runs through the inductors of the arrangement. The I_j and E_k represent the currents and voltages impressed on the storage elements by the rest of the circuit. Provided that the circuit elements are all modeled as linear devices, these (forcing terms from the point of view of the individual capacitors and inductors) are *linear* functions of the other variables and external source functions.

If these equations are written in vector form, the result is

$$\frac{d}{dt}\begin{bmatrix} \mathbf{v} \\ \mathbf{i} \end{bmatrix} = \mathbf{A}\begin{bmatrix} \mathbf{v} \\ \mathbf{i} \end{bmatrix} + \mathbf{f}(t),$$

a vector equation for the voltage-current vector of the circuit.

The usual n-th order linear equation can be written in the form

$$\frac{d}{dt}\begin{bmatrix} y(t) \\ \frac{dy}{dt} \\ \frac{d^2y}{dt^2} \\ \vdots \\ \frac{d^{n-1}y}{dt^{n-1}} \end{bmatrix} = \begin{bmatrix} 0 & 1 & 0 & \cdots & 0 \\ 0 & 0 & 1 & \cdots & 0 \\ \vdots & & & & \vdots \\ -\frac{a_0(t)}{a_n(t)} & -\frac{a_1(t)}{a_n(t)} & \cdots & \cdots & -\frac{a_{n-1}(t)}{a_n(t)} \end{bmatrix} \begin{bmatrix} y(t) \\ \frac{dy}{dt} \\ \frac{d^2y}{dt^2} \\ \vdots \\ \frac{d^{n-1}y}{dt^{n-1}} \end{bmatrix},$$

which in vector notation could be compactly written as

$$\frac{d}{dt}\mathbf{x} = \mathbf{A}(t)\mathbf{x}(t),$$

where the vector quantity corresponds to the vector of successive derivatives of the scalar n-th order equation. If a constant coefficient problem that is inhomogeneous is put in this form, then what results is

$$\frac{d}{dt}\mathbf{x} = \mathbf{A}\mathbf{x}(t) + \mathbf{f}(t),$$

which as a vector equation is indistinguishable from the physical examples above.

7.2 Bases and Eigenvectors

The treatment of constant coefficient systems of differential equations relies on the ability to convert such problems to a standard form through a suitable change of basis. This section reviews the fundamental facts of this process in the context of finite dimensional matrix problems.

Our convention (already adopted in previous examples of systems of differential equations) is to treat vectors as coordinate column arrays. Then linear mappings operate by the usual matrix multiplication from the left side.

In this arena, changes of coordinates are represented by multiplication by a nonsingular matrix; such a matrix has linearly independent columns, and so the columns form a basis for the vector space (either R^n of C^n). Such a change of basis matrix is often denoted by $\mathbf{P}$. It is useful to single out the columns of $\mathbf{P}$ for use as a basis set:

$$\mathbf{P} = \left[\mathbf{p}_1 \,|\, \mathbf{p}_2 \,|\, \dots \,|\, \mathbf{p}_n\right].$$

Such a change of basis matrix is invertible, so for any arbitrary vector $\mathbf{x}$ we have the obvious identity

$$\mathbf{x} = \mathbf{P}\,\mathbf{P}^{-1}\mathbf{x} = \left[\mathbf{p}_1 \,|\, \mathbf{p}_2 \,|\, \dots \,|\, \mathbf{p}_n\right]\left(\mathbf{P}^{-1}\mathbf{x}\right).$$

The combination

$$\left(\mathbf{P}^{-1}\mathbf{x}\right) = \begin{bmatrix} z_1 \\ z_2 \\ \vdots \\ z_n \end{bmatrix}$$

is a column vector multiplied by a matrix, hence a column. When this is multiplied by $\mathbf{P}$, the result is

$$\mathbf{x} = \left[\mathbf{p}_1 \,|\, \mathbf{p}_2 \,|\, \dots \,|\, \mathbf{p}_n\right] \begin{bmatrix} z_1 \\ z_2 \\ \vdots \\ z_n \end{bmatrix} = z_1\,\mathbf{p}_1 + z_2\,\mathbf{p}_2 + \dots + z_n\,\mathbf{p}_n.$$

This shows the arbitrary vector $\mathbf{x}$ expanded as a linear combination of the columns of $\mathbf{P}$. Since the columns form a basis, and expansion coefficients with respect to a basis are unique, the $\mathbf{z}$ vector is a column of the components of the vector $\mathbf{x}$ with respect to the new ($\mathbf{P}$ column) basis. Multiplication by $\mathbf{P}^{-1}$ converts the original to the new coordinates.

A fertile hunting ground for basis vectors is the set of eigenvectors of a matrix. To get a hint that eigenvectors might be involved with solutions of systems of differential equations, suppose that we are confronted with the vector-matrix system

$$\frac{d}{dt}\mathbf{x} = \mathbf{A}\,\mathbf{x},$$

where the coefficient matrix has constant coefficients. Since we are used to finding exponential functions among the solutions of constant coefficient differential equations, we might be tempted to try to find a solution in the form of an unknown vector multiplied by an exponentail function

$$\mathbf{x}(t) = \begin{bmatrix} a_1 \\ a_2 \\ \vdots \\ a_n \end{bmatrix} e^{\lambda t}.$$

Substituting this trial solution in the equation, we see that the trial solution works when

$$\lambda \begin{bmatrix} a_1 \\ a_2 \\ \vdots \\ a_n \end{bmatrix} e^{\lambda t} = \mathbf{A} \begin{bmatrix} a_1 \\ a_2 \\ \vdots \\ a_n \end{bmatrix} e^{\lambda t},$$

or in more familiar form,

$$\mathbf{A} \begin{bmatrix} a_1 \\ a_2 \\ \vdots \\ a_n \end{bmatrix} = \lambda \begin{bmatrix} a_1 \\ a_2 \\ \vdots \\ a_n \end{bmatrix}.$$

The conclusion is that eigenvectors multiplied by exponential functions, with the corresponding eigenvalues as the time constant, solve the vector matrix equation. There are as many such solutions as there are eigenvectors. A systematic investigation of this leads to change of basis problems, and this is explored in the following section.

7.3 Diagonalization

Some vector differential equations are particularly easy to solve. If the equation

$$\frac{d}{dt}\mathbf{x} = \mathbf{A}\mathbf{x}(t)$$

happens to take the form

$$\frac{d}{dt} \begin{bmatrix} x_1(t) \\ x_2(t) \\ \vdots \\ x_n(t) \end{bmatrix} = \begin{bmatrix} \lambda_1 & 0 & \dots & 0 \\ 0 & \lambda_2 & 0\dots & \\ \vdots & \vdots & \ddots & \vdots \\ 0 & 0 & \dots & \lambda_n \end{bmatrix} \begin{bmatrix} x_1(t) \\ x_2(t) \\ \vdots \\ x_n(t) \end{bmatrix},$$

then the equations for the vector components are decoupled: only x_1 affects the rate of change of x_1, and so on. The equations may be integrated separately to give the solution

$$\begin{bmatrix} x_1(t) \\ x_2(t) \\ \vdots \\ x_n(t) \end{bmatrix} = \begin{bmatrix} e^{\lambda_1 t} x_1(0) \\ e^{\lambda_2 t} x_2(0) \\ \vdots \\ e^{\lambda_n t} x_n(0) \end{bmatrix}$$

in short order.

The homogeneous vector-matrix system can be handled virtually as easily, provided that the coefficient matrix is diagonalizable. Suppose that we have

$$\frac{d}{dt}\mathbf{x} = \mathbf{A}\mathbf{x}(t)$$

as a constant coefficient system, and that the coefficient matrix $\mathbf{A}$ can be put into diagonalized form by a similarity transformation. This assumption means that there is an invertible matrix $\mathbf{P}$ such that

$$\mathbf{P}^{-1}\mathbf{A}\mathbf{P} = \begin{bmatrix} \lambda_1 & 0 & \dots & 0 \\ 0 & \lambda_2 & 0\dots & \\ \vdots & \vdots & \ddots & \vdots \\ 0 & 0 & \dots & \lambda_n \end{bmatrix}.$$

Then from

$$\frac{d}{dt}\mathbf{x} = \mathbf{A}\mathbf{x}(t)$$

we obtain

$$\frac{d}{dt}\mathbf{P}^{-1}\mathbf{x} = \mathbf{P}^{-1}\mathbf{A}\mathbf{x}(t),$$

and then ("multiplying by 1")

$$\frac{d}{dt}\mathbf{P}^{-1}\mathbf{x} = \mathbf{P}^{-1}\mathbf{A}\mathbf{P}\mathbf{P}^{-1}\mathbf{x}(t).$$

If we name the vector

$$\mathbf{z} = \mathbf{P}^{-1}\mathbf{x},$$

the above is

$$\begin{aligned} \frac{d}{dt}\mathbf{z} &= \left(\mathbf{P}^{-1}\mathbf{A}\mathbf{P}\right)\mathbf{z}(t) \\ &= \begin{bmatrix} \lambda_1 & 0 & \dots & 0 \\ 0 & \lambda_2 & 0\dots & \\ \vdots & \vdots & \ddots & \vdots \\ 0 & 0 & \dots & \lambda_n \end{bmatrix}\mathbf{z}. \end{aligned}$$

The equation in terms of $\mathbf{z}$ is diagonalized, so the solution in those terms can be written down directly. It is

$$\begin{bmatrix} z_1(t) \\ z_2(t) \\ \vdots \\ z_n(t) \end{bmatrix} = \begin{bmatrix} e^{\lambda_1 t} z_1(0) \\ e^{\lambda_2 t} z_2(0) \\ \vdots \\ e^{\lambda_n t} z_n(0) \end{bmatrix}.$$

Since

$$\mathbf{x} = \mathbf{P}\mathbf{z},$$

the solution in terms of the "original" variables is simply

$$\mathbf{x}(t) = \mathbf{P}\begin{bmatrix} e^{\lambda_1 t} z_1(0) \\ e^{\lambda_2 t} z_2(0) \\ \vdots \\ e^{\lambda_n t} z_n(0) \end{bmatrix}.$$

If a matrix **A** can be diagonalized by a similarity transform as assumed above, then there is a basis (say, for R^n) consisting of eigenvectors of **A**. In terms of the discussion above, the diagonalization equation

$$\mathbf{P}^{-1}\,\mathbf{A}\,\mathbf{P} = \begin{bmatrix} \lambda_1 & 0 & \dots & 0 \\ 0 & \lambda_2 & 0\dots & \\ \vdots & \vdots & \ddots & \vdots \\ 0 & 0 & \dots & \lambda_n \end{bmatrix}.$$

can be written in the form

$$\mathbf{A}\,\mathbf{P} = \mathbf{P}\begin{bmatrix} \lambda_1 & 0 & \dots & 0 \\ 0 & \lambda_2 & 0\dots & \\ \vdots & \vdots & \ddots & \vdots \\ 0 & 0 & \dots & \lambda_n \end{bmatrix}.$$

If **P** is written in terms of its columns, then the above equation takes the form

$$\mathbf{A}\left[\mathbf{p}_1 \,|\, \mathbf{p}_2 \,|\, \dots \,|\, \mathbf{p}_n\right] = \left[\mathbf{p}_1 \,|\, \mathbf{p}_2 \,|\, \dots \,|\, \mathbf{p}_n\right]\begin{bmatrix} \lambda_1 & 0 & \dots & 0 \\ 0 & \lambda_2 & 0\dots & \\ \vdots & \vdots & \ddots & \vdots \\ 0 & 0 & \dots & \lambda_n \end{bmatrix}$$

or

$$\left[\mathbf{A}\mathbf{p}_1 \,|\, \mathbf{A}\mathbf{p}_2 \,|\, \dots \,|\, \mathbf{A}\mathbf{p}_n\right] = \left[\lambda_1\mathbf{p}_1 \,|\, \lambda_2\mathbf{p}_2 \,|\, \dots \,|\, \lambda_n\mathbf{p}_n\right].$$

Matching columns in the matrix equation, this amounts to writing that the columns of **P** are eigenvectors of **A**, with the $\{\lambda_1, \dots, \lambda_n\}$ as eigenvalues. Because the equation

$$\mathbf{x} = \mathbf{P}\mathbf{P}^{-1}\mathbf{x}$$

writes **x** as a linear combination of the columns of **P**, and basis vector expansions are unique, the quantity

$$\mathbf{P}^{-1}\mathbf{x}$$

represents a column vector consisting of the components of $\mathbf{x}$ with respect to the eigenvector basis. Since this is exactly what has been called $\mathbf{z}$ above, the calculations of the $\mathbf{z}$ solution show that it is the equations for the eigenvector components that arc in simple form.

Example
A three dimensional system example is

$$\frac{d}{dt}\mathbf{x} = \begin{bmatrix} 0 & 1 & 0 \\ 0 & 0 & 1 \\ -6 & -11 & -6 \end{bmatrix} \mathbf{x}.$$

This corresponds to the vector representation of the scalar third order equation

$$\frac{d^3y}{dt^3} + 6\frac{d^2y}{dt^2} + 11\frac{dy}{dt} + 6y = 0,$$

but we wish to produce solutions by vector methods.

The first problem is to calculate the eigenvalues of the coefficient matrix. The characteristic equation of the coefficient matrix is

$$\det\left(s\,\mathbf{I} - \begin{bmatrix} 0 & 1 & 0 \\ 0 & 0 & 1 \\ -6 & -11 & -6 \end{bmatrix}\right) = (s+1)(s+2)(s+3),$$

so that the eigenvalues are $\{-1, -2, -3\}$. Since these are distinct, the eigenvectors corresponding to these eigenvalues will provide a basis which diagonalizes the matrix.

The matrix of eigenvectors can be calculated as

$$\mathbf{P} = \begin{bmatrix} 3 & -3 & 1 \\ -3 & 6 & -3 \\ 3 & -12 & 9 \end{bmatrix},$$

and verified by checking that

$$\begin{bmatrix} 0 & 1 & 0 \\ 0 & 0 & 1 \\ -6 & -11 & -6 \end{bmatrix} \begin{bmatrix} 3 & -3 & 1 \\ -3 & 6 & -3 \\ 3 & -12 & 9 \end{bmatrix} = \begin{bmatrix} -3 & 6 & -3 \\ 3 & -12 & 9 \\ -3 & 24 & -27 \end{bmatrix}.$$

The inverse of the eigenvector matrix evaluates to

$$\mathbf{P}^{-1} = \begin{bmatrix} 1 & \frac{5}{6} & \frac{1}{6} \\ 1 & \frac{4}{3} & \frac{1}{3} \\ 1 & \frac{3}{2} & \frac{1}{2} \end{bmatrix},$$

and one can then verify the expected diagonalization

$$\begin{bmatrix} 1 & \frac{5}{6} & \frac{1}{6} \\ 1 & \frac{4}{3} & \frac{1}{3} \\ 1 & \frac{3}{2} & \frac{1}{2} \end{bmatrix} \begin{bmatrix} 0 & 1 & 0 \\ 0 & 0 & 1 \\ -6 & -11 & -6 \end{bmatrix} \begin{bmatrix} 3 & -3 & 1 \\ -3 & 6 & -3 \\ 3 & -12 & 9 \end{bmatrix} = \begin{bmatrix} -1 & 0 & 0 \\ 0 & -2 & 0 \\ 0 & 0 & -3 \end{bmatrix}.$$

Solving the system of differential equations by diagonalization proceeds by introducing the new improved coordinate system with

$$\mathbf{z} = \mathbf{P}^{-1}\mathbf{x}.$$

This turns the original system into

$$\frac{d}{dt}\mathbf{P}^{-1}\mathbf{x} = \mathbf{P}^{-1} \begin{bmatrix} 0 & 1 & 0 \\ 0 & 0 & 1 \\ -6 & -11 & -6 \end{bmatrix} \mathbf{P}\,\mathbf{P}^{-1}\mathbf{x},$$

$$\frac{d}{dt}\mathbf{z} = \begin{bmatrix} -1 & 0 & 0 \\ 0 & -2 & 0 \\ 0 & 0 & -3 \end{bmatrix} \mathbf{z}.$$

In the $\mathbf{z}$ coordinates the equation is readily solved to give

$$\mathbf{z}(t) = \begin{bmatrix} e^{-1t} & 0 & 0 \\ 0 & e^{-2t} & 0 \\ 0 & 0 & e^{-3t} \end{bmatrix} \mathbf{z}(0).$$

To get the solution in terms of the original variables, just introduce the relations between $\mathbf{z}$ and $\mathbf{x}$ into the solution equation. Then

$$\mathbf{x}(t) = \mathbf{P}\mathbf{z}(t) = \mathbf{P} \begin{bmatrix} e^{-1t} & 0 & 0 \\ 0 & e^{-2t} & 0 \\ 0 & 0 & e^{-3t} \end{bmatrix} \mathbf{P}^{-1}\mathbf{x}(0).$$

The temptation to multiply this out is resisted as the resulting formula appears in a later example, obtained by another method.

Maple connection **Page:** 312

7.3.1 *Using Maple*

For systems of size bigger than 2×2, hand calculation of eigenvectors is tedious at best. The linear algebra routines of the Maple `linalg` package include eigenvector routines that are useful for such problems. Examples are provided in the Maple exercises part of this text.

Maple connection **Page:** 307

7.4 Jordan Forms

While things are simple when matrices can be diagonalized, an unfortunate fact is that not all matrices are in this category. Examples that illustrate the difficulty are not difficult to construct. The innocent matrix

$$\mathbf{J} = \begin{bmatrix} 1 & 1 \\ 0 & 1 \end{bmatrix}$$

cannot be put into diagonal form. The problem arises when an attempt to find the eigenvectors is made. The characteristic equation is

$$\det(\lambda \mathbf{I} - \mathbf{J}) = (\lambda - 1)^2,$$

so that there is only a single eigenvalue $\lambda = 1$. The eigenvector equation for $\lambda = 1$ is

$$\begin{bmatrix} 0 & -1 \\ 0 & 0 \end{bmatrix} \begin{bmatrix} x_1 \\ x_2 \end{bmatrix} = \begin{bmatrix} 0 \\ 0 \end{bmatrix}.$$

Unfortunately, this system only produces one eigenvector

$$\mathbf{u_1} = \begin{bmatrix} 1 \\ 0 \end{bmatrix}$$

instead of the required pair. The conclusion is that there does not exist a similarity transformation (change of basis) which will diagonalize the above matrix **J**.

The question which then arises is whether there is another standard form for matrices, which can always be attained by a change of basis. It turns out that there is such a form; the troublesome example matrix **J** turns out to *already* be in the best available standard form, called the Jordan canonical form.

The Jordan canonical form theorem is tricky to prove, but can be readily described. In essence, the content of the theorem is that there exists a change of basis that puts the matrix into *nearly* diagonal form. More precisely, the form consists of a sequence of blocks, each of which is nearly diagonal.

The blocks are in a form called a *basic Jordan block*, consisting of an eigenvalue along the diagonal, the number 1 along the first super diagonal, and zeroes elsewhere. A basic Jordan block associated with the eigenvalue λ_1 is

$$\begin{bmatrix} \lambda_1 & 1 & 0 & \dots \\ 0 & \lambda_1 & 1 & \dots \\ \vdots & & \ddots & \ddots \\ 0 & 0 & \dots & \lambda_1 \end{bmatrix}.$$

The nondiagonalizable example J above is actually in this terminology a 2×2 basic Jordan block associated with the eigenvalue 1.

The Jordan canonical form of a matrix consists of a number of basic Jordan blocks, associated with the same or different eigenvalues, arrayed along the main diagonal of the canonical form matrix. For a case in which there are two eigenvalues, each having only a single Jordan block associated with it, the Jordan canonical form would look like

$$\mathbf{M} = \begin{bmatrix} \begin{bmatrix} \lambda_1 & 1 & 0 & \dots \\ 0 & \lambda_1 & 1 & \dots \\ \vdots & & \ddots & \ddots \\ 0 & 0 & \dots & \lambda_1 \end{bmatrix} & \mathbf{0} \\ \mathbf{0} & \begin{bmatrix} \lambda_2 & 1 & 0 & \dots \\ 0 & \lambda_2 & 1 & \dots \\ \vdots & & \ddots & \ddots \\ 0 & 0 & \dots & \lambda_2 \end{bmatrix} \end{bmatrix}.$$

The importance of this form in the context of systems of differential equations is that it reveals that the vector component differential equations split into two independent systems of equations. If we have the system of differential equations

$$\frac{d}{dt}\mathbf{y} = \mathbf{M}\,\mathbf{y}$$

with $\mathbf{M}$ in the form above, then the independent systems come from dividing $\mathbf{y}$ into into two sub-vectors, with the sizes of the λ_1 and λ_2 Jordan blocks

$$\mathbf{y} = \begin{bmatrix} \mathbf{y_1} \\ \mathbf{y_2} \end{bmatrix}.$$

Because of the zero blocks in the canonical form, the differential equation system is really a pair of equations

$$\begin{aligned} \frac{d}{dt}\mathbf{y_1} &= \begin{bmatrix} \lambda_1 & 1 & 0 & \dots \\ 0 & \lambda_1 & 1 & \dots \\ \vdots & & \ddots & \ddots \\ 0 & 0 & \dots & \lambda_1 \end{bmatrix} \mathbf{y_1}, \\ \frac{d}{dt}\mathbf{y_2} &= \begin{bmatrix} \lambda_2 & 1 & 0 & \dots \\ 0 & \lambda_2 & 1 & \dots \\ \vdots & & \ddots & \ddots \\ 0 & 0 & \dots & \lambda_2 \end{bmatrix} \mathbf{y_2}. \end{aligned}$$

This state of affairs is the true one, generalizing the (generally impossible) case in which the coefficient matrix can be put into diagonal form, in which case the system of equations is equivalent to a set of scalar first order equations. Here the complicated coupled set of original equations breaks into a set of decoupled subsystems, where the coupling within each subsystem is of the special form given by the basic Jordan block.

This splitting into decoupled subsystems continues in the general case. The Jordan form theorem asserts that given a real or complex matrix $\mathbf{A}$, there exists a similarity transformation (i.e., a change of basis) that changes $\mathbf{A}$ into Jordan form: with respect to the new basis the matrix consists of basic Jordan blocks arrayed along the diagonal.

A case with two eigenvalues and three blocks will have the form

$$\begin{bmatrix} \begin{bmatrix} \lambda_1 & 1 & 0 & \cdots \\ 0 & \lambda_1 & 1 & \cdots \\ \vdots & & \ddots & \ddots \\ 0 & 0 & \cdots & \lambda_1 \end{bmatrix} & \mathbf{0} & \mathbf{0} \\ \mathbf{0} & \begin{bmatrix} \lambda_1 & 1 & 0 & \cdots \\ 0 & \lambda_1 & 1 & \cdots \\ \vdots & & \ddots & \ddots \\ 0 & 0 & \cdots & \lambda_1 \end{bmatrix} & \mathbf{0} \\ \mathbf{0} & \mathbf{0} & \begin{bmatrix} \lambda_2 & 1 & 0 & \cdots \\ 0 & \lambda_2 & 1 & \cdots \\ \vdots & & \ddots & \ddots \\ 0 & 0 & \cdots & \lambda_2 \end{bmatrix} \end{bmatrix}.$$

A system of equations with this as coefficient matrix decouples into three independent Jordan block subsystems.

In order to understand the solutions of systems of constant coefficient differential equations, it is only necessary to solve the problem for the case in which the coefficient matrix is a single Jordan block. In view of the Jordan form theorem, the solution of the entire problem will be linear combinations of such solutions, as they arise by "unchanging the basis" back to the original one from the Jordan form basis.

The Jordan system

$$\frac{d}{dt}\mathbf{y} = \begin{bmatrix} \lambda & 1 & 0 & \cdots \\ 0 & \lambda & 1 & \cdots \\ \vdots & & \ddots & \ddots \\ 0 & 0 & \cdots & \lambda \end{bmatrix} \mathbf{y}$$

can be solved most easily by Laplace transforms. Applying the Laplace transform to both sides of the equation, we get

$$s\mathbf{Y}(s) - \mathbf{y}(0) = \begin{bmatrix} \lambda & 1 & 0 & \cdots \\ 0 & \lambda & 1 & \cdots \\ \vdots & & \ddots & \ddots \\ 0 & 0 & \cdots & \lambda \end{bmatrix} \mathbf{Y}(s),$$

$$\left[s\,\mathbf{I} - \begin{bmatrix} \lambda & 1 & 0 & \dots \\ 0 & \lambda & 1 & \dots \\ \vdots & & \ddots & \ddots \\ 0 & 0 & \dots & \lambda \end{bmatrix} \right] \mathbf{Y}(s) = \mathbf{y}(0),$$

$$\mathbf{Y}(s) = \left[s\,\mathbf{I} - \begin{bmatrix} \lambda & 1 & 0 & \dots \\ 0 & \lambda & 1 & \dots \\ \vdots & & \ddots & \ddots \\ 0 & 0 & \dots & \lambda \end{bmatrix} \right]^{-1} \mathbf{y}(0).$$

The simple form of the Jordan block makes it possible to explicitly calculate the inverse above, essentially by a process of "back substitution". The result is

$$\left[s\,\mathbf{I} - \begin{bmatrix} \lambda & 1 & 0 & \dots \\ 0 & \lambda & 1 & \dots \\ \vdots & & \ddots & \ddots \\ 0 & 0 & \dots & \lambda \end{bmatrix} \right]^{-1}$$
$$= \begin{bmatrix} \frac{1}{(s-\lambda)} & \frac{1}{(s-\lambda)^2} & \dots & \frac{1}{(s-\lambda)^{n-1}} & \frac{1}{(s-\lambda)^n} \\ 0 & \frac{1}{(s-\lambda)} & \frac{1}{(s-\lambda)^2} & \dots & \frac{1}{(s-\lambda)^{n-1}} \\ 0 & 0 & \ddots & \dots & \frac{1}{(s-\lambda)^{n-2}} \\ \vdots & \vdots & \ddots & \ddots & \dots \\ 0 & 0 & 0 & \frac{1}{(s-\lambda)} & \frac{1}{(s-\lambda)^2} \\ 0 & 0 & 0 & 0 & \frac{1}{(s-\lambda)} \end{bmatrix}$$

Inverse transforming to get the solution in the time domain, we get

$$\mathbf{y}(t) = \begin{bmatrix} e^{\lambda t} & t\,e^{\lambda t} & \dots & \frac{t^{n-1}}{(n-1)!}\,e^{\lambda t} \\ 0 & e^{\lambda t} & \dots & \frac{t^{n-2}}{(n-2)!}\,e^{\lambda t} \\ \vdots & \ddots & \ddots & \vdots \\ 0 & 0 & 0 & e^{\lambda t} \end{bmatrix} \mathbf{y}(0).$$

Maple connection Page: 310

The conclusion to be drawn from this is concerns the character of all solutions to systems of constant coefficient linear homogeneous equations of the form

$$\frac{d}{dt}\,\mathbf{y}(t) = \mathbf{A}\mathbf{y}(t).$$

Each component of each solution of such an equation consists of a linear combination of exponential functions (including complex exponentials) or polynomial multiples of exponential functions.

The proof of this statement is simply that systems in Jordan form have that character, and returning to the original coordinates produces linear combinations of the Jordan form solutions.

Example

The system to be analyzed is

$$\frac{d}{dt}\mathbf{x} = \mathbf{M}\mathbf{x},$$

where $\mathbf{M}$ is the 8×8 matrix

$$\mathbf{M} = \begin{bmatrix} \frac{-17}{9} & \frac{16}{9} & \frac{-2}{3} & \frac{5}{9} & \frac{-4}{9} & \frac{1}{3} & \frac{-2}{9} & \frac{1}{9} \\ \frac{5}{9} & \frac{-19}{9} & \frac{8}{3} & \frac{-20}{9} & \frac{16}{9} & \frac{-4}{3} & \frac{8}{9} & \frac{-4}{9} \\ \frac{10}{9} & \frac{-20}{9} & \frac{7}{3} & \frac{-31}{9} & \frac{32}{9} & \frac{-8}{3} & \frac{16}{9} & \frac{-8}{9} \\ \frac{8}{9} & \frac{-16}{9} & \frac{8}{3} & \frac{-41}{9} & \frac{40}{9} & \frac{-10}{3} & \frac{20}{9} & \frac{-10}{9} \\ \frac{-1}{3} & \frac{2}{3} & -1 & \frac{4}{3} & \frac{-8}{3} & 3 & \frac{-4}{3} & \frac{2}{3} \\ \frac{-8}{9} & \frac{16}{9} & \frac{-8}{3} & \frac{32}{9} & \frac{40}{9} & \frac{13}{3} & \frac{-2}{9} & \frac{1}{9} \\ \frac{2}{3} & \frac{-4}{3} & 2 & \frac{-8}{3} & \frac{10}{3} & -4 & \frac{20}{3} & \frac{-13}{3} \\ \frac{8}{9} & \frac{-16}{9} & \frac{8}{3} & \frac{-32}{9} & \frac{40}{9} & \frac{-16}{3} & \frac{56}{9} & \frac{-46}{9} \end{bmatrix}.$$

The matrix $\mathbf{M}$ happens not to be diagonalizable, but it can be put into Jordan canonical form. The Jordan normal form of $\mathbf{M}$ is

$$\mathbf{J_M} = \begin{bmatrix} -1 & 1 & 0 & 0 & 0 & 0 & 0 & 0 \\ 0 & -1 & 1 & 0 & 0 & 0 & 0 & 0 \\ 0 & 0 & -1 & 0 & 0 & 0 & 0 & 0 \\ 0 & 0 & 0 & -1 & 1 & 0 & 0 & 0 \\ 0 & 0 & 0 & 0 & -1 & 0 & 0 & 0 \\ 0 & 0 & 0 & 0 & 0 & 2 & 1 & 0 \\ 0 & 0 & 0 & 0 & 0 & 0 & 2 & 0 \\ 0 & 0 & 0 & 0 & 0 & 0 & 0 & -2 \end{bmatrix}.$$

In the course of determining the Jordan form of the matrix, the change of basis matrix representing the change of coordinates is determined (by Maple, for example). For this case the matrices are given by (according to our convention, $\mathbf{P}$

provides the basis, while $\mathbf{P}^{-1}$ converts coordinates)

$$\mathbf{P}^{-1} = \begin{bmatrix} \frac{8}{9} & \frac{-7}{9} & \frac{2}{3} & \frac{-5}{9} & \frac{4}{9} & \frac{-1}{3} & \frac{2}{9} & \frac{-1}{9} \\ \frac{-7}{9} & \frac{14}{9} & \frac{-4}{3} & \frac{10}{9} & \frac{-8}{9} & \frac{2}{3} & \frac{-4}{9} & \frac{2}{9} \\ \frac{2}{3} & \frac{-4}{3} & 2 & \frac{-5}{3} & \frac{4}{3} & -1 & \frac{2}{3} & \frac{-1}{3} \\ \frac{-5}{9} & \frac{10}{9} & \frac{-5}{3} & \frac{20}{9} & \frac{-16}{9} & \frac{4}{3} & \frac{-8}{9} & \frac{4}{9} \\ \frac{4}{9} & \frac{-8}{9} & \frac{4}{3} & \frac{-16}{9} & \frac{20}{9} & \frac{-5}{3} & \frac{10}{9} & \frac{-5}{9} \\ \frac{-1}{3} & \frac{2}{3} & -1 & \frac{4}{3} & \frac{-5}{3} & 2 & \frac{-4}{3} & \frac{2}{3} \\ \frac{2}{9} & \frac{-4}{9} & \frac{2}{3} & \frac{-8}{9} & \frac{10}{9} & \frac{-4}{3} & \frac{14}{9} & \frac{-7}{9} \\ \frac{-1}{9} & \frac{2}{9} & \frac{-1}{3} & \frac{4}{9} & \frac{-5}{9} & \frac{2}{3} & \frac{-7}{9} & \frac{8}{9} \end{bmatrix}$$

and

$$\mathbf{P} = \begin{bmatrix} 2 & 1 & 0 & 0 & 0 & 0 & 0 & 0 \\ 1 & 2 & 1 & 0 & 0 & 0 & 0 & 0 \\ 0 & 1 & 2 & 1 & 0 & 0 & 0 & 0 \\ 0 & 0 & 1 & 2 & 1 & 0 & 0 & 0 \\ 0 & 0 & 0 & 1 & 2 & 1 & 0 & 0 \\ 0 & 0 & 0 & 0 & 1 & 2 & 1 & 0 \\ 0 & 0 & 0 & 0 & 0 & 1 & 2 & 1 \\ 0 & 0 & 0 & 0 & 0 & 0 & 1 & 2 \end{bmatrix}.$$

The transformations of the system of equations goes according to

$$\begin{aligned} \frac{d}{dt}\mathbf{x} &= \mathbf{M}\,\mathbf{x}, \\ \frac{d}{dt}\mathbf{P}^{-1}\mathbf{x} &= \mathbf{P}^{-1}\,\mathbf{M}\,\mathbf{P}\,\mathbf{P}^{-1}\,\mathbf{x}, \\ \frac{d}{dt}\mathbf{z} &= \mathbf{J_M}\,\mathbf{z} \end{aligned}$$

where

$$\mathbf{J_M} = \mathbf{P}^{-1}\,\mathbf{M}\,\mathbf{P}$$

is the "Jordan-ized" (as opposed to diagonalized) form of the original matrix.

The solution in terms of the **z** coordinates involves multiplication by the Jordan block solutions for each of the independent subsystems. This is represented by

$$\begin{bmatrix} e^{(-t)} & te^{(-t)} & \frac{1}{2}t^2e^{(-t)} & 0 & 0 & 0 & 0 & 0 \\ 0 & e^{(-t)} & te^{(-t)} & 0 & 0 & 0 & 0 & 0 \\ 0 & 0 & e^{(-t)} & 0 & 0 & 0 & 0 & 0 \\ 0 & 0 & 0 & e^{(-t)} & te^{(-t)} & 0 & 0 & 0 \\ 0 & 0 & 0 & 0 & e^{(-t)} & 0 & 0 & 0 \\ 0 & 0 & 0 & 0 & 0 & e^{(2t)} & te^{(2t)} & 0 \\ 0 & 0 & 0 & 0 & 0 & 0 & e^{(2t)} & 0 \\ 0 & 0 & 0 & 0 & 0 & 0 & 0 & e^{(-2t)} \end{bmatrix}$$

To get the answer back in terms of the original **x** coordinates, the above must be multiplied by **P** on one side, and $\mathbf{P}^{-1}$ on the other. The result is substantially more complex in terms of the original coordinate choice. For reasons of display space, the result is formatted into separate rows.

$$\Big[e^{(-t)} - \frac{8}{9}te^{(-t)} + \frac{2}{3}t^2e^{(-t)}, \frac{16}{9}te^{(-t)} - \frac{4}{3}t^2e^{(-t)}, -\frac{2}{3}te^{(-t)} + 2t^2e^{(-t)},$$
$$\frac{5}{9}te^{(-t)} - \frac{5}{3}t^2e^{(-t)}, -\frac{4}{9}te^{(-t)} + \frac{4}{3}t^2e^{(-t)}, \frac{1}{3}te^{(-t)} - t^2e^{(-t)},$$
$$-\frac{2}{9}te^{(-t)} + \frac{2}{3}t^2e^{(-t)}, \frac{1}{9}te^{(-t)} - \frac{1}{3}t^2e^{(-t)} \Big]$$

...

...

$$\Big[\frac{2}{9}e^{(2t)} - \frac{2}{9}e^{(-2t)}, -\frac{4}{9}e^{(2t)} + \frac{4}{9}e^{(-2t)}, \frac{2}{3}e^{(2t)} - \frac{2}{3}e^{(-2t)},$$
$$-\frac{8}{9}e^{(2t)} + \frac{8}{9}e^{(-2t)}, \frac{10}{9}e^{(2t)} - \frac{10}{9}e^{(-2t)}, -\frac{4}{3}e^{(2t)} + \frac{4}{3}e^{(-2t)},$$
$$\frac{14}{9}e^{(2t)} - \frac{14}{9}e^{(-2t)}, -\frac{7}{9}e^{(2t)} + \frac{16}{9}e^{(-2t)} \Big].$$

7.4.1 Using Maple

The Maple `linalg` package can compute eigenvalues and eigenvectors, and diagonalize matrices with distinct eigenvalues. Hence, the solution by diagonalization described in Section 7.3 can be carried out in particular cases that happen to satisfy the diagonalizability condition.

Maple connection **Page:** 307

In the general case, diagonalization is not guaranteed, and all that can be hoped for is a transformation into Jordan form. Maple has a Jordan form computation subroutine, which will return the change of basis matrix as well. With this method, explicit solutions can be calculated for any system of constant coefficient differential equations.

7.5 Matrix Exponentials

The scalar first order differential equation

$$\frac{dy}{dt} = a\, y(t)$$

has the solution

$$y(t) = e^{at}\, y(0).$$

In the previous sections it was suggested that the way to handle the vector version of this problem

$$\frac{d}{dt}\mathbf{y} = \mathbf{A}\,\mathbf{y}(t)$$

is to find a change of basis matrix **P** that puts **A** either into diagonal or Jordan form, from which point solutions can be written down directly. An alternative and less laborious way to solve the problems is to say the solution is simply

$$\mathbf{y}(t) = e^{\mathbf{A}\,t}\,\mathbf{y}(0).$$

The rest of this section is devoted to explaining what the above ought to mean, and showing that it really is the solution.

One way to "invent" the exponential of a matrix is to look in detail at what emerges when the Picard iteration procedure is applied to the constant coefficient system of differential equations. The iteration for such a system is

$$\mathbf{y}_n(t) = \mathbf{y}(0) + \int_0^t \mathbf{A}\,\mathbf{y}_{n-1}(t)\,dt.$$

If this is carried out, the generated sequence is

$$\begin{aligned}
\mathbf{y}_0 &= \mathbf{y}(0) \\
\mathbf{y}_1 &= \mathbf{y}(0) + \int_0^t \mathbf{A}\,\mathbf{y}(0)\,dt \\
\mathbf{y}_1 &= \mathbf{y}(0) + \mathbf{A}\,\mathbf{y}(0)\,t \\
\mathbf{y}_2 &= \mathbf{y}(0) + \int_0^t \mathbf{A}\,\{\mathbf{y}(0) + \mathbf{A}\,\mathbf{y}(0)\,t\}\,dt \\
\mathbf{y}_2 &= \mathbf{y}(0) + \mathbf{A}\,t\,\mathbf{y}(0) + \mathbf{A}^2\,\frac{t^2}{2!}\mathbf{y}(0) \\
\mathbf{y}_3 &= \mathbf{y}(0) + \mathbf{A}\,t\,\mathbf{y}(0) + \mathbf{A}^2\,\frac{t^2}{2!}\,\mathbf{y}(0) + \mathbf{A}^3\,\frac{t^3}{3!}\,\mathbf{y}(0) \\
\vdots &= \vdots \\
\mathbf{y}_n &= \mathbf{y}(0) + \mathbf{A}\,t\,\mathbf{y}(0) + \mathbf{A}^2\,\frac{t^2}{2!}\,\mathbf{y}(0) + \mathbf{A}^3\,\frac{t^3}{3!}\,\mathbf{y}(0) + \ldots + \mathbf{A}^n\,\frac{t^n}{n!}\,\mathbf{y}(0)
\end{aligned}$$

This Picard iteration should converge to the solution of the differential equation in the limit, and the sequence takes a familiar form. We can write down the nth term in the sequence calculated in closed form as

$$\mathbf{y}_n = \mathbf{y}(0) + \mathbf{A}\,t\,\mathbf{y}(0) + \mathbf{A}^2\,\frac{t^2}{2!}\,\mathbf{y}(0) + \mathbf{A}^3\,\frac{t^3}{3!}\,\mathbf{y}(0) + \ldots + \mathbf{A}^n\,\frac{t^n}{n!}\,\mathbf{y}(0),$$

$$\mathbf{y}_n = \left(\mathbf{I} + \mathbf{A}\,t\, + \mathbf{A}^2\,\frac{t^2}{2!}\, + \mathbf{A}^3\,\frac{t^3}{3!}\, + \ldots + \mathbf{A}^n\,\frac{t^n}{n!}\right)\,\mathbf{y}(0)$$

and hence

$$\mathbf{y}(t) = \lim_{n\to\infty}\left(\mathbf{I} + \mathbf{A}\,t\, + \mathbf{A}^2\,\frac{t^2}{2!}\, + \mathbf{A}^3\,\frac{t^3}{3!}\, + \ldots + \mathbf{A}^n\,\frac{t^n}{n!}\right)\,\mathbf{y}(0).$$

The terms above look exactly like the terms of the Taylor series for the exponential function, except that they have the matrix quantity $\mathbf{A}\,t$ substituted where one would expect to see the (scalar) argument. If we are going to fish about for an appropriate *name* for the limit in the above expression, there is only one candidate.

Definition
The matrix exponential is defined as

$$e^{\mathbf{A}\,t} = \lim_{n\to\infty}\left(\mathbf{I} + \mathbf{A}\,t\, + \mathbf{A}^2\,\frac{t^2}{2!}\, + \mathbf{A}^3\,\frac{t^3}{3!}\, + \ldots + \mathbf{A}^n\,\frac{t^n}{n!}\right).$$

Although the matrix exponential is defined as the limit of an infinite series, there is no implication that attempting to sum the series "by hand" is an advisable computational technique. The preferred methods follow the techniques already used, that is, basis change and Laplace transform methods.

Example
If the matrix $\mathbf{A}$ is actually in diagonal form, then it is easy to calculate the matrix exponential function. Given

$$\mathbf{A} = \begin{bmatrix} \lambda_1 & 0 & \dots & 0 \\ 0 & \lambda_2 & 0 & \dots \\ 0 & 0 & \dots & 0 \\ 0 & 0 & \dots & \lambda_n \end{bmatrix},$$

it is easy to see that the Picard iteration procedure will only produce entries along the diagonal of the matrix sequence, and also that the resulting diagonal entries will be identical in form to those of the scalar case. (The problem in effect is n simultaneous uncoupled first order problems). The final result is

$$e^{\mathbf{A}t} = \begin{bmatrix} e^{\lambda_1 t} & 0 & \dots & 0 \\ 0 & e^{\lambda_2 t} & 0 & \dots \\ 0 & 0 & \dots & 0 \\ 0 & 0 & \dots & e^{\lambda_n t} \end{bmatrix}.$$

Example
If the matrix in question is not in diagonal form, then on a lucky day a change of basis will put it there, and one can proceed (in the new coordinates, of course) as the previous example advocates. Of course, an arbitrary matrix need not be diagonalizable, and the problem of computing the matrix exponential of a matrix in Jordan form then arises.

For this problem (and practically speaking most others) Laplace transform methods work, as discussed in the next Section.

7.5.1 Laplace Transforms

If the constant coefficient system

$$\frac{d}{dt}\mathbf{x}(t) = \mathbf{A}\,\mathbf{x}(t)$$

is treated by Laplace transforms, then we obtain

$$s\mathcal{L}\{\mathbf{x}\} - \mathbf{x}(0) = \mathbf{A}\,\mathcal{L}\{\mathbf{x}\}.$$

Note that this actually requires some effort. In the first place, one must *define* the Laplace transform of a vector function by

$$\mathcal{L}\{\mathbf{x}(t)\} = \begin{bmatrix} \mathcal{L}\{x_1(t)\} \\ \mathcal{L}\{x_2(t)\} \\ \mathcal{L}\{x_3(t)\} \\ \vdots \\ \mathcal{L}\{x_n(t)\} \end{bmatrix},$$

that is, the vector of Laplace transforms of the components. Then the rule for transforming a derivative works out as before, and one can also verify that the transform operator can be pulled inside the matrix-vector multiplication on the left of the equation. See the problems below.

If the equation above is solved for the solution transform,

$$\begin{aligned}(s\,\mathbf{I}-\mathbf{A})\,\mathcal{L}\{\mathbf{x}(t)\} &= \mathbf{x}(0),\\ \mathcal{L}\{\mathbf{x}(t)\} &= (s\,\mathbf{I}-\mathbf{A})^{-1}\,\mathbf{x}(0).\end{aligned}$$

Because the matrix exponential representation of the solution is

$$\mathbf{x}(t) = e^{\mathbf{A}t}\,\mathbf{x}(0),$$

we can conclude that

$$\mathbf{x}(t) = e^{\mathbf{A}t}\,\mathbf{x}(0) = \mathcal{L}^{-1}\{(s\,\mathbf{I}-\mathbf{A})^{-1}\}\,\mathbf{x}(0),$$

so that an alternative way to compute the matrix exponential must be

$$e^{\mathbf{A}t} = \mathcal{L}^{-1}\left\{(s\,\mathbf{I}-\mathbf{A})^{-1}\right\}.$$

Maple can readily evaluate the required symbolic matrix inverses, and apply the inverse transform to each element of the result.

7.5.2 *Using Maple*

Maple connection **Page:** 313

Proceed to calculate the matrix exponential corresponding to the standard harmonic oscillator model.

```
> A := matrix(2,2, [[0,1],[-\omega_n^2, -2*zeta*\omega_n]]);
```

$$A := \begin{bmatrix} 0 & 1 \\ -\omega_n{}^2 & -2\zeta\,\omega_n \end{bmatrix}$$

```
> evalm(s*Id - A);
```

$$\begin{bmatrix} s & -1 \\ \omega_n{}^2 & 2\zeta\,\omega_n + s \end{bmatrix}$$

Now invert the matrix

$$(\mathbf{I}\,s-\mathbf{A})\,.$$

The result is called the resolvent function of the matrix **A**.

$$(\mathbf{I}s - \mathbf{A})^{-1} = \begin{bmatrix} \dfrac{2\zeta\omega_n + s}{s^2 + 2\zeta\omega_n s + \omega_n{}^2}, & \dfrac{1}{s^2 + 2\zeta\omega_n s + \omega_n{}^2} \\ -\dfrac{\omega_n{}^2}{s^2 + 2\zeta\omega_n s + \omega_n{}^2} & \dfrac{s}{s^2 + 2\zeta\omega_n s + \omega_n{}^2} \end{bmatrix}.$$

Using Maple it is simple to invert each element of a matrix of transforms to produce the output:

$$e^{\mathbf{A}t} = \begin{bmatrix} \dfrac{\zeta e^{(-\zeta\omega_n t)}\sin\left(\sqrt{1-\zeta^2}\omega_n t\right)}{\sqrt{1-\zeta^2}} + e^{(-\zeta\omega_n t)}\cos\left(\sqrt{1-\zeta^2}\omega_n t\right), & \dfrac{e^{(-\zeta\omega_n t)}\sin\left(\sqrt{1-\zeta^2}\omega_n t\right)}{\sqrt{1-\zeta^2}\omega_n} \\ -\dfrac{\omega_n e^{(-\zeta\omega_n t)}\sin\left(\sqrt{1-\zeta^2}\omega_n t\right)}{\sqrt{1-\zeta^2}}, & -\dfrac{\zeta e^{(-\zeta\omega_n t)}\sin\left(\sqrt{1-\zeta^2}\omega_n t\right)}{\sqrt{1-\zeta^2}} + e^{(-\zeta\omega_n t)}\cos\left(\sqrt{1-\zeta^2}\omega_n t\right) \end{bmatrix}$$

Try the same calculations for a 3 × 3 example. Take a case with distinct eigenvalues, starting from a given characteristic polynomial.

```
> p(s) := expand((s-1)*(s-2)*(s-3));
```

$$p(s) := s^3 - 6s^2 + 11s - 6.$$

A "companion matrix" constructed from the polynomial will have the given eigenvalues.

```
> B := matrix(3,3, [[0,1,0],[0,0,1],[6,-11, 6]]);
```

$$B := \begin{bmatrix} 0 & 1 & 0 \\ 0 & 0 & 1 \\ 6 & -11 & 6 \end{bmatrix}$$

The Laplace transform calculation of the matrix exponential requires

$$(s\mathbf{I} - \mathbf{B})^{-1} =$$

$$\begin{bmatrix} \dfrac{-6s + s^2 + 11}{s^3 - 6s^2 + 11s - 6}, & \dfrac{-6 + s}{s^3 - 6s^2 + 11s - 6}, & \dfrac{1}{s^3 - 6s^2 + 11s - 6} \\ 6\dfrac{1}{s^3 - 6s^2 + 11s - 6}, & \dfrac{s(-6+s)}{s^3 - 6s^2 + 11s - 6}, & \dfrac{s}{s^3 - 6s^2 + 11s - 6} \\ 6\dfrac{s}{s^3 - 6s^2 + 11s - 6}, & -\dfrac{11s - 6}{s^3 - 6s^2 + 11s - 6}, & \dfrac{s^2}{s^3 - 6s^2 + 11s - 6} \end{bmatrix}.$$

Inverting the the Laplace transform term-by-term gives an expression for the matrix exponential.

$$e^{\mathbf{B}t} = \begin{bmatrix} 3e^t - 3e^{(2t)} + e^{(3t)}, -\frac{5}{2}e^t + 4e^{(2t)} - \frac{3}{2}e^{(3t)}, \frac{1}{2}e^t - e^{(2t)} + \frac{1}{2}e^{(3t)} \\ 3e^t - 6e^{(2t)} + 3e^{(3t)}, -\frac{5}{2}e^t + 8e^{(2t)} - \frac{9}{2}e^{(3t)}, \frac{1}{2}e^t - 2e^{(2t)} + \frac{3}{2}e^{(3t)} \\ 3e^t - 12e^{(2t)} + 9e^{(3t)}, -\frac{5}{2}e^t + 16e^{(2t)} - \frac{27}{2}e^{(3t)}, \frac{1}{2}e^t - 4e^{(2t)} + \frac{9}{2}e^{(3t)} \end{bmatrix}$$

One can verify also that the matrix exponential expressions, complicated though they may appear, actually evaluate to an identity matrix when $t = 0$. This property, of course, is what guarantees that the matrix exponential form of the solution meets the initial condition prescribed at $t = 0$. The evaluation is easily done using Maple.

7.6 Problems

1. Find the general solution of the system

$$\frac{d}{dt}\mathbf{x} = \begin{bmatrix} 0 & 1 \\ -2 & -3 \end{bmatrix}\mathbf{x}$$

by diagonalizing the coefficient matrix.

2. Find the general solution of the system

$$\frac{d}{dt}\mathbf{x} = \begin{bmatrix} 0 & 1 \\ 0 & -1 \end{bmatrix}\mathbf{x}$$

also by diagonalizing the coefficient matrix.

3. Under what condition on the entries does the matrix

$$\begin{bmatrix} a & b \\ c & d \end{bmatrix}$$

have distinct eigenvalues? It is useful to check to see if you are guaranteed diagonalizability before you launch into the attempt.

4. A coupled harmonic oscillator (with no damping) is described by

$$\begin{aligned} M_2 \frac{d^2 x_2}{dt^2} &= -k_2\,(x_2(t) - x_1(t)), \\ M_1 \frac{d^2 x_1}{dt^2} &= -k_1\, x_1(t) + k_2\,(x_2(t) - x_1(t)). \end{aligned}$$

The solutions to this coupled system consist of a mixture of oscillations at two oscillation frequencies. Use Laplace transform methods to find the frequencies, without actually solving the whole problem.

5. Pack a lunch, and then find the general solution of

$$\frac{d}{dt}\mathbf{x} = \begin{bmatrix} 0 & 1 \\ -\omega_n^2 & -2\zeta\omega_n \end{bmatrix}\mathbf{x}$$

by means of diagonalization and change of basis. In the "usual" situation the eigenvalues and eigenvectors will be complex quantities, so this enterprise is not for the faint of heart.

6. Calculate the resolvent of a 3×3 Jordan block:

$$\left(s\mathbf{I} - \begin{bmatrix} \lambda & 1 & 0 \\ 0 & \lambda & 1 \\ 0 & 0 & \lambda \end{bmatrix} \right)^{-1}.$$

7. Evaluate

$$e^{\begin{bmatrix} \lambda & 1 & 0 \\ 0 & \lambda & 1 \\ 0 & 0 & \lambda \end{bmatrix} t}.$$

8. Show that

$$\left[s\mathbf{I} - \begin{bmatrix} \lambda & 1 & 0 & \cdots \\ 0 & \lambda & 1 & \cdots \\ \vdots & & \ddots & \ddots \\ 0 & 0 & \cdots & \lambda \end{bmatrix} \right]^{-1}$$

$$= \begin{bmatrix} \frac{1}{(s-\lambda)} & \frac{1}{(s-\lambda)^2} & \cdots & \frac{1}{(s-\lambda)^{n-1}} & \frac{1}{(s-\lambda)^n} \\ 0 & \frac{1}{(s-\lambda)} & \frac{1}{(s-\lambda)^2} & \cdots & \frac{1}{(s-\lambda)^{n-1}} \\ 0 & 0 & \ddots & \cdots & \frac{1}{(s-\lambda)^{n-2}} \\ \vdots & \vdots & \ddots & \ddots & \cdots \\ 0 & 0 & 0 & \frac{1}{(s-\lambda)} & \frac{1}{(s-\lambda)^2} \\ 0 & 0 & 0 & 0 & \frac{1}{(s-\lambda)} \end{bmatrix}$$

9. Using Problem 8 show that

$$e^{\begin{bmatrix} \lambda & 1 & 0 & \cdots \\ 0 & \lambda & 1 & \cdots \\ \vdots & & \ddots & \ddots \\ 0 & 0 & \cdots & \lambda \end{bmatrix} t} = \begin{bmatrix} e^{\lambda t} & t\,e^{\lambda t} & \cdots & \frac{t^{n-1}}{(n-1)!}\, e^{\lambda t} \\ 0 & e^{\lambda t} & \cdots & \frac{t^{n-2}}{(n-2)!}\, e^{\lambda t} \\ \vdots & \ddots & \ddots & \vdots \\ 0 & 0 & 0 & e^{\lambda t} \end{bmatrix}$$

10. If the Laplace transform of a vector function of time is defined by the only plausible recipe

$$\mathcal{L}\{\begin{bmatrix} x_1(t) \\ x_2(t) \\ \vdots \\ x_n(t) \end{bmatrix}\} = \begin{bmatrix} X_1(s) \\ X_2(s) \\ \vdots \\ X_n(s) \end{bmatrix},$$

prove that

$$\mathcal{L}\{\frac{d}{dt}\begin{bmatrix} x_1(t) \\ x_2(t) \\ \vdots \\ x_n(t) \end{bmatrix}\} = s\begin{bmatrix} X_1(s) \\ X_2(s) \\ \vdots \\ X_n(s) \end{bmatrix} - \begin{bmatrix} x_1(0) \\ x_2(0) \\ \vdots \\ x_n(0) \end{bmatrix}.$$

11. Prove that the Laplace transformation of vectors commutes with multiplication by a constant matrix **A**. That is,

$$\mathcal{L}\left\{\mathbf{A}\begin{bmatrix} x_1(t) \\ x_2(t) \\ \vdots \\ x_n(t) \end{bmatrix}\right\} = \mathbf{A}\,\mathcal{L}\left\{\begin{bmatrix} x_1(t) \\ x_2(t) \\ \vdots \\ x_n(t) \end{bmatrix}\right\}.$$

12. Use Maple to find the general solution to the following system of equations:

$$\frac{d}{dt}\begin{bmatrix} x_1(t) \\ x_2(t) \\ x_3(t) \end{bmatrix} = \begin{bmatrix} 1 & 2 & 3 \\ 4 & 5 & 6 \\ 7 & 8 & 9 \end{bmatrix}\begin{bmatrix} x_1(t) \\ x_2(t) \\ x_3(t) \end{bmatrix}.$$

13. Suppose that the matrix **A** has distinct eigenvalues, so that

$$\det\ (\mathbf{I}s - \mathbf{A}) = p(s) = (s - p_1)(s - p_2)\ldots(s - p_n).$$

Show that by partial fractions we can express the resolvent of **A** in the form

$$(\mathbf{I}s - \mathbf{A})^{-1} = \frac{\mathbf{A}_1}{s - p_1} + \frac{\mathbf{A}_2}{s - p_2} + \ldots + \frac{\mathbf{A}_n}{s - p_n},$$

which is essentially a partial fraction expansion with matrices for coefficients.

14. Show that with the matrix **A** as in Problem 13, the matrix exponential can be written as

$$e^{\mathbf{A}t} = \mathbf{A}_1\, e^{p_1 t} + \mathbf{A}_2\, e^{p_2 t} + \ldots + \mathbf{A}_n\, e^{p_n t}.$$

15. Show that the expansion matrices of Problem 13 satisfy

$$\mathbf{A}_1 + \mathbf{A}_2 + \ldots + \mathbf{A}_n = \mathbf{I}.$$

16. Use vector equation methods and Maple to find the general solution to the scalar higher order equation

$$\frac{d^3x}{dt^3} + 3\frac{d^2x}{dt^2} + 3\frac{dx}{dt} + x = 0.$$

8
Stability

The stability of differential equations is a topic of very wide interest and paramount importance in many subject areas. Many technological systems incorporate control devices of various sorts, and lives literally depend on the stability properties of such designs. Analysis of such stability problems began with Maxwell's investigation of steam engine governors, and has continued to develop along with the increasing sophistication of control systems. Some historical account of these issues is in [8].

A problem of continuing interest (and still unresolved after centuries of effort) concerns the stability of planetary orbits. It would be comforting to know that the earth is not about to fly off its orbit and crash into the sun, but the stability problem for the solar system is so complex that nothing approaching a complete solution is known. One of the original discussions of these issues is [11], and a large amount of literature has followed.

Stability problems are usually discussed first in terms of low order cases for which clear geometric insight is available. This approach is taken in the discussion below.

8.1 Second Order Problems

We have seen that second order differential equations arise naturally in problems of particle mechanics, and so are of interest for that reason. If they are formulated as vector systems (as in previous chapters) the vectors are only of dimension two, so the situation is much more easily visualized than it is in the general case of higher dimensions.

The appropriate visualization is in the notion of a "phase plane". The terminology originates with single degree of freedom nonlinear mechanical systems. The equation of motion of such a model is

$$M\frac{d^2x}{dt^2} = f(x(t), \frac{dx}{dt}).$$

If the mechanical system is represented in the vector form

$$\begin{aligned} \frac{d}{dt}x(t) &= v(t) \\ \frac{d}{dt}v(t) &= \frac{1}{M}\, f(x(t), v(t)) \end{aligned}$$

with position and velocity as the natural component variables, then the (vector) solution $(x(t), v(t))$ can be visualized as a parameterized curve in the plane. The plane is referred to as the "phase plane", and is usually drawn with the position axis horizontal (as abscissa) with the velocity vertical (as ordinate).

In this simple two dimensional world, closed curves correspond to periodic solutions of the differential equations. Stable and unstable solutions have characteristic appearances, at least locally in the case of non-linear systems. For linear second order systems, the problem (with a change of parameter names) comes down to another re-examination of the harmonic oscillator.

8.2 Harmonic Oscillator Again

In the vector system formulation, the harmonic oscillator takes the form

$$\frac{d}{dt}\mathbf{x}(t) = \begin{bmatrix} 0 & 1 \\ -\omega_n{}^2 & -2\zeta\,\omega_n \end{bmatrix} \mathbf{x}(t),$$

with initial condition

$$\mathbf{x}(0) = \mathbf{x}_0.$$

The general solution of this model can be represented in terms of the exponential of the coefficient matrix, and takes the form

$$\mathbf{x}(t) = \begin{bmatrix} \frac{\zeta e^{(-\zeta\omega_n t)}\sin\left(\sqrt{1-\zeta^2}\omega_n t\right)}{\sqrt{1-\zeta^2}} + e^{(-\zeta\omega_n t)}\cos\left(\sqrt{1-\zeta^2}\omega_n t\right), & \frac{e^{(-\zeta\omega_n t)}\sin\left(\sqrt{1-\zeta^2}\omega_n t\right)}{\sqrt{1-\zeta^2}\omega_n} \\ -\frac{\omega_n e^{(-\zeta\omega_n t)}\sin\left(\sqrt{1-\zeta^2}\omega_n t\right)}{\sqrt{1-\zeta^2}}, & -\frac{\zeta e^{(-\zeta\omega_n t)}\sin\left(\sqrt{1-\zeta^2}\omega_n t\right)}{\sqrt{1-\zeta^2}} + e^{(-\zeta\omega_n t)}\cos\left(\sqrt{1-\zeta^2}\omega_n t\right) \end{bmatrix} \mathbf{x}_0.$$

The components of the state vector $\mathbf{x}$ are identified with the position and velocity of the "original" (originality in the eye of the beholder) second order equation

$$\frac{d^2y}{dt^2} + 2\,\zeta\,\omega_n\,\frac{dy}{dt} + \omega_n^2\,y = 0.$$

If one were to construct a "phase plane" with the original scalar problem position and velocity as coordinates, then the vector solution formula above can be visualized as a mapping of the phase plane into itself. The mapping is linear, and parameterized by t, the time of evolution of the system. The mapping sends the initial point $\mathbf{x}_0$ to its position after t seconds of motion according to

$$\mathbf{x}_0 \mapsto \begin{bmatrix} \frac{\zeta e^{(-\zeta\omega_n t)}\sin\left(\sqrt{1-\zeta^2}\omega_n t\right)}{\sqrt{1-\zeta^2}} + e^{(-\zeta\omega_n t)}\cos\left(\sqrt{1-\zeta^2}\omega_n t\right), & \frac{e^{(-\zeta\omega_n t)}\sin\left(\sqrt{1-\zeta^2}\omega_n t\right)}{\sqrt{1-\zeta^2}\omega_n} \\ -\frac{\omega_n e^{(-\zeta\omega_n t)}\sin\left(\sqrt{1-\zeta^2}\omega_n t\right)}{\sqrt{1-\zeta^2}}, & -\frac{\zeta e^{(-\zeta\omega_n t)}\sin\left(\sqrt{1-\zeta^2}\omega_n t\right)}{\sqrt{1-\zeta^2}} + e^{(-\zeta\omega_n t)}\cos\left(\sqrt{1-\zeta^2}\omega_n t\right) \end{bmatrix} \mathbf{x}_0.$$

As mentioned above the expressions above are (at least for $\zeta^2 \neq 1$) valid, if a bit misleading, for all ranges of the damping and frequency parameters. One case, forming the boundary between stable and unstable behavior, occurs when the damping coefficient ζ vanishes. If $\zeta = 0$, then energy is preserved in the system. This is reflected in the calculation that

$$\frac{d}{dt}\frac{1}{2}\left(\omega_n^2\, y(t)^2 + \frac{dy}{dt}^2\right) = 0.$$

This means that in this case, the solutions move (periodically at frequency ω_n) along the contours of an ellipse in the phase plane.

For other values of the parameters the geometric (phase plane) view of the solutions distinguishes strongly between oscillatory, exponential, stable and unstable cases.

8.3 Nodes and Spirals

The solutions favored by the notation used above arise in the case in which $\zeta^2 < 1$. Then the solutions are oscillatory with vibration frequency

$$\sqrt{1-\zeta^2}\,\omega_n.$$

The effect of the damping term is to decrease the frequency of vibration below that of the undamped case. If $0 < \zeta < 1$, then the oscillation is modulated by a negative exponential, and solutions all tend to the origin as t approaches infinity. This case is referred to as a "stable spiral" in the phase plane, and is illustrated below in Figure 8.1.

Maple connection **Page: 315**

As ζ tends to 1, the oscillation frequency goes to zero, and the resulting diagram moves from a stable spiral to what is referred to as a "stable node". This case

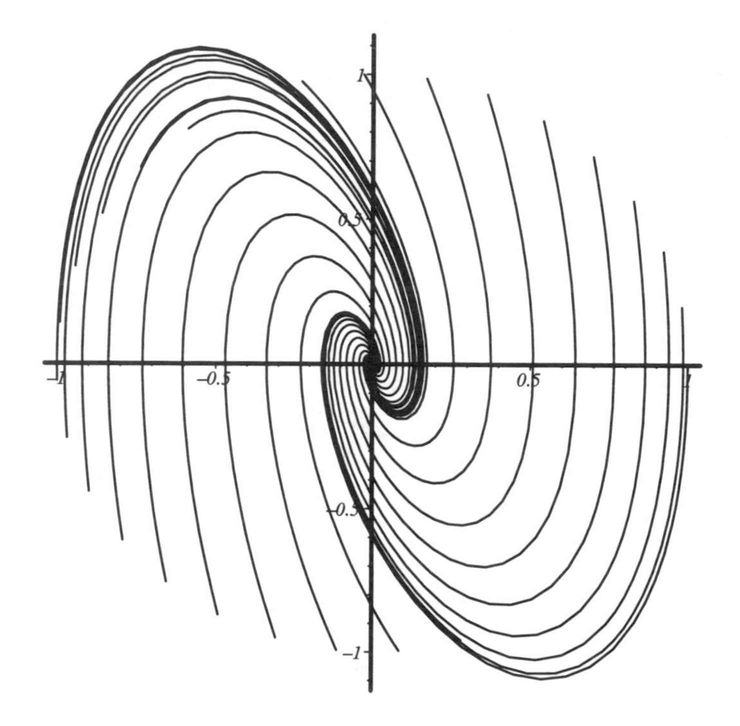

FIGURE 8.1. A Stable Spiral

corresponds to a double real root in the characteristic equation. The corresponding plot is in Figure 8.2.

Increasing the damping beyond this point leads to an "over–damped" system, where the characteristic equation has a pair of real (and negative, for positive ζ) roots. The phase plane category is "over–damped system", illustrated in Figure 8.3.

The over–damped case shows an evident tendency for trajectories to head for the origin along a particular line. This phenomenon can be understood in terms of the eigenvector discussion of the previous chapter. The natural situation to consider in this case is a 2×2 homogeneous system

$$\frac{d}{dt}\mathbf{x} = \mathbf{A}\,\mathbf{x},$$

assuming that $\mathbf{A}$ has two real distinct eigenvalues. This is slightly more general than the harmonic oscillator equation, for which it is impossible (due to the parameterization of the coefficients) to have two real eigenvalues of opposite sign.

With these assumptions, the coefficient matrix may be diagonalized by changing to the eigenvector basis. With

$$\mathbf{P} = [\mathbf{p}_1 \,|\, \mathbf{p}_2],$$

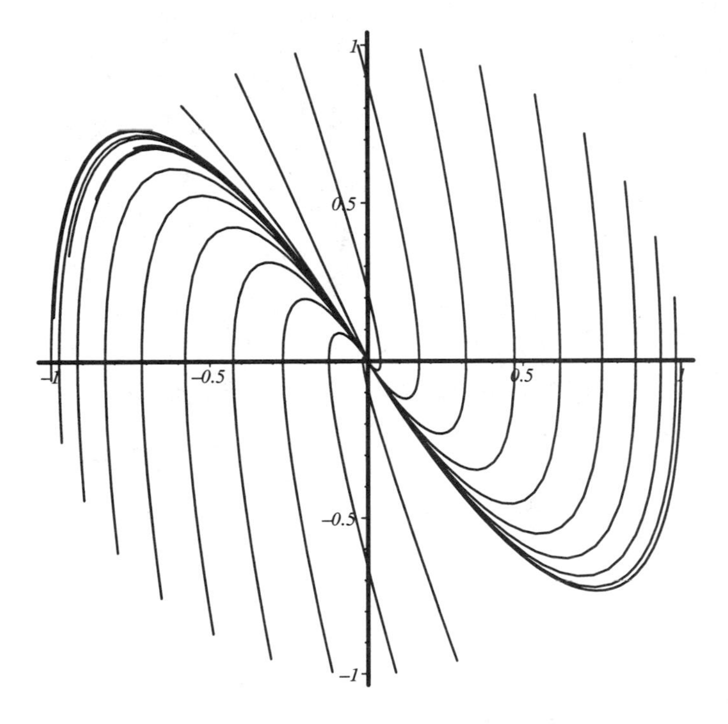

FIGURE 8.2. A Stable Node

we will have

$$\begin{aligned} \mathbf{A}\,\mathbf{p}_1 &= \lambda_1\,\mathbf{p}_1, \\ \mathbf{A}\,\mathbf{p}_2 &= \lambda_2\,\mathbf{p}_2, \\ \mathbf{P}^{-1}\,\mathbf{A}\,\mathbf{P} &= \begin{bmatrix} \lambda_1 & 0 \\ 0 & \lambda_2 \end{bmatrix}. \end{aligned}$$

The solution to the differential equation system will then be simply

$$\mathbf{x}(t) = z_1\,\mathbf{p}_1\,e^{\lambda_1 t} + z_2\,\mathbf{p}_2\,e^{\lambda_2 t},$$

where the z_i are the coefficients for expanding the initial state in the eigenvector basis:

$$\begin{bmatrix} z_1 \\ z_2 \end{bmatrix} = \mathbf{P}^{-1}\,\mathbf{x}(0).$$

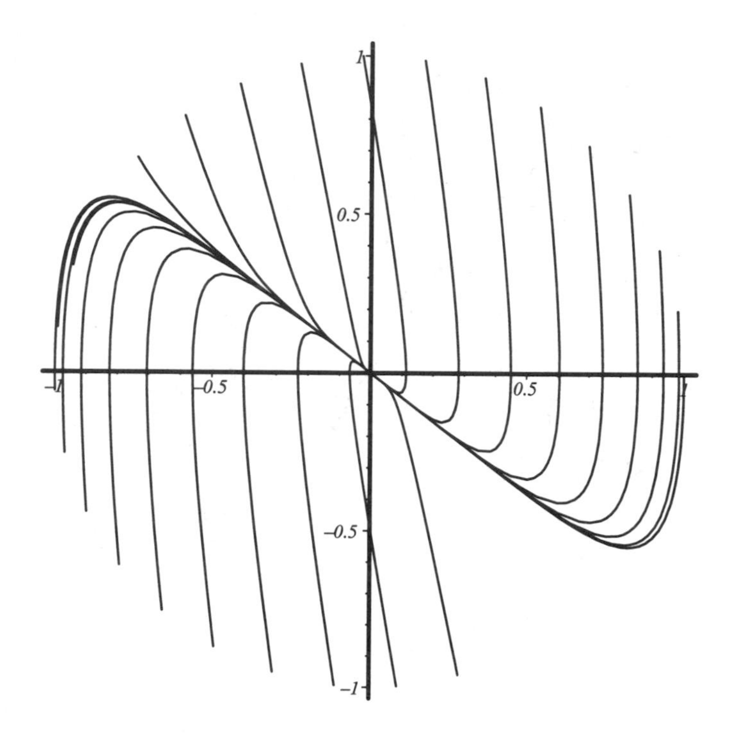

FIGURE 8.3. Over–damped Case

For the harmonic oscillator problem in the over–damped case, the formulas for the eigenvalues are

$$\lambda_1 = -\zeta\,\omega_n + \sqrt{\zeta^2 - 1}\,\omega_n$$

$$\lambda_2 = -\zeta\,\omega_n - \sqrt{\zeta^2 - 1}\,\omega_n.$$

The second eigenvalue is larger in magnitude, and as time evolves, the term

$$z_2\,\mathbf{p}_2\,e^{\lambda_2 t}$$

in the solution disappears faster than the solution as a whole. The solution is approximated by

$$x(t) = z_1\,\mathbf{p}_1\,e^{\lambda_1 t}.$$

In the phase plane plot of the solution, the apparent direction of the final approach to the origin is therefore the orientation of the eigenvector associated with the slower of the two real time constants. The direction of "rapid approach" is associated with the eigenvector of the eigenvalue with the large (negative) real part.

Although not attainable with the harmonic oscillator model, systems with one positive and one negative eigenvalue are possible. The phase portrait of such a system is referred to as a "saddle".

Moving to negative values of the damping coefficient in the harmonic oscillator model is mathematically equivalent to reversing the direction of time in the previous discussion. The eigenvalues in this case will have positive real parts, so that solutions grow without bound for $\zeta < 0$. As far as the phase portrait of the solutions is concerned, there are unstable spirals and unstable nodes that are the counterparts of the stable case.

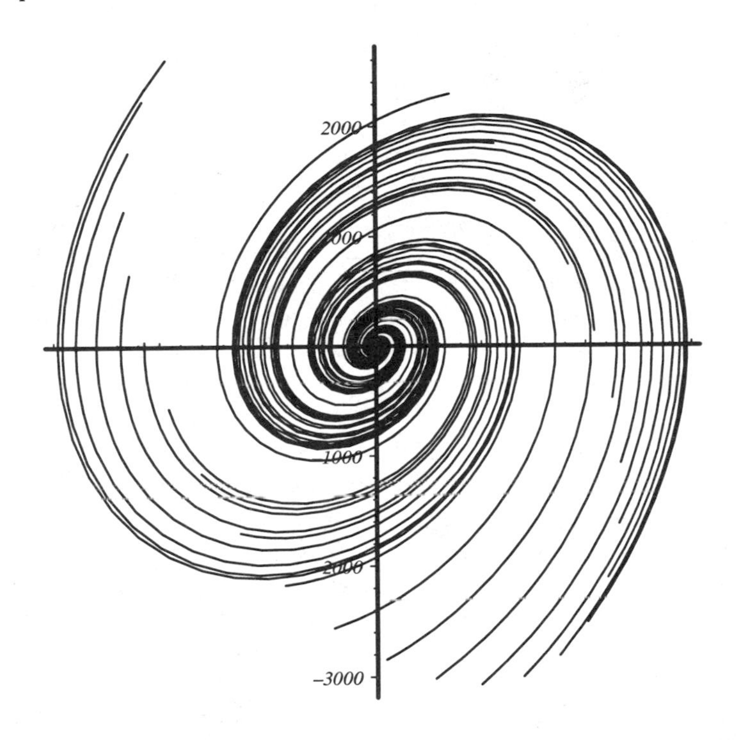

FIGURE 8.4. An Unstable Spiral

8.4 Higher Order Problems

The ability to draw phase portraits of the solutions is only available for two dimensional systems, although one could stretch the notion to three dimensional systems by generating perspective views of the trajectories in three dimensional space. Producing pictures of four dimensional solutions would then require a movie of perspective views, ...

The nature of solutions of higher order linear constant coefficient systems can in any case be discussed (as far as stability issues are concerned) by recourse to appropriate basis changes.

8.4.1 The Easy Cases

The problem is to determine conditions that guarantee that all solutions of the system

$$\frac{d}{dt}\mathbf{x} = \mathbf{A}\,\mathbf{x}$$

tend to zero as time tends to infinity. This is referred to as "asymptotic stability", or "asymptotic stability in the large". A related notion is "stability", which is taken to mean that all solutions remain bounded as time tends to infinity. Asymptotically stable systems are thus stable, but the inclusion does not work the other way.

The easy case is defined as the one in which the coefficient matrix **A** can be diagonalized by a change of basis. From the discussion in Section 7.3, the solution to the differential equation system can be written in terms of the eigenvalues and eigenvectors of **A** as

$$\mathbf{x}(t) = z_1\,\mathbf{p}_1\,e^{\lambda_1 t} + z_2\,\mathbf{p}_2\,e^{\lambda_2 t} + \ldots + z_n\,\mathbf{p}_n\,e^{\lambda_n t}.$$

The fact that the change of basis result shows every solution to be of this form allows the conditions to be stated succinctly. If the real parts of all of the eigenvalues of **A** are negative, then solutions of the system are *asymptotically stable.*

If any of the eigenvalues of **A** has a positive real part, then there is the possibility of solutions that are unbounded as time tends to infinity. To avoid this outcome it would have to happen that the initial condition vector was chosen in such a way that the (z coordinate) components associated with all positive real part eigenvalues vanish. Assuming that the problem models a real system, this could never be guaranteed, and so the system must be declared *unstable* in this case.

The boundary case (literally) occurs when some of the eigenvalues of **A** are purely imaginary (have zero real part). Then solutions that are purely sinusoidal occur. The system is therefore *stable* according to the above definition since all solutions are bounded.[1] An alternative terminology is to call the system *marginally stable* in this situation. This terminology has a slightly pejorative ring to it, probably reflecting the fact that small perturbations in the entries of the **A** coefficients could put the system in the unstable category.

---

[1]This conclusion depends on the assumption that the coefficient matrix is diagonalizable. It is possible to have unstable solutions from purely imaginary eigenvalues in a case where the change of basis results in a Jordan form.

8.4.2 *Jordan Form*

In the general case and in the face of Murphy's Law, it cannot be assumed that the coefficient matrix **A** of the system

$$\frac{d}{dt}\mathbf{x} = \mathbf{A}\mathbf{x}$$

is diagonalizable. The most that can be assumed is that there exists a change of basis that reduces to coefficient matrix to Jordan canonical form. The given system is then equivalent to a set of decoupled systems of lower dimension, with each in the standard form

$$\frac{d}{dt}\mathbf{z} = \begin{bmatrix} \lambda & 1 & 0 & \cdots \\ 0 & \lambda & 1 & \cdots \\ \vdots & & \ddots & \ddots \\ 0 & 0 & \cdots & \lambda \end{bmatrix} \mathbf{z}.$$

This system has the explicit solution (see page 184 in Section 7.4)

$$\mathbf{z}(t) = \begin{bmatrix} e^{\lambda t} & t\,e^{\lambda t} & \cdots & \frac{t^{n-1}}{n-1!}\,e^{\lambda t} \\ 0 & e^{\lambda t} & \cdots & \frac{t^{n-2}}{n-2!}\,e^{\lambda t} \\ \vdots & \ddots & \ddots & \vdots \\ 0 & 0 & \cdots & e^{\lambda t} \end{bmatrix} \mathbf{z}(0).$$

This explicit form allows us to draw conclusions parallel to those of the diagonalizable case.

If the real parts of all of the eigenvalues of **A** are strictly less than zero, then the system is asymptotically stable. (A negative exponential will beat any power of t.) If the real part of any eigenvalue is strictly positive, then the system will be unstable, with unbounded solutions for all but specially chosen initial conditions.

The case of purely imaginary eigenvalues is in this case more delicate than when the coefficient matrix is diagonalizable.

If any purely imaginary eigenvalue is associated with a Jordan block of dimension two or greater, then the system has solutions of the form

$$t\ \sin(\omega t)$$

in addition to purely sinusoidal solutions. Such solutions become unbounded as time tends to infinity. The conclusion is that the system is marginally stable provided that the real part of all of the eigenvalues are *less than or equal to zero*, and further, that all of the Jordan blocks associated with imaginary eigenvalues are of size 1×1. If all of the eigenvalues are less than or equal to zero and any of the Jordan blocks associated with purely imaginary eigenvalues is of dimension two or larger, then the system is unstable.

8.5 Linearization

In view of the apparent fact that the world is nonlinear, one might wonder if the linear analysis of the previous Section was of any relevance. It turns out that it is, and the connection hinges on the process of linearizing a nonlinear system.

Linearization involves choice of a base point. This is generally a particular solution of a given system of differential equations. What is then sought is an (approximate) equation for the deviation between a general solution and the particular one chosen. For an introduction we discuss linearization using an equilibrium point as the base, although it is both possible and not unusual to linearize about other particular solutions. Periodic solutions are of particular interest in this regard.

An *autonomous* (or time invariant) nonlinear system can be written in the form

$$\frac{d}{dt}\mathbf{x}(t) = \mathbf{f}(\mathbf{x}(t)),$$

where $\mathbf{f}(\cdot)$ represents a vector valued function of a vector argument. If this vector equation were written in terms of components, then the appearance would be that of the set of equations

$$\begin{aligned}
\frac{d}{dt}x_1 &= f_1(x_1, x_2, \ldots, x_n) \\
\frac{d}{dt}x_2 &= f_2(x_1, x_2, \ldots, x_n) \\
&\vdots \\
\frac{d}{dt}x_{n-1} &= f_{n-1}(x_1, x_2, \ldots, x_n) \\
\frac{d}{dt}x_n &= f_n(x_1, x_2, \ldots, x_n).
\end{aligned}$$

A *stationary point* or *equilibrium point* of such a system is a vector $\mathbf{x}_0$ such that

$$\mathbf{f}(\mathbf{x}_0) = \begin{bmatrix} 0 \\ 0 \\ \vdots \\ 0 \end{bmatrix}.$$

This means that the constant function $\mathbf{x}(t) = \mathbf{x}_0$ is a solution of the governing differential equations, or more simply, that the system stays at that point if it is initially placed (exactly) there.

A question that immediately arises about equilibrium points is whether the situation they represent is a stable one. That is, if the system is disturbed slightly from the equilibrium point, does it return there or wander off for parts unknown?

To attempt to answer that question, we obtain an equation for the deviation of a solution from the equilibrium point.

Since

$$\frac{d}{dt}(\mathbf{x}(t) - \mathbf{x}_0) = \frac{d}{dt}\mathbf{x}(t)$$

and (since $\mathbf{x}_0$ is an equilibrium point)

$$\mathbf{f}(\mathbf{x}(t)) = \mathbf{f}(\mathbf{x}(t)) - 0 = \mathbf{f}(\mathbf{x}(t)) - \mathbf{f}(\mathbf{x}_0),$$

we have

$$\frac{d}{dt}(\mathbf{x}(t) - \mathbf{x}_0) = \mathbf{f}(\mathbf{x}(t)) - \mathbf{f}(\mathbf{x}_0).$$

Expecting our interest to be in "small deviations" from the equilibrium, it is natural to expand the right hand side of the equation above in the form of a (multivariable) Taylor series. This takes the form

$$\mathbf{f}(\mathbf{x}(t)) - \mathbf{f}(\mathbf{x}_0) = \frac{\partial \mathbf{f}}{\partial \mathbf{x}}(\mathbf{x}_0)\ (\mathbf{x}(t) - \mathbf{x}_0) + O(\|\mathbf{x}(t) - \mathbf{x}_0\|^2),$$

where the linear term involves the Jacobian matrix of the slope function from the system equations evaluated at the base point $\mathbf{x}_0$. That is,

$$\frac{\partial \mathbf{f}}{\partial \mathbf{x}}(\mathbf{x}_0) = \begin{bmatrix} \frac{\partial f_1}{\partial x_1} & \frac{\partial f_1}{\partial x_2} & \cdots & \frac{\partial f_1}{\partial x_n} \\ \frac{\partial f_2}{\partial x_1} & \frac{\partial f_2}{\partial x_2} & \cdots & \frac{\partial f_2}{\partial x_n} \\ \vdots & \vdots & \vdots & \vdots \\ \frac{\partial f_n}{\partial x_1} & \frac{\partial f_n}{\partial x_2} & \cdots & \frac{\partial f_n}{\partial x_n} \end{bmatrix}(\mathbf{x}_0).$$

Note that this is a *constant* matrix. If the quadratic error term in the Taylor expansion above is neglected, what is obtained is a linear, constant coefficient system of differential equations for the deviation from equilibrium. This takes the form

$$\frac{d}{dt}(\mathbf{x}(t) - \mathbf{x}_0) = \mathbf{A}\ (\mathbf{x}(t) - \mathbf{x}_0),$$

where $\mathbf{A}$ is the Jacobian matrix

$$\mathbf{A} = \frac{\partial \mathbf{f}}{\partial \mathbf{x}}(\mathbf{x}_0).$$

On the basis of the linear analysis carried out above, we expect that the equilibrium is stable provided that all of the eigenvalues of the coefficient matrix $\mathbf{A} = \frac{\partial \mathbf{f}}{\partial \mathbf{x}}(\mathbf{x}_0)$ have strictly negative real parts. Conversely, if any of the eigenvalues has a positive real part, then the equilibrium can be expected to be unstable. If the linear analysis indicates marginal stability, then the problem is too delicate for this blunt approach. Of course, in order to support the claims of the linear analysis, something must be done to justify the validity of ignoring the quadratic error terms in the above. Such results are available by means of Lyapunov methods, introduced below.

Example

Introductory solid mechanics is largely concerned with analyzing the deflections of transversally loaded beams, and such problems have been treated earlier. Beams are also used as supporting columns with a direct axial loading. Such arrangements are sometimes subject to catastrophic failure when the axially loaded member fails by buckling.

Analysis of this problem is beyond the range of the shear and bending moment methods used for transverse deflection problems. The buckling phenomenon really is an example of a problem of stability. The condition in which the column supports the load is a condition of equilibrium, but for sufficiently high axial loads the equilibrium is an unstable one. Small perturbations from the equilibrium are amplified by the instability, and failure results.

To fully analyze this issue requires modeling the column in terms of a system of partial differential equations, but the essence of the phenomenon becomes clear with the analysis of simpler mechanical models crafted to (hopefully) share the basic physical mechanisms that act in the more complicated genuine column case.

If a beam is bent transversally, then potential energy is stored in the elastic deformation. The beam also has mass distributed along its length, so that there are inertial effects involved as well. The simplest mechanical analog that can be imagined might be the arrangement illustrated in Figure 8.5. This consists of a pair of mass-less rods of length l, fixed to slide along a frictionless track, and hinged in the middle. The column axial load is modeled as axial forces pushing along the track.

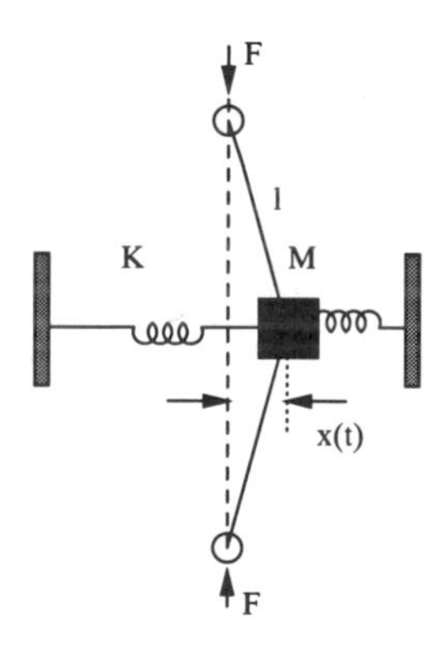

FIGURE 8.5. Simple Column Analog

To model the inertial and elastic properties of the real beam, we attach a mass at the hinge point, and restrain it with a pair of springs. The equation of motion for this system involves calculating the net horizontal forces acting on the mass. In the first place there is a restoring force of

$$-2\,K\,x(t),$$

acting due to the deformation of the springs. With the assumption that the connecting rods are massless, the forces acting on each end of the rods are equal and opposite. To calculate the effect on the mass, we first calculate the tensile force T in the rod according to

$$T\,\frac{\sqrt{l^2 - x^2}}{l} = F,$$

keeping the vertical component of the T equal to the axial load F. The forcing acting on the mass is the horizontal component of the tensile stress, and hence is

$$\frac{x(t)}{l}\,T = F\,\frac{x}{\sqrt{l^2 - x^2}}.$$

The equation of motion of the mass becomes

$$\frac{d^2x}{dt^2} = \frac{1}{M}\left(-2\,K\,x(t) + 2\,F\,\frac{x(t)}{\sqrt{l^2 - x(t)^2}}\right).$$

If this is represented as a vector system, the system of equations is

$$\begin{aligned}
\frac{d}{dt}x(t) &= v(t),\\
\frac{d}{dt}v(t) &= \frac{1}{M}\left(-2\,K\,x(t) + 2\,F\,\frac{x(t)}{\sqrt{l^2 - x(t)^2}}\right).
\end{aligned}$$

Of course, this is a nonlinear system of equations, so that one can attempt to analyze the stability of the equilibrium by linearizing the equation of motion. The Jacobian of the "velocity function" is

$$\frac{\partial \mathbf{f}}{\partial \mathbf{x}} = \begin{bmatrix} 0 & 1 \\ \frac{1}{M}\left(-2\,K\ + 2\,F\,(\frac{1}{\sqrt{l^2 - x(t)^2}} + \frac{x^2}{(l^2 - x^2)^{\frac{3}{2}}})\right) & 0 \end{bmatrix},$$

and evaluating this at the equilibrium point $x = 0,\ v = 0$ gives

$$\mathbf{A} = \begin{bmatrix} 0 & 1 \\ \frac{1}{M}(-2\,K + 2\,\frac{F}{l}) & 0 \end{bmatrix}.$$

To check the stability of the equilibrium, compute the characteristic polynomial of the coefficient matrix and check the eigenvalues. The characteristic polynomial is

$$p(s) = s^2 + \left(2K - 2\frac{F}{l}\right).$$

At the critical point where

$$\left(2K - 2\frac{F}{l}\right) = 0,$$

the eigenvalues change from purely imaginary values to a pair of real eigenvalues of opposite signs. If

$$2\frac{F}{l} > 2K,$$

the linearized system is unstable, and the equilibrium position is unstable as well.

The simple model therefore displays the effect observed with the column: a transition to an unstable equilibrium when the axial load reaches a critical value.

A perhaps more realistic model could take into account the mass of the connecting rods, and the fact that a column is typically loaded from above, and fixed at the base. If this is done, the governing differential equation becomes substantially more complicated. This case is discussed in Problem 2 below.

8.6 Introduction to Lyapunov Theory

In stability problems, it is relatively rare that explicit closed form calculations of solutions can be made. This is the case (in principle) for linear constant coefficient systems, but many system models include non-linear terms and preclude the possibilities of explicit formulas. In addition, for many systems of interest, the models are really not completely known. Parameter values may drift over time, or be manufactured out of tolerance. One would like to be able to predict that the system will actually remain stable in the face of such variations, and in order to do this an analysis method that allows for some uncertainty in the model is required.

Lyapunov methods are a useful approach in this regard. The basic idea is geometrical in origin, and does not require anything like calculations of explicit solutions. (Historically, this is referred to as "Lyapunov's second method", the "first method" was explicit solution-bound.)

The basic idea of the method can be succinctly stated: if one could find some sort of function that measured the distance of a system from equilibrium, and determine that this measure decreased over time, then the conclusion should be that the equilibrium was a stable point. This idea can be illustrated with the familiar example of a damped harmonic oscillator. A spring mass system with linear damping is modeled by

$$M\frac{d^2y}{dt^2} + c\,\frac{dy}{dt} + k\,y = 0.$$

The total energy of such a system is represented by

$$V\left(y(t), \frac{dy}{dt}\right) = \frac{1}{2}\,k\,y^2(t) + \frac{1}{2}\,M\left(\frac{dy}{dt}\right)^2.$$

In the phase plane, the curves of constant V represent "concentric" ellipses surrounding the equilibrium point at $y = 0$, $\frac{dy}{dt} = 0$. If the time rate of change of V along the motion of the system is calculated, then the result is

$$\begin{aligned}
\frac{d}{dt}V\left(y(t), \frac{dy}{dt}\right) &= \frac{d}{dt}\left(\frac{1}{2}\,k\,y^2(t) + \frac{1}{2}\,M\left(\frac{dy}{dt}\right)^2\right) \\
&= k\,y(t)\,\frac{dy}{dt} + \frac{dy}{dt}\,M\,\frac{d^2y}{dt^2} \\
&= k\,y(t)\,\frac{dy}{dt} - \frac{dy}{dt}\left(c\,\frac{dy}{dt} + k\,y\right) \\
&= -c\left(\frac{dy}{dt}\right)^2.
\end{aligned}$$

This means that the system trajectories are moving inward across the equal energy contours represented by the ellipses. On this basis stability can be concluded, without actually calculating any explicit solutions or auxiliary information like eigenvalues.

The utility of such methods becomes evident if one considers a variant of the above in which the damping is a nonlinear function of the velocity. The model for such a system is

$$M\frac{d^2y}{dt^2} + f\left(\frac{dy}{dt}\right) + k\,y = 0.$$

If the same V as the above is chosen as the *Lyapunov function* for the problem, then

$$\begin{aligned}
\frac{d}{dt} V\left(y(t), \frac{dy}{dt}\right) &= \frac{d}{dt}\left(\frac{1}{2} k\, y^2(t) + \frac{1}{2} M \left(\frac{dy}{dt}\right)^2\right) \\
&= k\, y(t) \frac{dy}{dt} + \frac{dy}{dt} M \frac{d^2y}{dt^2} \\
&= k\, y(t) \frac{dy}{dt} - \frac{dy}{dt}\left(f(\frac{dy}{dt}) + k\, y\right) \\
&= -f\left(\frac{dy}{dt}\right) \frac{dy}{dt}.
\end{aligned}$$

The natural assumption one would make about the damping term is that it opposes the motion, so that the graph of f should lie in the first and third quadrants. But under those circumstances, $\frac{dy}{dt}$ and $f(\frac{dy}{dt})$ always have the same algebraic sign, so that

$$f\left(\frac{dy}{dt}\right) \frac{dy}{dt} > 0,$$

making the time derivative of V negative in this (nonlinear) situation. This would indicate that the harmonic oscillator would be stable with a rather arbitrary nonlinear damping mechanism, rather than just in the linear case where explicit calculations are available.

8.6.1 Linear Problems

Lyapunov methods are an incredibly powerful approach to showing that certain systems of differential equations are stable, but the power comes at a price. The truth of the situation is that there is no recipe for construction of a Lyapunov function. It is basically a matter of clever guesswork, and experience and physical intuition are the only guides available for the search.

If searches for Lyapunov functions are to be undertaken, the prudent search party must consider what the prospects for finding such a function are. Unless there are some assurances that there actually exist Lyapunov functions for systems that are "really" stable, a lot of effort could be expended trying to invent energy-like functions that are never going to work anyway.

So-called converse Lyapunov theorems do exist (see books devoted to Lyapunov methods, such as [9]), although they are of little help in guiding guesses in particular problems. Another mine for usable Lyapunov functions is in "nearby problems". For problems that are slightly nonlinear, or only slowly time-varying, constant coefficient linear systems offer themselves as "nearby" candidates that might provide Lyapunov functions. Linear constant coefficient systems also are a benchmark for the possibilities of construction of Lyapunov functions appropriate to systems known to be stable. In principle, constant coefficient systems can be completely analyzed by linear algebraic methods, and it would be reassuring to know that Lyapunov methods also offer a sufficient tool.

The question to consider then is, given the asymptotically stable constant coefficient system

$$\frac{d}{dt}\mathbf{x}(t) = \mathbf{A}\,\mathbf{x}(t),$$

is it possible to discover a function V of the state vector $\mathbf{x}(t)$ such that

$$\frac{d}{dt}V(\mathbf{x}(t)) < 0$$

along trajectories of the constant coefficient differential equation system.

It turns out that for systems of the above form, it is not necessary to seek *arbitrary* functions V. For stable systems, Lyapunov functions that establish the stability can be found in the form of a quadratic function of the state variable components. In short, the sort of Lyapunov function that worked for the harmonic oscillator example suffices for the most general sort of constant coefficient system of differential equations. We therefore seek a Lyapunov function in the form

$$V(\mathbf{x}(t)) = \mathbf{x}'(t)\mathbf{Q}\mathbf{x}(t).$$

Then

$$\begin{aligned}
\frac{d}{dt}V(\mathbf{x}(t)) &= \frac{d}{dt}\mathbf{x}'(t)\mathbf{Q}\mathbf{x}(t) + \mathbf{x}'(t)\mathbf{Q}\frac{d}{dt}\mathbf{x}(t) \\
&= \mathbf{x}'(t)\mathbf{A}'\mathbf{Q}\mathbf{x}(t) + \mathbf{x}'(t)\mathbf{Q}\mathbf{A}\mathbf{x}(t), \\
&= \mathbf{x}'(t)\left[\mathbf{A}'\mathbf{Q} + \mathbf{Q}\mathbf{A}\right]\mathbf{x}(t).
\end{aligned}$$

This quantity can be made negative if $\mathbf{Q}$ can be chosen in such a way that it satisfies (the Lyapunov equation)

$$\mathbf{A}'\mathbf{Q} + \mathbf{Q}\mathbf{A} = -\mathbf{I}.$$

Then our quadratic Lyapunov function would satisfy

$$\frac{d}{dt}V(\mathbf{x}(t)) = -\|\mathbf{x}(t)\|^2.$$

Provided that the level surfaces of $\mathbf{x}'\mathbf{Q}\mathbf{x}$ are energy-like (multidimensional ellipsoids), this would then verify stability by Lyapunov methods. The problem is to know that the above equation for $\mathbf{Q}$ has a solution with the required properties.

This is not as difficult as one might fear. The system of equations above actually represents a linear system of equations for the elements of the unknown matrix $\mathbf{Q}$, but row reduction is not the approach to take. The solution to the system of equations can be written explicitly in terms of matrix exponentials.

The form of solution is the integral (convergent by the stability assumption)

$$\mathbf{Q} = \int_0^\infty e^{\mathbf{A}'t}\, e^{\mathbf{A}t}\, dt.$$

That this is a solution can be verified by substituting it into the equation. Then we see

$$\begin{aligned}
\mathbf{A}' \int_0^\infty e^{\mathbf{A}'t}\, e^{\mathbf{A}t}\, dt + \int_0^\infty e^{\mathbf{A}'t}\, e^{\mathbf{A}t}\, dt\, A &= \int_0^\infty \frac{d}{dt}\left(e^{\mathbf{A}'t}\, e^{\mathbf{A}t}\right) dt \\
&= e^{\mathbf{A}'t}\, e^{\mathbf{A}t} \Big|_0^\infty \\
&= \mathbf{0} - \mathbf{I}.
\end{aligned}$$

The fact that the $\mathbf{Q}$ defined this way provides level surfaces of the appropriate sort follows from the observation that the form of solution makes $\mathbf{Q}$ a positive definite symmetric matrix. Symmetry is built into the integral formula, and if we test for positive definiteness by constructing the quadratic form $\mathbf{x}'\mathbf{Q}\mathbf{x}$, the result is

$$\begin{aligned}
\mathbf{x}'\mathbf{Q}\mathbf{x} &= \int_0^\infty \mathbf{x}'\, e^{\mathbf{A}'t}\, e^{\mathbf{A}t}\mathbf{x}\, dt \\
&= \int_0^\infty \|e^{\mathbf{A}t}\mathbf{x}\|^2\, dt \\
&= \geq 0.
\end{aligned}$$

This shows that $\mathbf{Q}$ is at least positive semidefinite. Positive definiteness follows from the observation that the expression can only vanish for $\mathbf{x} = \mathbf{0}$, as the exponential is always invertible.

The conclusion then is that Lyapunov methods suffice for showing stability of constant coefficient systems, and moreover that there are explicit formulas for the Lyapunov functions appropriate to such problems.

8.6.2 Applications

An immediate use for the linear Lyapunov results is as justification for conclusions drawn from linearizing non-linear differential equations about a point of equilibrium point. The tactic is surprisingly direct: just try on the Lyapunov function appropriate to the linearized model as a candidate Lyapunov function for the original non-linear system of equations. The situation is of an *autonomous* (or time invariant) nonlinear system in the form

$$\frac{d}{dt}\mathbf{x}(t) = \mathbf{f}(\mathbf{x}(t)),$$

of a vector system of equations, or in component form as

$$\begin{aligned}
\frac{d}{dt}x_1 &= f_1(x_1, x_2, \ldots, x_n) \\
\frac{d}{dt}x_2 &= f_2(x_1, x_2, \ldots, x_n) \\
&\vdots \\
\frac{d}{dt}x_{n-1} &= f_{n-1}(x_1, x_2, \ldots, x_n) \\
\frac{d}{dt}x_n &= f_n(x_1, x_2, \ldots, x_n).
\end{aligned}$$

$\mathbf{x}_0$ is an equilibrium point such that

$$\mathbf{f}(\mathbf{x}_0) = \begin{bmatrix} 0 \\ 0 \\ \vdots \\ 0 \end{bmatrix}.$$

The linearization of this system about the equilibrium point is

$$\frac{d}{dt}(\mathbf{x}(t) - \mathbf{x}_0) = \mathbf{A}\,(\mathbf{x}(t) - \mathbf{x}_0),$$

where the constant coefficient matrix $\mathbf{A}$ is the Jacobian matrix

$$\mathbf{A} = \frac{\partial \mathbf{f}}{\partial \mathbf{x}}(\mathbf{x}_0)$$

evaluated at the equilibrium point. We would like to believe that if the linearized system is asymptotically stable, then initial values close to the equilibrium point will lead to solutions that go to the equilibrium point as t tends to infinity. To attempt to show this, we use the linearized system coefficient matrix to find a $\mathbf{Q}$ satisfying

$$\mathbf{A}'\mathbf{Q} + \mathbf{Q}\mathbf{A} = -\mathbf{I}.$$

Then we choose

$$V(\mathbf{x}(t)) = (\mathbf{x}(\mathbf{t}) - \mathbf{x}_0)'\mathbf{Q}(\mathbf{x}(\mathbf{t}) - \mathbf{x}_0)$$

as a possible Lyapunov function. The beauty of Lyapunov methods is that the problem reduces to a mechanical calculation once a Lyapunov function has been selected. Either the chosen function has negative derivatives in a neighborhood of the equilibrium point, or a new guess is in order.

The rate of change of our chosen V is computed as

$$\frac{d}{dt} V(\mathbf{x}(t)) = \frac{d}{dt} (\mathbf{x(t)} - \mathbf{x}_0)'\mathbf{Q}(\mathbf{x(t)} - \mathbf{x}_0)$$

$$= \left(\frac{d}{dt} (\mathbf{x(t)} - \mathbf{x}_0)\right)' \mathbf{Q}\,(\mathbf{x(t)} - \mathbf{x}_0) + (\mathbf{x(t)} - \mathbf{x}_0)'\,\mathbf{Q} \left(\frac{d}{dt}(\mathbf{x(t)} - \mathbf{x}_0)\right)$$

$$= (\mathbf{f}(\mathbf{x(t)}) - \mathbf{f}(\mathbf{x}_0))'\,\mathbf{Q}\,(\mathbf{x(t)} - \mathbf{x}_0) + (\mathbf{x(t)} - \mathbf{x}_0)'\,\mathbf{Q}\,(\mathbf{f}(\mathbf{x(t)}) - \mathbf{f}(\mathbf{x}_0))$$

$$= (\mathbf{A}(\mathbf{x(t)} - \mathbf{x}_0) + O(\|\mathbf{x(t)} - \mathbf{x}_0\|^2))'\mathbf{Q}(\mathbf{x(t)} - \mathbf{x}_0)$$

$$+ (\mathbf{x(t)} - \mathbf{x}_0)'\mathbf{Q}(\mathbf{A}(\mathbf{x(t)} - \mathbf{x}_0) + O(\|\mathbf{x(t)} - \mathbf{x}_0\|^2))$$

$$= (\mathbf{x(t)} - \mathbf{x}_0)' \left(\mathbf{A}'\mathbf{Q} + \mathbf{Q}\mathbf{A}\right) (\mathbf{x(t)} - \mathbf{x}_0) + O(\|\mathbf{x(t)} - \mathbf{x}_0\|^3)$$

$$= -\|\mathbf{x(t)} - \mathbf{x}_0\|^2 + O(\|\mathbf{x(t)} - \mathbf{x}_0\|^3).$$

For $\|\mathbf{x(t)}-\mathbf{x}_0\|$ sufficiently small, this quantity is negative, leading to the conclusion that the equilibrium of the nonlinear system would be at least locally stable. Note that this Lyapunov based style of argument would not in this instance lead to the conclusion that *all* solutions of the system head for the equilibrium position, but only that those starting from initial conditions sufficiently close to the equilibrium do.

8.7 Problems

1. The model of a "weight on a string" pendulum is

$$m\,l^2 \frac{d^2\theta}{dt^2} + m\,g\,l\,\sin(\theta) = 0.$$

   Derive the linearized vector version of this model.

2. The equation of a more serious spring-loaded hinge (Figure 8.6) uses a mass as an axial load, and includes the mass and hence inertial effects of the legs of the hinge. In this case, the motion cannot be treated as a simple one dimensional one, and the model is probably best derived by using the methods of Lagrangian mechanics. The natural variable then is the angle of

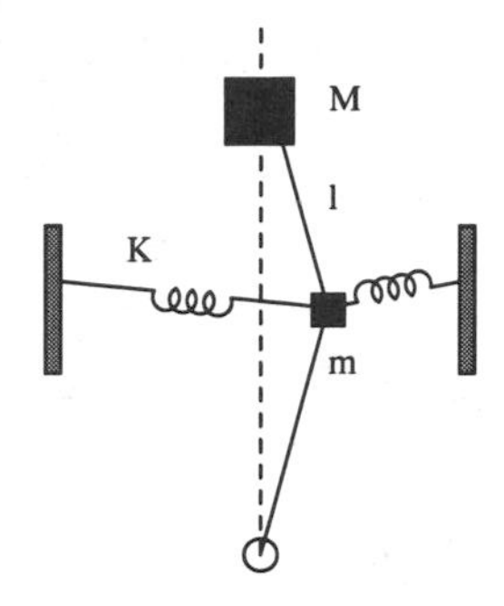

FIGURE 8.6. Loaded Model

the column base rod from the vertical, and the resulting equation of motion turns out to be the nonlinear one:

$$
\begin{aligned}
&2\Big(4\,M\,l^2\sin(\theta(t))\cos(\theta(t))\left(\frac{\partial}{\partial t}\theta(t)\right) \\
&\qquad - \mu l^2\cos(\theta(t))\sin(\theta(t))\left(\frac{\partial}{\partial t}\theta(t)\right)\Big)\left(\frac{\partial}{\partial t}\theta(t)\right) \\
&+2\left(\frac{1}{3}\rho l^3 + 2M\,l^2\sin(\theta(t))^2 + \frac{1}{2}\mu l^2\cos(\theta(t))^2\right)\left(\frac{\partial^2}{\partial t^2}\theta(t)\right) - \\
&\left(4\,M\,l^2\sin(\theta(t))\cos(\theta(t)) - \mu l^2\cos(\theta(t))\sin(\theta(t))\right)\left(\frac{\partial}{\partial t}\theta(t)\right)^2 \\
&\qquad - (2Mgl + 2\rho l^2 g)\sin(\theta(t)) + 2Kl^2\sin(\theta(t))\cos(\theta(t)) = 0.
\end{aligned}
$$

Find a linearized vector version, about the equilibrium rest point.

3. Characterize each of the following differential equations as either asymptotically stable, marignally stable, or unstable. For each of the asymptotically stable cases, classify the situation as overdamped, critically damped, or underdamped.

(a)

$$\frac{d^2y}{dt^2} + 3\frac{dy}{dt} + 2y = 0$$

(b)

$$\frac{d^2y}{dt^2} + 2\frac{dy}{dt} + 5y = 0$$

(c)

$$\frac{d^2y}{dt^2} + -y = 0$$

(d)

$$\frac{d^2y}{dt^2} + 4y = 0$$

(e)

$$\frac{d^2y}{dt^2} + 2\frac{dy}{dt} + y = 0$$

(f)

$$\frac{d^2y}{dt^2} + 5\frac{dy}{dt} + 6.25y = 0$$

(g)

$$\frac{d^2y}{dt^2} - \frac{dy}{dt} - 6y = 0$$

(h)

$$\frac{d^2y}{dt^2} - 2\frac{dy}{dt} + 10y = 0$$

4. Is the vector equation

$$\frac{d}{dt}\mathbf{x} = \begin{bmatrix} 0 & 1 \\ -1 & -1 \end{bmatrix} \mathbf{x}$$

a stable one?

5. Is the system of equations

$$\frac{d}{dt}\mathbf{x} = \begin{bmatrix} 0 & 1 & 0 \\ 0 & 0 & 1 \\ -4 & 0 & 3 \end{bmatrix} \mathbf{x}$$

stable or unstable?

6. What stability category applies to

$$\frac{d}{dt}\mathbf{x} = \begin{bmatrix} 0 & 1 \\ 0 & 0 \end{bmatrix} \mathbf{x}?$$

7. For each of the stable examples of Question 3, find a Lyapunov equation that establishes stability.

8. Show that

$$\frac{d}{dt}\frac{1}{2}\left(\omega_n^2\, y(t)^2 + \left(\frac{dy}{dt}\right)^2\right) = 0$$

for the undamped harmonic oscillator model

$$\frac{d^2x}{dt^2} + \omega_n^2 x = 0.$$

9. Find a Lyapunov equation for the system of Question 4 by solving the matrix equation

$$\begin{bmatrix} 0 & 1 \\ -1 & -1 \end{bmatrix}' \mathbf{Q} + \mathbf{Q} \begin{bmatrix} 0 & 1 \\ -1 & -1 \end{bmatrix} = \begin{bmatrix} -1 & 0 \\ 0 & -1 \end{bmatrix}.$$

10. A model for a damped harmonic oscillator with a non-linear spring is

$$\frac{d^2y}{dt^2} + c\frac{dy}{dt} + f(y) = 0,$$

where the nonlinear spring characteristic is assumed to satisfy

$$\zeta\, f(\zeta) \geq \alpha\zeta^2,\ \alpha > 0.$$

Using the work

$$F(y) = \int_0^y f(\zeta)\, d\zeta$$

done in stretching the spring, find a Lyapunov function that shows that the nonlinear damped harmonic oscillator is stable.

11. Suppose that the model of Question 10 has a nonlinear damping mechanism as well as a nonlinear spring. That is,

$$\frac{d^2y}{dt^2} + c\left(\frac{dy}{dt}\right) + f(y) = 0,$$

where the damping is a first and third quadrant function, so that

$$\zeta\, c(\zeta) \geq \beta\zeta^2,\ \beta > 0.$$

Show that this model is also stable, by finding the "obvious" Lyapunov function.

9

Periodic Problems

Many physical problems of interest involve periodic phenomena. In a sense, the widely studied constant coefficient cases are a somewhat extreme example in which the coefficients in the model are periodic with any period one cares to choose. Other problems with periodic coefficients arise from investigating the stability of the oscillator circuits underlying many (all ?) communication systems. The long-standing issues of the stability of planetary orbits also can be approached through analysis of a linearized system with periodic coefficients.

These examples involve the periodicity of the coefficient functions in the governing differential equations. Other problems of interest involve constant coefficient differential equations (or systems of such) that have periodic forcing functions. In this case what is wanted is what has earlier been called a particular (or forced) solution of the differential equation. In this case, it is a periodic solution which is sought.

9.1 Periodic Inputs

A periodic input problem can take the form of a single higher order differential equation (with constant coefficients)

$$a_n \frac{d^n y}{dt^n} + a_{n-1} \frac{d^{n-1} y}{dt^{n-1}} + a_{n-2} \frac{d^{n-2} y}{dt^{n-2}} + \ldots + a_0\, y(t) = f(t),$$

where the waveform of the forcing function is periodic, that is

$$f(t+T) = f(t)$$

for some period T. For certain cases of this problem, particular solutions have been calculated either by means of undetermined coefficients or Laplace transforms. For

$$a_n \frac{d^n y}{dt^n} + a_{n-1} \frac{d^{n-1} y}{dt^{n-1}} + a_{n-2} \frac{d^{n-2} y}{dt^{n-2}} + \ldots + a_0\, y(t) = \cos(\omega t)$$

the period of the forcing function is

$$T = \frac{2\pi}{\omega},$$

and particular (periodic) solutions can (often, barring resonance) be found in the form

$$y(t) = A\ \cos(\omega t) + B\ \sin(\omega t).$$

Constant coefficient vector systems of differential equations with periodic forcing functions can also be considered. In vector notation these take the form

$$\frac{d}{dt}\mathbf{x}(t) = \mathbf{A}\,\mathbf{x}(t) + \mathbf{f}(t),$$

where the forcing function now is a periodic vector function

$$\mathbf{f}(t+T) = \mathbf{f}(t).$$

As usual, this way of considering the problem is of more general use than the single equation of order n, since the single equation can always be written in vector form as

$$\frac{d}{dt}\mathbf{x} = \begin{bmatrix} 0 & 1 & 0 & \cdots \\ 0 & 0 & 1 & \cdots \\ \vdots & \vdots & \vdots & \vdots \\ 0 & 0 & \cdots & 1 \\ -\frac{a_0}{a_n} & -\frac{a_1}{a_n} & \cdots & -\frac{a_{n-1}}{a_n} \end{bmatrix} \mathbf{x} + \begin{bmatrix} 0 \\ 0 \\ \vdots \\ -\frac{f(t)}{a_n} \end{bmatrix}.$$

9.2 Phasors

Solving constant coefficient problems with sinusoidal forcing functions can be made easier than undetermined coefficients or Laplace transform methods might indicate in the case (most common) in which the differential equation coefficients are real valued. Then Euler's formula can be used to do the required calculations in terms of complex arithmetic. The method is easily illustrated with a scalar first order example.

We consider the forced equation

$$\frac{d}{dt}x(t) = -a\,x(t) + e^{i\,\omega t},$$

assuming that a is a real number. Then undetermined coefficients will lead to the periodic, forced, particular solution

$$x(t) = \frac{1}{i\omega + a} e^{i\,\omega\,t}.$$

This solution is complex valued, and hence has a real and an imaginary component. If we use the notation

$$x(t) = x_R(t) + i\,x_I(t),$$

then the two components can be determined just by splitting the solution formula into real and imaginary parts. Using the polar form

$$i\omega + a = \sqrt{\omega^2 + a^2}\, e^{i\ \arctan(\frac{\omega}{a})}$$

we can rewrite the solution as

$$\begin{aligned}
x(t) &= \frac{1}{i\omega + a} e^{i\,\omega\,t} \\
&= \frac{1}{\sqrt{\omega^2 + a^2}} e^{-i\ \arctan(\frac{\omega}{a})}\, e^{i\,\omega\,t} \\
&= \frac{1}{\sqrt{\omega^2 + a^2}} e^{-i\,(\omega t - \arctan(\frac{\omega}{a}))} \\
&= \frac{1}{\sqrt{\omega^2 + a^2}} \left(\cos(\omega t - \arctan(\frac{\omega}{a})) + i\ \sin(\omega t - \arctan(\frac{\omega}{a}))\right).
\end{aligned}$$

The real and imaginary parts of the solution are then

$$\begin{aligned}
x_R(t) &= \frac{1}{\sqrt{\omega^2 + a^2}} \cos(\omega t - \arctan(\frac{\omega}{a})), \\
x_I(t) &= \frac{1}{\sqrt{\omega^2 + a^2}} \sin(\omega t - \arctan(\frac{\omega}{a})).
\end{aligned}$$

The point of this arises from writing the original equation in terms of real and imaginary parts.

$$\begin{aligned}
\frac{d}{dt}\,(x_R(t) + i x_I(t)) &= -a\ (x_R(t) + i x_I(t)) + e^{i\,\omega\,t}, \\
\frac{d}{dt}\,x_R(t) &= -a\,x_R(t) + \cos(\,\omega\,t), \\
\frac{d}{dt}\,x_I(t) &= -a\,x_I(t) + \sin(\,\omega\,t).
\end{aligned}$$

With this information the particular solution to

$$\frac{d}{dt} x(t) = -a\, x(t) + A \cos(\omega t) + B \sin(\omega t)$$

can be written down immediately by superposition of the particular solutions computed by complex means to get

$$x(t) = A\, x_R(t) + B\, x_I(t).$$

This calculation can be carried out in like fashion for an arbitrary linear, constant real coefficient differential equation.

The complex forcing function

$$e^{i\,\omega t}$$

is a complex number of magnitude 1 for all t, rotating counterclockwise in the complex plane with angular velocity ω. This is referred to as a *phasor*, and a particular solution calculated with $e^{i\,\omega t}$ is called the *phasor response* of the system.

General phasor response calculations may be made for single differential equations of arbitrary order. If the forced equation of order n is given, the problem of calculating the phasor response is to find a particular periodic solution of

$$a_n \frac{d^n y}{dt^n} + a_{n-1} \frac{d^{n-1} y}{dt^{n-1}} + a_{n-2} \frac{d^{n-2} y}{dt^{n-2}} + \ldots + a_0\, y(t) = e^{i\,\omega t}.$$

Using the method of undetermined coefficients, the solution is (assuming that $i\omega$ is not a root of the characteristic polynomial)

$$x(t) = \frac{1}{a_n (i\omega)^n + a_{n-1} (i\omega)^{n-1} + \ldots + a_0} e^{i\,\omega t}.$$

The argument from this point proceeds exactly as in the simple example, with the exception that the complex arithmetic becomes more involved with additional terms.[1]

The complex fraction appearing in the answer above defines the phase and magnitude of the response of the system to a phasor input as function of the frequency of the input (phasor) signal. For this reason

$$G(i\,\omega) = \frac{1}{a_n (i\omega)^n + a_{n-1} (i\omega)^{n-1} + \ldots + a_0}$$

is referred to as the *frequency response* of the differential equation.

[1] If the arithmetic becomes too tedious, there is always Maple ...

9.3 Fourier Series

Periodic functions are often represented as a sum of trigonometric functions, and such forms lend themselves to calculations of periodic solutions in the same form. The basic idea starts with a forced differential equation

$$a_n \frac{d^n y}{dt^n} + a_{n-1} \frac{d^{n-1} y}{dt^{n-1}} + a_{n-2} \frac{d^{n-2} y}{dt^{n-2}} + \ldots + a_0 \, y(t) = f(t),$$

where $f(t)$ is a "phasor sum"

$$f(t) = \Sigma_{n=-N}^{N} \, c_n \, e^{i \, n \, \omega_0 t}.$$

The terms in this sum are all oscillating at integer multiples of a fundamental frequency ω_0, so that the entire sum is periodic of period $T = \frac{2\pi}{\omega_0}$.

Although c_n in the above expansion is generally a complex quantity, the entire sum will emerge real valued provided that replacing n by $-n$ in the coefficient produces the complex conjugate quantity

$$c_{-n} = \overline{c_n}.$$

Since each of the terms in the sum is a phasor quantity, the particular solution corresponding to the sum forcing term is a superposition of the individual responses, with scaling to match the input terms. Hence the solution takes the form

$$x(t) = \Sigma_{n=-N}^{N} \, \frac{1}{p(i \, n \, \omega_0)} \, c_n \, e^{i \, n \, \omega_0 t},$$

where $p(\cdot)$ is the characteristic polynomial of the differential equation. The interpretation of this solution is that it simply multiplies the input phasor components by the corresponding frequency response.

Example
The (periodic particular) solution to the periodically forced harmonic oscillator

$$\frac{d^y}{dt^2} + 2 \, \zeta \, \omega_n \, \frac{dy}{dt} + \omega_n^2 \, y = \Sigma_{n=-N}^{N} \, c_n \, e^{i \, n \, \omega_0 t}$$

is

$$x(t) = \Sigma_{n=-N}^{N} \, \frac{1}{(i \, n \, \omega_0)^2 + 2 \, \zeta \, \omega_n \, (i \, n \, \omega_0) + \omega_n^2} \, c_n \, e^{i \, n \, \omega_0 t}.$$

Note that the magnitude effect of the terms in the above can be visualized from the resonance response curves calculated earlier in Figure 5.2.

The representation of periodic functions by an expansion of the form appearing above is called a Fourier Series representation of the function. The expression above is what is called a finite Fourier series, as only a finite number of terms are involved in the sum. The form actually is an instance of an expansion in terms of a set of mutually orthogonal vectors. The relationship that reveals this fact is that for $n \neq m$

$$\begin{aligned} \frac{1}{T}\int_0^T e^{i\,n\,\omega_0\,t}\,\overline{e^{i\,m\,\omega_0\,t}}\,dt &= \frac{1}{T}\int_0^T e^{i\,(n-m)\,\omega_0\,t}\,dt \\ &= \frac{1}{(n-m)\,\omega_0\,T} e^{i\,(n-m)\,\omega_0\,t}\Big|_0^T \\ &= \frac{1}{(n-m)\,\omega_0\,T}\left(e^{i\,(n-m)\,2\pi} - 1\right) \\ &= 0. \end{aligned}$$

At the same time

$$\begin{aligned} \frac{1}{T}\int_0^T e^{i\,n\,\omega_0\,t}\,\overline{e^{i\,n\,\omega_0\,t}}\,dt &= \frac{1}{T}\int_0^T 1\,dt \\ &= 1. \end{aligned}$$

These relations lead to the Fourier coefficient formula

$$c_n = \frac{1}{T}\int_0^T f(t)\,e^{-i\,n\,\omega_0\,t}\,dt$$

for evaluating the expansion coefficient from the periodic function, and in turn to the representation

$$f(t) = \Sigma_{n=-\infty}^{\infty}\, c_n\, e^{i\,n\,\omega_0 t}$$

for the Fourier Series expansion of a periodic function f.

The formal solution of the periodically forced differential equation

$$\begin{aligned} a_n\,\frac{d^n y}{dt^n} &+ a_{n-1}\,\frac{d^{n-1} y}{dt^{n-1}} + a_{n-2}\,\frac{d^{n-2} y}{dt^{n-2}} + \ldots + a_0\,y(t) = f(t) \\ &= \Sigma_{n=-\infty}^{\infty}\, c_n\, e^{i\,n\,\omega_0 t} \end{aligned}$$

is then

$$x(t) = \Sigma_{-\infty}^{\infty}\, c_n\, \frac{1}{a_n(i\omega_0)^n + a_{n-1}(i\omega_0)^{n-1} + \ldots + a_0} e^{i\,n\,\omega_0 t}.$$

The interpretation of this expression is simply that each phasor component of the forcing function appears scaled by the frequency response in the periodic particular solution.

9.4 Time Domain Methods

While the Fourier Series approach to periodic solutions outlined above is the classical one, it has certain drawbacks. One involves the effort in computing the Fourier Series expansion of the forcing function. For all but the simplest examples the required integration may prove tedious.

A more serious problem has to do with the nature of convergence of the Fourier series. Because of the fact that it represents an orthonormal expansion the series naturally converges in the mean square sense, which means that

$$\lim_{N\to\infty} \frac{1}{T} \int_0^T |f(t) - \Sigma_{n=-N}^{N} c_n\, e^{i\,n\,\omega_0 t}|^2\, dt = 0.$$

The meaning of this is that the average squared error is going to zero, but what typically happens is that the convergence is slow with a lot of overshoot appearing.

This makes it awkward to determine what the periodic solution waveform actually looks like in a particular example by means of Fourier Series method since the convergence may be slow.

An alternative approach is to attempt a direct "time domain" calculation of the required periodic solution. If this succeeds, then the product will be explicit expressions for the solution curves as functions of time, without any need for dealing with the problems of summing an infinite series. The time domain approach also involves tedious calculations, but they are finite in number and in many examples easily carried out by use of Maple facilities.

The periodic solution problem is actually much more easily treated in the vector equation format. Hence we consider the system

$$\frac{d}{dt}\mathbf{x}(t) = \mathbf{A}\,\mathbf{x}(t) + \mathbf{f}(t),$$

representing a forced constant coefficient system of differential equations. We seek a solution of the above equation with the property that it returns to its initial value after a period of T seconds.

If such a magic solution can be found, then it will provide a periodic solution of the differential equation. For, if

$$\mathbf{x}(T) = \mathbf{x}(0),$$

and the forcing function then repeats its values over the time interval $T \mapsto 2\,T$, then the solution over the second interval will repeat the trajectory obtained over the first. The system there starts from the same "initial" condition, and is subject to the same "input values" over the second interval, and hence produces the same response.

To solve the problem this way, simply consider solving for the required initial condition. The problem

$$\frac{d}{dt}\mathbf{x}(t) = \mathbf{A}\,\mathbf{x}(t) + \mathbf{f}(t),$$

can be solved by Laplace transforms (since it is a constant coefficient problem). Taking this tack,

$$\begin{aligned}
\mathcal{L}\left\{\frac{d}{dt}\mathbf{x}(t)\right\} &= \mathbf{A}\mathcal{L}\{\mathbf{x}(t)\} + \mathcal{L}\{\mathbf{f}(t)\} \\
(\mathbf{I}s - \mathbf{A})\,\mathcal{L}\{\mathbf{x}(t)\} &= \mathbf{x}(0) + \mathcal{L}\{\mathbf{f}(t)\} \\
\mathcal{L}\{\mathbf{x}(t)\} &= (\mathbf{I}s - \mathbf{A})^{-1}\,\mathbf{x}(0) + (\mathbf{I}s - \mathbf{A})^{-1}\,\mathcal{L}\{\mathbf{f}(t)\} \\
\mathbf{x}(t) &= e^{\mathbf{A}\,t}\,\mathbf{x}(0) + \mathbf{G}(t)
\end{aligned}$$

where this last term represents the forced response

$$\mathbf{G}(t) = \mathcal{L}^{-1}\{(\mathbf{I}s - \mathbf{A})^{-1}\,\mathcal{L}\{\mathbf{f}(t)\}\}.$$

The expression above represents the general solution of the forced differential equation, starting from the arbitrary initial condition $\mathbf{x}(0)$ with the forcing function $\mathbf{f}(t)$. If we evaluate this at the end of the first period, we obtain

$$\mathbf{x}(T) = e^{\mathbf{A}\,T}\,\mathbf{x}(0) + \mathbf{G}(T).$$

In order for the solution to be periodic of period T, the calculated value at T must be the same as at the initial time. The initial condition that produces a periodic solution must satisfy

$$\begin{aligned}
\mathbf{x}(0) &= e^{\mathbf{A}\,T}\,\mathbf{x}(0) + \mathbf{G}(T) \\
\left(\mathbf{I} - e^{\mathbf{A}\,T}\right)\mathbf{x}(0) &= \mathbf{G}(T).
\end{aligned}$$

This represents a set of linear equations to be solved for the initial condition. Determining the initial condition that results in a periodic solution of the equation requires solving the above linear algebra problem.

The first question to consider is whether the coefficient matrix of the system is an invertible one. If it is *not* invertible, this means that there is a nonzero vector $\mathbf{x}_0$ (an element of the null space of the coefficient matrix) satisfying

$$\left(\mathbf{I} - e^{\mathbf{A}\,T}\right)\mathbf{x}(0) = \mathbf{0}.$$

But this equation amounts to the condition for a periodic solution, but applied in the absence of the forcing function. If there is such a nontrivial solution, then the homogeneous equation

$$\frac{d}{dt}\mathbf{x} = \mathbf{A}\,\mathbf{x}(t)$$

would have a periodic solution of period T. The situation in which the coefficient matrix is not invertible then corresponds to seeking a periodic forced solution response in the case in which the homogeneous equation has solutions "with a time constant" that may be present in the forcing function. The terminology used to describe this situation is one of *resonance*. In fact, in a situation of resonance, the system of equations may fail to have a periodic solution, and instead may grow without bound.

The operative word is *may*, because the crux of the matter is whether or not the equation

$$\left(\mathbf{I} - e^{\mathbf{A}T}\right)\mathbf{x}(0) = \mathbf{G}(T)$$

determining the periodic solution initial condition has a solution for $\mathbf{x}(0)$.

If the homogeneous equation

$$\frac{d}{dt}\mathbf{x} = \mathbf{A}\,\mathbf{x}(t)$$

has *no* solutions periodic of period T, then the coefficient matrix

$$\left(\mathbf{I} - e^{\mathbf{A}T}\right)$$

must be invertible, and there will be a unique initial condition

$$\mathbf{x}(0) = \left(\mathbf{I} - e^{\mathbf{A}T}\right)^{-1}\mathbf{G}(T)$$

giving rise to a periodic solution.

If the coefficient matrix

$$\left(\mathbf{I} - e^{\mathbf{A}T}\right)$$

is not invertible, it might still be the case that the system can be solved for $\mathbf{x}(0)$. In the language of row reduction, the augmented matrix must reduce in such a way as to produce the required zeroes in the bottom rows. This imposes a requirement on the forcing function which can be interpreted as an absence of terms in the forcing function that are *in resonance* with solutions of the homogeneous equation.

Example

The problem is to find the periodic solution of period 3 for

$$\frac{dx}{dt} + x = f(t),$$

where

$$f(t) = \begin{cases} \sin(2t) & 0 < t < 3 \\ \sin(2(t-3)) & \text{otherwise.} \end{cases}$$

First solve the differential equation

$$\frac{dx}{dt} + x = \sin(2\,t)$$

over the first cycle, treating the unknown initial condition $x(0)$ as a free parameter. The result of this calculation is

$$x(t) = \frac{2}{5}\,e^{(-t)} + e^{(-t)}\,x(0) - \frac{2}{5}\cos(2\,t) + \frac{1}{5}\sin(2\,t).$$

Evaluate this at the end of the first period to get

$$x(3) = \frac{2}{5}\,e^{(-3)} + e^{(-3)}\,x(0) - \frac{2}{5}\cos(6) + \frac{1}{5}\sin(6).$$

Equate this to $x(0)$ and solve for the initial condition

$$x(0) = -\frac{1}{5}\,\frac{2\,e^{(-3)} - 2\cos(6) + \sin(6)}{e^{(-3)} - 1}.$$

Finally, if an expression for the actual solution is desired, the initial condition can be substituted back in the solution formula.

$$x(t) = \frac{2}{5}\,e^{(-t)} - e^{(-t)}\,\frac{1}{5}\,\frac{2\,e^{(-3)} - 2\cos(6) + \sin(6)}{e^{(-3)} - 1} - \frac{2}{5}\cos(2\,t) + \frac{1}{5}\sin(2\,t).$$

Example

Maple connection Page: 317

The problem is to compute the periodic solution of the system

$$\frac{d^2y}{dt^2} + 2\,\zeta\,\omega_n\,\frac{dy}{dt} + \omega_n^2\,y = f(t),$$

where $f(t)$ is a periodic ramp function (see Figure 9.1) defined by

$$f(t) = \begin{cases} t & \text{for } 0 \le t < T \\ f(t+T) & \text{otherwise} \end{cases}$$

If this problem is represented in vector form, it takes the appearance

$$\frac{d}{dt}\,\mathbf{x} = \mathbf{A}\,\mathbf{x} + \mathbf{g}(t)$$

where the coefficient matrix is

$$\mathbf{A} = \begin{bmatrix} 0 & 1 \\ -{\omega_n}^2 & -2\,\zeta\,\omega_n \end{bmatrix},$$

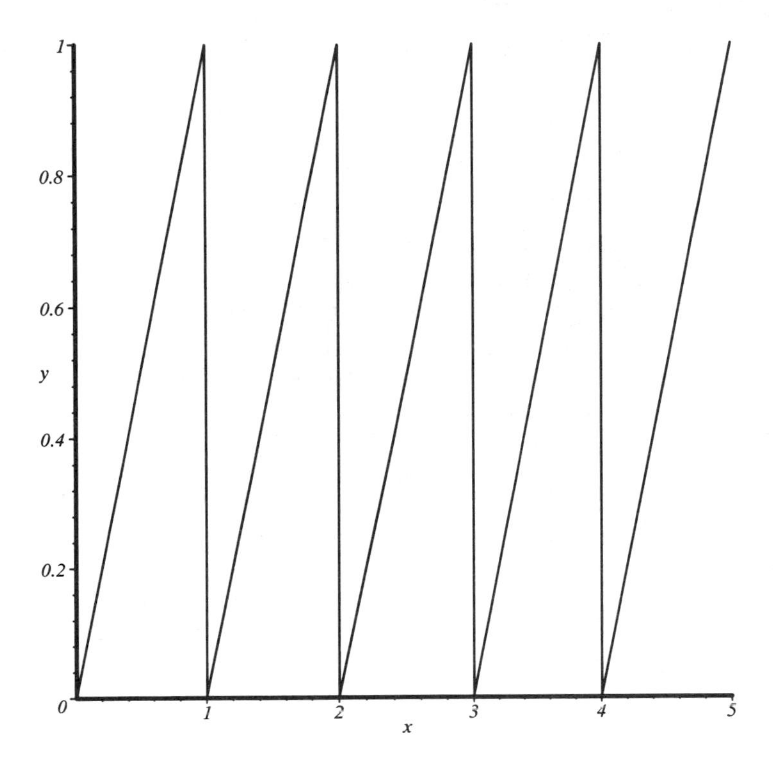

FIGURE 9.1. Periodic ramp function

and the forcing vector is (in a form valid for all values of time)

$$\mathbf{g}(t) = \begin{bmatrix} 0 \\ f(t) \end{bmatrix}.$$

The resolvent of the coefficient matrix is

$$(\mathbf{I}s - \mathbf{A})^{-1} = \begin{bmatrix} \dfrac{2\,\zeta\,\omega_n + s}{2\,s\,\zeta\,\omega_n + s^2 + {\omega_n}^2} & \dfrac{1}{2\,s\,\zeta\,\omega_n + s^2 + {\omega_n}^2} \\ -\dfrac{{\omega_n}^2}{2\,s\,\zeta\,\omega_n + s^2 + {\omega_n}^2} & \dfrac{s}{2\,s\,\zeta\,\omega_n + s^2 + {\omega_n}^2} \end{bmatrix},$$

and a term-by-term inversion gives the exponential of the matrix

$$e^{\mathbf{A}t} = \begin{bmatrix} \dfrac{\zeta e^{(-\zeta\omega_n t)}\sin\left(\sqrt{1-\zeta^2}\omega_n t\right)}{\sqrt{1-\zeta^2}} + e^{(-\zeta\omega_n t)}\cos\left(\sqrt{1-\zeta^2}\omega_n t\right), & \dfrac{e^{(-\zeta\omega_n t)}\sin\left(\sqrt{1-\zeta^2}\omega_n t\right)}{\sqrt{1-\zeta^2}\omega_n} \\ -\dfrac{\omega_n e^{(-\zeta\omega_n t)}\sin\left(\sqrt{1-\zeta^2}\omega_n t\right)}{\sqrt{1-\zeta^2}}, & -\dfrac{\zeta e^{(-\zeta\omega_n t)}\sin\left(\sqrt{1-\zeta^2}\omega_n t\right)}{\sqrt{1-\zeta^2}} + e^{(-\zeta\omega_n t)}\cos\left(\sqrt{1-\zeta^2}\omega_n t\right) \end{bmatrix}.$$

The right hand side of the equation for the periodic solution initial condition involves calculation of the response of the system to the forcing function over the "first period" of the solution. Over this period the forcing function can be taken to be just

$$\mathbf{g}(t) = \begin{bmatrix} 0 \\ t \end{bmatrix}.$$

Since the equation does not realize that the input is periodic until the first period expires, the response to the forcing function can be calculated *as though* the input were going to be $g(t) = t$ for all time. Since only the value of the forced response at the end of the first cycle is required, calculation are enormously simplified by pretending that the straight line input will persist forever. The response then can be evaluated through a Laplace transform calculation. The calculation is

$$(s\,\mathbf{I} - \mathbf{A})^{-1}\ \mathcal{L}\{\begin{bmatrix} 0 \\ t \end{bmatrix}\} = \begin{bmatrix} \dfrac{1}{(2\,s\,\zeta\,\omega_n + s^2 + {\omega_n}^2)\,s^2} \\ \dfrac{1}{s\,(2\,s\,\zeta\,\omega_n + s^2 + {\omega_n}^2)} \end{bmatrix}$$

The forced response term for this example must be calculated by inverting the above Laplace transform expression. That is,

$$\mathbf{G}(t) = \mathcal{L}^{-1}\left\{\begin{bmatrix} \dfrac{1}{(2\,s\,\zeta\,\omega_n + s^2 + {\omega_n}^2)\,s^2} \\ \dfrac{1}{s\,(2\,s\,\zeta\,\omega_n + s^2 + {\omega_n}^2)} \end{bmatrix}\right\}.$$

To determine the initial condition leading to a periodic solution, we need the value of the forced response at the end of the first period $t = T$. Evaluating the expression gives the surprisingly compact expression

$$\mathbf{G}(T) = \begin{bmatrix} -\dfrac{e^{(-\zeta\omega_n T)}\sin\left(\sqrt{1-\zeta^2}\,\omega_n T\right)}{{\omega_n}^3\sqrt{1-\zeta^2}} + 2\dfrac{\zeta^2 e^{(-\zeta\omega_n T)}\sin\left(\sqrt{1-\zeta^2}\,\omega_n T\right)}{{\omega_n}^3\sqrt{1-\zeta^2}} \\ +2\dfrac{\zeta e^{(-\zeta\omega_n T)}\cos\left(\sqrt{1-\zeta^2}\,\omega_n T\right)}{{\omega_n}^3} - 2\dfrac{\zeta}{{\omega_n}^3} + \dfrac{T}{{\omega_n}^2} \\ \dfrac{1}{{\omega_n}^2} - \dfrac{\zeta e^{(-\zeta\omega_n T)}\sin\left(\sqrt{1-\zeta^2}\,\omega_n T\right)}{{\omega_n}^2\sqrt{1-\zeta^2}} - \dfrac{e^{(-\zeta\omega_n T)}\cos\left(\sqrt{1-\zeta^2}\,\omega_n T\right)}{{\omega_n}^2} \end{bmatrix}.$$

The initial condition equation also involves the valu of the system matrix exponential at the end of the first period. This is

$$eAT := \begin{bmatrix} \dfrac{\zeta e^{(-\zeta\omega_n T)}\sin\left(\sqrt{1-\zeta^2}\,\omega_n T\right)}{\sqrt{1-\zeta^2}} + e^{(-\zeta\omega_n T)}\cos\left(\sqrt{1-\zeta^2}\,\omega_n T\right), & \dfrac{e^{(-\zeta\omega_n T)}\sin\left(\sqrt{1-\zeta^2}\,\omega_n T\right)}{\sqrt{1-\zeta^2}\,\omega_n} \\ -\dfrac{\omega_n e^{(-\zeta\omega_n T)}\sin\left(\sqrt{1-\zeta^2}\,\omega_n T\right)}{\sqrt{1-\zeta^2}}, & -\dfrac{\zeta e^{(-\zeta\omega_n T)}\sin\left(\sqrt{1-\zeta^2}\,\omega_n T\right)}{\sqrt{1-\zeta^2}} + e^{(-\zeta\omega_n T)}\cos\left(\sqrt{1-\zeta^2}\,\omega_n T\right) \end{bmatrix}.$$

From this point the initial condition leading to the periodic solution can be calculated by solving for the initial condition $\mathbf{x}_0$[2] This gives the initial condition as

$$\mathbf{x}_0 = \left(\mathbf{I} - e^{\mathbf{A}T}\right)^{-1} \mathbf{G}(T).$$

The expression for the periodic solution over the first cycle can then be calculated as (a formula valid for the solution over the first period of length T)

$$\mathbf{x}(t) = e^{\mathbf{A}t}\mathbf{x}_0 + \mathbf{G}(t).$$

This expression is quite unweildy when written in terms in terms of all of the free parameters of the problem statement. As a particular numerical case, the components of the solution for parameter values $T = 4$, $\omega_n = 4$, $\zeta = .5$ can be calculated as

$$
\begin{aligned}
y(t) = {} & .01804219591 \Big(\\
& .50\sin(3.464101615t)(\mathrm{e}^{(-8.0)})^2\cos(13.85640646)^2 \\
& - 9.00\sin(3.464101615t)\mathrm{e}^{(-8.0)}\cos(13.85640646) \\
& + 8.50\sin(3.464101615t) \\
& + .50\sin(3.464101615t)(\mathrm{e}^{(-8.0)})^2\sin(13.85640646)^2 - \\
& .8660254038 \\
& \cos(3.464101615t)(\mathrm{e}^{(-8.0)})^2\sin(13.85640646)^2 \\
& + 12.99038106\cos(3.464101615t) - 12.12435565 \\
& \cos(3.464101615t)\mathrm{e}^{(-8.0)}\cos(13.85640646) - \\
& .8660254038 \\
& \cos(3.464101615t)(\mathrm{e}^{(-8.0)})^2\cos(13.85640646)^2 \\
& + 8.0\cos(3.464101615t)\mathrm{e}^{(-8.0)}\sin(13.85640646) - \\
& 13.85640646\sin(3.464101615t)\mathrm{e}^{(-8.0)}\sin(13.85640646) \Big) \\
& \mathrm{e}^{(-2.0t)} \Big/ \Big((\mathrm{e}^{(-8.0)})^2\sin(13.85640646)^2 \\
& + (\mathrm{e}^{(-8.0)})^2\cos(13.85640646)^2 - 2\mathrm{e}^{(-8.0)}\cos(13.85640646) \\
& + 1 \Big) - .009021097956\mathrm{e}^{(-2.0t)}\sin(3.464101615t) \\
& + .01562500000\mathrm{e}^{(-2.0t)}\cos(3.464101615t) \\
& - .01562500000 + \frac{1}{16}t
\end{aligned}
$$

---

[2]Strictly speaking, we should check that the coefficient matrix is invertible. The coefficent matrix is *always* invertible when the damping coefficient $\zeta \neq 0$. If $\zeta = 0$, the period T must not be a period of natural oscillation of the harmonic oscillator system.

and

$$\frac{dy}{dt} = -.07216878363 \Big(15.5\sin(3.464101615t)$$
$$- .5\sin(3.464101615t)(e^{(-8.0)})^2\cos(13.85640646)^2$$
$$- .5\sin(3.464101615t)(e^{(-8.0)})^2\sin(13.85640646)^2$$
$$- 15.0\sin(3.464101615t)e^{(-8.0)}\cos(13.85640646) -$$
$$.8660254038$$
$$\cos(3.464101615t)(e^{(-8.0)})^2\sin(13.85640646)^2 -$$
$$.8660254038$$
$$\cos(3.464101615t)(e^{(-8.0)})^2\cos(13.85640646)^2$$
$$+ 16\cos(3.464101615t)e^{(-8.0)}\sin(13.85640646)$$
$$- .8660254038\cos(3.464101615t) + 1.732050808$$
$$\cos(3.464101615t)e^{(-8.0)}\cos(13.85640646) \Big) e^{(-2.0t)} \Big/ \Big($$
$$(e^{(-8.0)})^2\sin(13.85640646)^2 + (e^{(-8.0)})^2\cos(13.85640646)^2$$
$$- 2e^{(-8.0)}\cos(13.85640646) + 1 \Big) + \frac{1}{16}$$
$$- .03608439181e^{(-2.0t)}\sin(3.464101615t)$$
$$- \frac{1}{16}e^{(-2.0t)}\cos(3.464101615t).$$

These curves are plotted in Figure 9.2 in a phase plane format. The axes represent “position” and “velocity”, and the periodic nature of the solution is evident from the fact that the phase plane plot is a closed curve. The corner on the phase plot appears because the forcing function has a jump discontinuity in the waveform.

The solution can also be plotted in other formats, as described in the Maple example for this problem.

Maple reference **Page:** 321

9.5 Periodic Coefficients

We have seen that stability questions involving systems of constant coefficient differential equations can be essentially completely answered by use of the methods of eigenvalue analysis from linear algebra. While many systems may be regarded as time invariant over the time scales of interest, many other problems involve regular, periodic time variation. Many biological and physical phenomena exhibit oscillations, and one question that arises is whether the oscillation observed is a

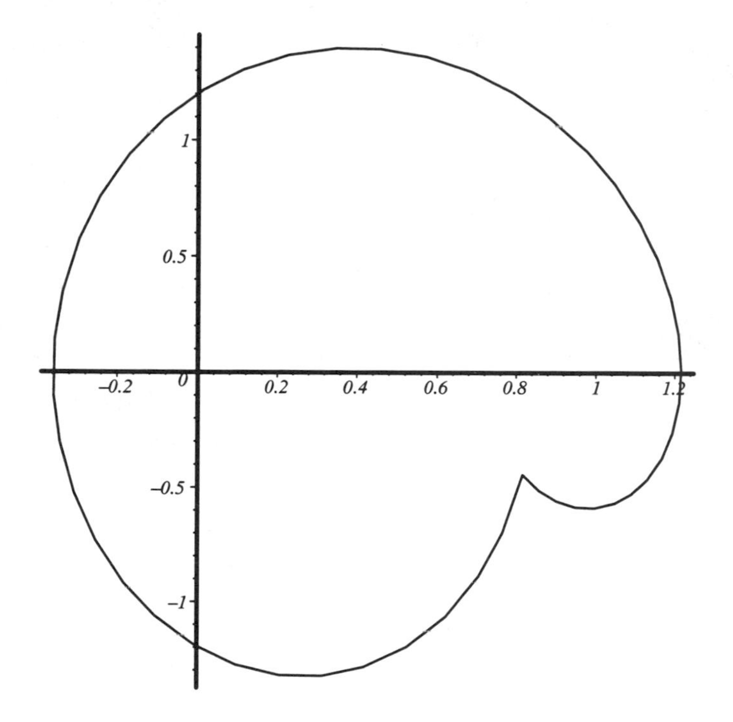

FIGURE 9.2. Periodic response phase plot

stable one. If a nonlinear system of differential equations

$$\frac{d}{dt}\mathbf{x} = \mathbf{F}(\mathbf{x}(t))$$

has a periodic solution

$$\mathbf{x}_p(t) = \mathbf{x}_p(t+T),$$

then

$$\frac{d}{dt}\mathbf{x}_p = \mathbf{F}(\mathbf{x}_p(t))$$

since the periodic solution satisfies the equation. If a solution deviates slightly from the periodic one,

$$\mathbf{x}(t) = \mathbf{x}_p(t) + \mathbf{e}(t),$$

where the error term $\mathbf{e}(t)$ is regarded as small, then expanding in a Taylor series about the periodic solution gives

$$
\begin{aligned}
\frac{d}{dt}\mathbf{x}(t) &= \frac{d}{dt}(\mathbf{x}_p(t)+\mathbf{e}(t)), \\
\frac{d}{dt}(\mathbf{x}_p(t)+\mathbf{e}(t)) &= \mathbf{F}(\mathbf{x}_p(t)+\mathbf{e}(t)) \\
&= \mathbf{F}(\mathbf{x}_p(t)) + \frac{\partial \mathbf{F}}{\partial \mathbf{x}}(\mathbf{x}_p(t))\,\mathbf{e}(t) + error.
\end{aligned}
$$

Subtracting the equation

$$\frac{d}{dt}\mathbf{x}_p = \mathbf{F}(\mathbf{x}_p(t))$$

from the above gives

$$\frac{d}{dt}\mathbf{e}(t) = \frac{\partial \mathbf{F}}{\partial \mathbf{x}}(\mathbf{x}_p(t))\,\mathbf{e}(t) + error,$$

and ignoring the "small" error term gives the linearized model for the deviation from the periodic solution

$$\frac{d}{dt}\mathbf{e}(t) = \frac{\partial \mathbf{F}}{\partial \mathbf{x}}(\mathbf{x}_p(t))\,\mathbf{e}(t).$$

Because the periodic solution $\mathbf{x}_p(t))$ is substituted into the Jacobian matrix, this equation is a system of linear equations with a coefficient matrix that is periodic rather than simply constant.

The historical interest in these problems arose from the question (still not satisfactorily resolved) of the stability of the solar system. It would be comforting to show that the motion of the earth in its orbit is a stable (nonlinear) oscillation. Motions of coupled electrical generators and oscillator circuits used for radio transmissions are subject to the same inquiry.

It turns out that there is a theoretically complete (although computationally burdensome) analysis of systems of linear differential equations with periodic coefficients.

9.6 Fundamental Matrices

In the case of time invariant systems of differential equations it was useful to consider matrix exponential functions, which essentially amount to the linear mapping that carries an initial condition vector into the solution vector at time t. The matrix exponential is entirely tied to time invariance, but there is an analog applicable to systems with time variable coefficient matrices. This replacement is called the *transition matrix* or *fundamental matrix* associated with a system of linear differential equations

$$\frac{d}{dt}\mathbf{x} = \mathbf{A}(t)\,\mathbf{x}.$$

The fundamental matrix was essentially introduced earlier (without announcement) in the Picard iteration discussion of existence of a fundamental set of solutions of a system of differential equations in Section 5.3.2.

The integral equation corresponding to the differential equation system is

$$\mathbf{x}(t) = \mathbf{x}(0) + \int_0^t \mathbf{A}(\sigma)\,\mathbf{x}(\sigma)\,d\sigma,$$

which can be solved by the *Picard iteration*

$$\mathbf{x}_n(t) = \mathbf{x}(0) + \int_0^t \mathbf{A}(t)\,\mathbf{x}_{n-1}(t)\,dt.$$

Since the equation is linear as long as the coefficient matrix is bounded, the iteration will converge to a solution of the vector equation for any initial condition $\mathbf{x}(0)$.

To generate a set of linearly independent solutions we choose initial conditions in the form of the standard basis elements of R^n.

The first solution is obtained by iterating the above Picard scheme with the initial condition

$$\mathbf{x}(0) = \begin{bmatrix} 1 \\ 0 \\ \vdots \\ 0 \end{bmatrix}.$$

The second basis solution comes from

$$\mathbf{x}(0) = \begin{bmatrix} 0 \\ 1 \\ \vdots \\ 0 \end{bmatrix},$$

and so on. Finally the last vector solution is obtained as the solution resulting from

$$\mathbf{x}(0) = \begin{bmatrix} 0 \\ 0 \\ \vdots \\ 1 \end{bmatrix}.$$

If these solutions are placed in the columns of a matrix

$$\Phi = [\mathbf{x}_1 \,|\, \mathbf{x}_2 \,|\, \ldots \,|\, \mathbf{x}_n]$$

then the solution of

$$\frac{d}{dt}\mathbf{x} = \mathbf{A}(t)\,\mathbf{x}$$

meeting the initial condition $\mathbf{x}(0) = \mathbf{x}_0$ is just

$$\mathbf{x}(t) = \Phi\,\mathbf{x}_0$$

(Φ turns into an identity matrix at the initial time.) This construction amounts to a time variable version of the matrix exponential, and really differs only in that closed form explicit calculation formulas are not generally available for the time varying case. The Picard iteration described above can also be described as being carried out in "parallel" in the form *Picard iteration*

$$\Phi_n(t) = \mathbf{I} + \int_0^t \mathbf{A}(t)\,\Phi_{n-1}(t)\,dt.$$

This in terms of column vectors amounts to

$$[\mathbf{x}_1 \,|\, \mathbf{x}_2 \,|\, \ldots \,|\, \mathbf{x}_n]_k\,(t) = \mathbf{I} + \int_0^t \mathbf{A}(\sigma)\;[\mathbf{x}_1 \,|\, \ldots \,|\, \mathbf{x}_n]_{k-1}\,(\sigma)\,d\sigma.$$

The discussion above is couched in terms of posing an initial condition problem at the initial time $t = 0$. In the case of systems of differential equations whose coefficient matrices are independent of time (i.e., time invariant problems) there is no loss of generality in doing this, since the initial value of time is of no consequence. However, if the coefficients in the problem are time varying (periodic for example) then the behavior of solutions does depend on what time the problem is "started" as well as the time of observation of the solution. In order to have a notation for this case, it is useful to regard the fundamental matrix Φ as a function of two variables: t, the "current time", and τ, the "initial time". Then the initial value problem

$$\begin{aligned}\frac{d}{dt}\mathbf{x} &= \mathbf{A}(t)\,\mathbf{x}(t)\\ \mathbf{x}(\tau) &= \mathbf{x}_\tau\end{aligned}$$

is solved by the expression

$$\mathbf{x}(t) = \Phi(t,\tau)\,\mathbf{x}_\tau,$$

while the coefficient matrix in the solution formula is the solution of the integral equation

$$\Phi(t,\tau) = \mathbf{I} + \int_\tau^t \mathbf{A}(\sigma)\,\Phi(t,\sigma)\,d\sigma$$

(solved by Picard iteration as above.)

$\Phi(t, \tau)$ is called the *fundamental matrix* or the *transition matrix* of the differential equation. This latter terminology comes from the fact that multiplication of the initial condition vector $\mathbf{x}_\tau$ by $\Phi(t, \tau)$ transforms the initial vector into the solution value at time t. This interpretation also leads to the fundamental composition law for transition matrices, namely

$$\Phi(t, \sigma)\, \Phi(\sigma, \tau) = \Phi(t, \tau).$$

This relation comes from stopping and immediately restarting the system at the "intermediate time" σ.

9.7 Stability

The stability of linear time invariant systems is studied (essentially to the point of complete analysis) by means of the theory of canonical forms for matrices. By a suitable change of basis, an equivalent system of equations is produced, and the stability properties are evident from the simplicity of the canonical form representation of the problem.

When the system of equations has variable coefficients, there is not in general an explicit description of what the solutions look like that goes beyond the "fundamental matrix" description outlined above. It is, however, possible to give a more complete description of the fundamental matrix (and hence the form of the general solution of the system of equations) in the case of periodic coefficients in the coefficient matrix. This representation is due to Floquet, and is referred to as Floquet theory.

The first observation leading to this theory is that the stability of the periodic system

$$\begin{aligned} \frac{d}{dt}\mathbf{x} &= \mathbf{A}(t)\,\mathbf{x}(t) \\ \mathbf{A}(t+T) &= \mathbf{A}(t) \\ \mathbf{x}(0) &= \mathbf{x}_0 \end{aligned}$$

is completely determined by the value of the transition matrix at the end of the "first cycle of operation". This follows from an application of the transition law discussed above. A general (final) time t can be represented as the sum of an integral number of periods plus the time into the "current cycle"

$$t = k * T + t'.$$

Then

$$\begin{aligned}\Phi(t,0) &= \Phi(k*T+t',0)\\ &= \Phi(k*T+t',k*T)\Phi(k*T,0)\\ &= \Phi(t',0)\Phi(k*T,0),\end{aligned}$$

where the last substitution is a consequence of the fact that the coefficient matrix takes the same values over the last (fractional) interval as it did over the original one. Similarly,

$$\Phi(k*T,0) = \Phi(k*T,(k-1)*T)\Phi((k-1)*T,(k-2)*T)\ldots\Phi(T,0)$$

$$= \Phi(T,0)^k.$$

This means that

$$\Phi(t,0) = \Phi(t',0)\,\Phi(T,0)^k,$$

so that the transition matrix is completely determined by its behavior over the first period of the coefficient variation.

The stability properties of the system may be described in terms of the Jordan canonical form of the one-period transition matrix $\Phi(T,0)$. If

$$\Phi(T,0) = \mathbf{P}\mathbf{J}\mathbf{P}^{-1}$$

is the Jordan canonical form representation, then

$$\Phi(T,0)^k = \mathbf{P}\mathbf{J}^k\mathbf{P}^{-1},$$

and all solutions of the system of equations tend to zero as time tends to infinity exactly in the case where

$$\mathbf{J}^k \rightarrow 0$$

as $k \rightarrow \infty$. By investigating the result of computing a power of a Jordan block matrix, it follows that what is required is that all of the eigenvalues of $\mathbf{J}$ be strictly less than 1 in magnitude. Further consideration leads to the conclusion that all solutions are bounded provided that no eigenvalues of $\mathbf{J}$ are greater than 1 in magnitude, and that any eigenvalues that are of magnitude 1 are associated with 1×1 Jordan blocks.

9.8 Floquet Representation

The argument leading to the stability analysis described in the previous section can be adapted to give the conclusion that a system of linear differential equations with periodic coefficients really amounts to a constant coefficient (that is, time-invariant) system with a time-periodic change of basis. This is referred to as the Floquet representation of the solution.

The representation hinges on a result from linear algebra to the effect that any nonsingular matrix can be written as a matrix exponential. This can be seen directly in the case in which the nonsingular matrix can be diagonalized, and the general fact follows from this by a "continuity argument". In the context of the periodic coefficient problem, we write

$$\Phi(T, 0) = e^{\mathbf{R}T},$$

where $\mathbf{R}$ is effectively the matrix of the general result divided by the constant period T. The Floquet result the follows from declaring

$$\mathbf{P}(t) = \Phi(t, 0)\, e^{-\mathbf{R}t}.$$

The definition of $\mathbf{R}$ and the properties of transition matrices then lead to

$$\begin{aligned}
\mathbf{P}(t+T) &= \Phi(t+T, 0)\, e^{-\mathbf{R}t+T} \\
&= \Phi(t+T, 0)\, e^{-\mathbf{R}T}\, e^{-\mathbf{R}t} \\
&= \Phi(t+T, T)\, \Phi(T, 0)\, e^{-\mathbf{R}T}\, e^{-\mathbf{R}t} \\
&= \Phi(t, 0)\, e^{-\mathbf{R}t} \\
&= \mathbf{P}(t),
\end{aligned}$$

so that the invertible matrix (by construction) $\mathbf{P}(t)$ can be thought of as a periodic change of basis.

If we introduce it as a change of basis associated with the periodic system

$$\begin{aligned}
\frac{d}{dt}\mathbf{x} &= \mathbf{A}(t)\, \mathbf{x}(t) \\
\mathbf{A}(t+T) &= \mathbf{A}(t) \\
\mathbf{x}(0) &= \mathbf{x}_0,
\end{aligned}$$

we define the new coordinates $\mathbf{z}(t)$ through

$$\mathbf{P}(t)\, \mathbf{z}(t) = \mathbf{x}(t).$$

To find the differential equation satisfied by $\mathbf{z}(t)$ simply differentiate the above equation relating $\mathbf{z}(t)$ and $\mathbf{x}(t)$. This gives

$$\frac{d}{dt}\mathbf{x} = \frac{d\mathbf{P}}{dt}\, \mathbf{z}(t) + \mathbf{P}\frac{d\mathbf{z}}{dt}(t).$$

Using the product rule the derivative of the change of basis matrix is

$$\frac{d\mathbf{P}}{dt} = \frac{d}{dt}\Phi(t,0)e^{-\mathbf{R}t} + \Phi(t,0)e^{-\mathbf{R}t}(-\mathbf{R})$$

$$= \mathbf{A}(t)\,\mathbf{P}(t) - \mathbf{P}(t)\mathbf{R}.$$

Expressing the fact that $\mathbf{x}$ satisfies the original equation in terms of $\mathbf{z}$,

$$\frac{d}{dt}\mathbf{x} = \mathbf{A}(t)\,\mathbf{x}(t)$$

$$= \mathbf{A}(t)\,\mathbf{P}(t)\,\mathbf{z}(t).$$

Combining the two expressions for $\frac{d}{dt}\mathbf{x}$ gives

$$\mathbf{P}(t)\,\frac{d\mathbf{z}}{dt} = \mathbf{P}(t)\,\mathbf{R}\,\mathbf{z}.$$

Since the $\mathbf{P}$ is invertible, the conclusion that $\mathbf{z}(t)$ satisfies

$$\frac{d\mathbf{z}}{dt} = \mathbf{R}\,\mathbf{z}$$

follows. In terms of the $\mathbf{z}$ coordinates the system description is a constant coefficient one. One form of the Floquet representation then comes from writing the $\mathbf{x}(t)$ solution in terms of the solution of the constant coefficient $\mathbf{z}(t)$ solution. This provides

$$\mathbf{x}(t) = \mathbf{P}(t)\,\mathbf{z}(t)$$

$$= \mathbf{P}(t)\,e^{\mathbf{R}t}\,\mathbf{z}(0)$$

$$= \mathbf{P}(t)\,e^{\mathbf{R}t}\,\mathbf{P}^{-1}(0)\,\mathbf{x}(0),$$

and because the solution can also be written in terms of the system transition matrix as

$$\mathbf{x}(t) = \Phi(t,0)\,\mathbf{x}(0),$$

this shows that the transition matrix of a linear system of periodic coefficient differential equations can be written in the form

$$\Phi(t,0) = \mathbf{P}(t)\,e^{\mathbf{R}t}\,\mathbf{P}^{-1}(0).$$

9.9 Problems

1. Use phasor methods to determine the solutions to

$$\frac{dx}{dt} + x = \cos(\omega t)$$

and

$$\frac{dx}{dt} + x = \sin(\omega t)$$

2. Use a phasor calculation to find the solution to

$$\frac{d^2x}{dt^2} + \frac{dx}{dt} + x = \cos(\omega t)$$

3. Would you rather solve

$$\frac{d^3x}{dt^3} + \frac{d^2x}{dt^2} + \frac{dx}{dt} + x = \sin(\omega t)$$

by the phasor approach, or by undetermined coefficients? If you have any doubt, try it both ways.

4. Define a periodic function by

$$f(t) = \begin{cases} 1 & \text{if } 0 < t < 1 \\ -1 & \text{if } 1 < t < 2 \\ f(t+2) & \text{otherwise} \end{cases} .$$

Find the complex Fourier series for $f(t)$.

5. Suppose that $f(t)$ is represented by the finite complex Fourier series

$$f(t) = \sum_{n=-N}^{N} c_n \, e^{in\omega_0 t}.$$

Show that by use of Euler's formula, the series can also be written in the "real form"

$$f(t) = a_0 + \sum_{n=1}^{N} a_n \cos(n\omega_0 t) + b_n \sin(n\omega_0 t).$$

6. Find the complex Fourier series expansion for the function

$$f(t) = |\sin(\omega_0 t)|.$$

What is the period of this function?

7. A periodic function is defined by

$$f(t) = \begin{cases} t & \text{if } 0 < t < T \\ f(t+T) & \text{otherwise} \end{cases} .$$

Compute the complex form of the Fourier series for $f(t)$.

8. Find the Fourier series form of the periodic solution of

$$\frac{dx}{dt} + x = f(t),$$

where $f(t)$ is as defined in the previous problem.

9. Solve the previous problem by Laplace transform and linear algebraic methods. You may find Maple useful for this, although the problem is certainly within the range of hand calculation.

10. The model for an RC-filtered, resistive-loaded, full wave rectified power supply is

$$RC\,\frac{dv}{dt} + (1 + \frac{R}{R_L})\,v = |\sin(\omega t)|.$$

Use time domain methods to calculate the periodic solution of this model. Is the periodic solution you calculate stable? That is, do all solutions tend to the periodic solution as time tends to infinity? Think general and particular solutions.

11. Consider the periodic coefficient system of equations

$$\frac{d\mathbf{x}}{dt} = \begin{bmatrix} 0 & 1 \\ f(t) & 0 \end{bmatrix} \mathbf{x},$$

where the sawtooth function $f(t)$ is as defined above. Explicitly write out the Picard iteration whose limit is the transition matrix associated with the system. Can you get Maple to evaluate and plot the first ten iterations ?

12. Look at the scalar valued periodic coefficient ordinary differential equation

$$\frac{dx}{dt} = a(t)\,x(t),$$

where $a(t) = a(t+T)$. What is the transition "matrix" of this system? Find the Floquet representation for this system, explicitly determining $\mathbf{R}$ and the change of basis "matrix" $\mathbf{P}(t)$ in terms of the coefficient function $a(t)$.

13. Consider the periodic system

$$\frac{d}{dt}\mathbf{x} = \begin{bmatrix} a(t) & 0 \\ 0 & a(2t) \end{bmatrix} \mathbf{x},$$

where $a(t) = a(t+T)$. Find the transition matrix, and the Floquet representation for this periodic system.

10

Impedances and Differential Equations

10.1 Introduction to Impedances

The term impedance is associated with algebraic methods for analyzing linear circuit problems. The methods can be (and have been) discussed in a vacuum as a set of rules, but they are in fact applications of the phasor and Laplace transform approaches to constant coefficient differential equation problems discussed above.

It is useful to realize that impedance methods are calculating either particular or in some cases general solutions of the governing differential equations. The problems that arise in some cases of impedance analysis can be understood on the basis of the relations between "solutions" calculated by undetermined coefficients, and Laplace transform solutions of the "same" problem. The essence of this issue is that making a calculation assuming a solution of a particular form exists, when in fact the solution has an entirely different character, is bound to lead to paradoxical calculations.

In principle, impedance methods could be adopted for subject areas other than circuit analysis, since the essence of the approach is solution methods for constant coefficient differential equations. In the event, however, the"interconnect-ability" of the basic circuit elements does not seem to have complete natural analogues in other subject areas.[1]

Within circuit analysis, impedance methods take two related forms. The first is AC analysis, based on sinusoidal particular solutions calculated by phasor methods. The terminology "transient analysis" also appears in the impedance context, and

[1]"Lumped" heat transfer models and basic hydraulics are analogues of RC circuits, for example.

this corresponds to solving the corresponding differential equations by Laplace transform calculations.

10.2 AC Circuits

The basic R, L, and C type circuit elements have governing equations that were illustrated in Figure 2.3. The resistance relation is an algebraic one, while the other two (energy storage) elements are governed by differential equations.

Suppose that we are interested in calculating the particular solution of the linear circuit problem symbolically illustrated in Figure 10.1. The linear circuit contains an interconnection of R, L, and C elements, as well as possibly linear amplifier and transformer components. Since this is a system of constant coefficient differ-

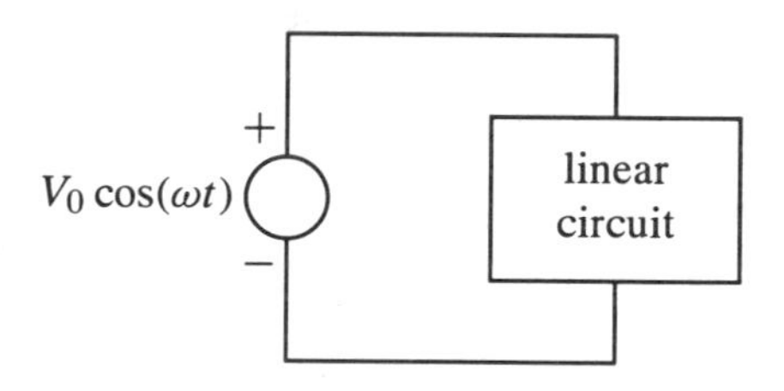

FIGURE 10.1. AC Circuit

ential (and occasional algebraic) equations, a particular solution corresponding to the $V_0 \cos(\omega t)$ forcing function could (assuming a sinusoidal solution exists) be calculated by assuming a general sinusoidal form and evaluating the undetermined coefficients.

A more computationally efficient approach is to "use phasors" and consider the (complex coefficient) circuit of Figure 10.2. Assuming circuit parameters are all

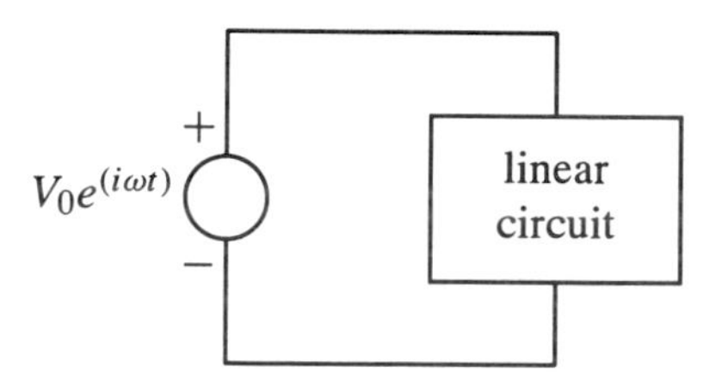

FIGURE 10.2. Phasor circuit

real-valued, the phasor circuit can be solved by (undetermined coefficients). This entails assuming that all voltages in the circuit are of the form

$$V\, e^{i\omega t}$$

with the corresponding current expressible by

$$I\, e^{i\omega t},$$

where the coefficients V, I are referred to as the (complex) phasor amplitudes. Substitution of the assumed forms into the component laws of Figure 2.3 results in the conclusion that the phasor amplitudes are related by

$$\begin{aligned} V &= R\, I \text{ for a resistor,} \\ V &= i\omega\, L \text{ for an inductor,} \\ V &= \frac{1}{i\omega C} I \text{ for the capacitor case.} \end{aligned}$$

These equations are represented by the diagrams in Figure 10.3. The form of

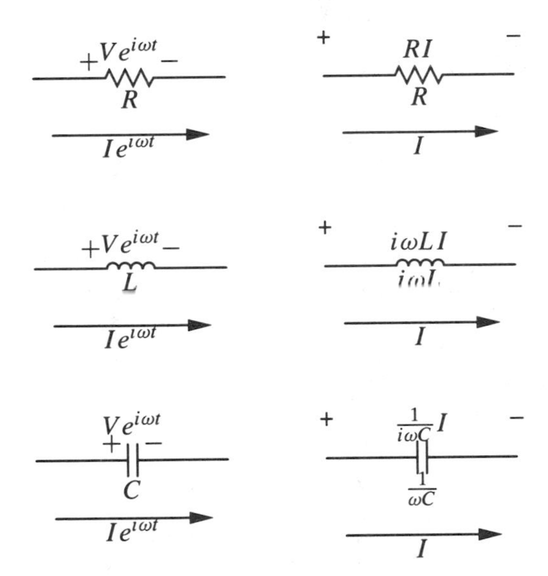

FIGURE 10.3. Phasor equivalent circuits

these equations is exactly that of Ohm's law except that the formulas involve the assumed phasor amplitudes of the circuit voltages and current, and the associated "resistance" is a complex (purely imaginary) value in the case of the capacitor and inductor. The factors appearing in the equations are referred to as the *impedances* of the standard circuit elements. The impedance of a resistor is R, of an inductor is $i\omega\, L$, and that of a capacitor is $\frac{1}{i\omega C}$. Ohm's law then can be revised to state that the voltage (phasor amplitude) is equal to the *impedance* times the current (phasor amplitude).

Beyond the basic element relations, circuits are governed by Kirchoff's laws. The first is Kirchoff's voltage law (KVL) to the effect that the sum of the voltage drops around a closed loop vanishes. [2] Such an equation generally takes the form (in the time domain)

$$v_1(t) + v_2(t) + \ldots + v_{\text{loop}}(t) = 0.$$

If we are assuming a particular solution in phasor form, then $v_1(t) = V_1\, e^{i\omega t}$, $v_2(t) = V_2\, e^{i\omega t} \ldots$, so that

$$\begin{aligned} V_1 e^{i\omega t} + V_2 e^{i\omega t} + \ldots + V_{\text{loop}} e^{i\omega t} &= 0, \\ V_1 + v_2 + \ldots + V_{\text{loop}} &= 0. \end{aligned}$$

Also the Kirchoff current law

$$i_1(t) + i_2(t) + \ldots + i_{\text{node}}(t) = 0$$

in the phasor assumption case amounts to

$$\begin{aligned} I_1 e^{i\omega t} + I_2 e^{i\omega t} + \ldots + I_{\text{node}} e^{i\omega t} &= 0, \\ I_1 + I_2 + \ldots + I_{\text{node}} &= 0. \end{aligned}$$

The conclusion of these calculations is that the Kirchoff circuit laws hold for the phasor amplitudes, as well as in the time domain. Combining this observation with the fact that Ohm's law holds for phasor amplitudes and the component impedances, the procedure for calculating a phasor form particular solution of the circuit can be described as follows:

- Redraw the original circuit, but with (upper case, by convention) phasor variables in place of the original time functions.
- Treat the circuit as though it was a resistive one, but instead of labelling the components with resistances, use the corresponding impedances for each circuit element.
- Write out Kirchoff's laws as appropriate, and solve the resulting equations for the phasor variables.

Example

The problem is the series RLC circuit of Figure 2.18, with the input source voltage $v(t) = V_0 \cos(\omega t)$. The phasor form of the circuit is Figure 10.4. The phasor loop

---

[2] Kirchoff's laws are a consequence of Maxwell's equations of electromagnetism, and assumptions about the localization of the fields in the circuit case.

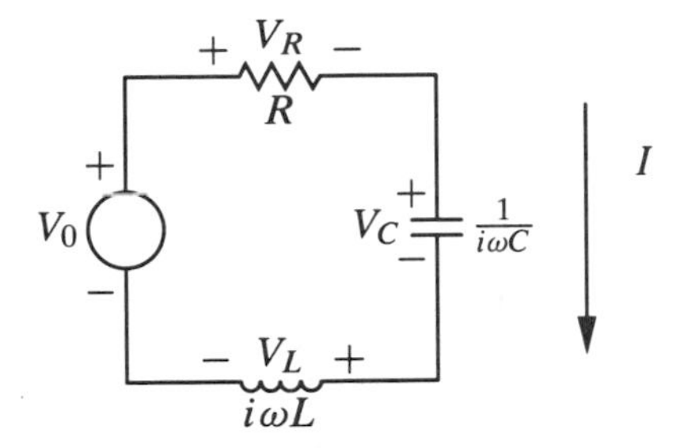

FIGURE 10.4. Phasor version RLC circuit

current can be calculated simply as

$$I = \frac{V_0}{R + i\omega L + \frac{1}{i\omega C}},$$

thinking of it as the source ("battery") voltage divided by the total series impedance. Impedances add for exactly the same reason that resistances do in resistive circuits.

The capacitor voltage can be calculated as

$$V_c = \frac{1}{i\omega C} I = \frac{V_0 \frac{1}{i\omega C}}{R + i\omega L + \frac{1}{i\omega C}},$$

using Ohm's impedance law, or alternatively, by the "voltage divider rule". This latter approach works, simply because Ohm's law, and Kirchoff's laws hold in terms of phasor amplitudes and impedances, and the "voltage divider rule" is a consequence of those. The final expression for the capacitor phasor amplitude is

$$V_C = \frac{V_0}{i\omega RC - \omega^2 LC + 1}.$$

Example

(Not!) If the resistance is removed from the previous example, the phasor circuit diagram appears as in Figure 10.5. The loop current phasor can be calculated as

$$I = \frac{V_0}{i\omega L + \frac{1}{i\omega C}} = \frac{V_0 i\omega C}{1 - \omega^2 LC}.$$

Before cheering, suppose that the driving frequency is $\omega = 2$, but the component values are such that $LC = \frac{1}{4}$?

Evidently something is wrong, since this formula is dividing by zero in this case.

The explanation is simply that for that choice of parameters, the circuit in question has no particular solution in the form of a phasor at the frequency $\omega = 2$. The AC impedance calculation *assumes that it does*, and so runs into an impasse. This is simply another example of what happens when the "wrong assumed form" is substituted for an undetermined coefficient calculation, since the latter is exactly what an AC impedance calculation amounts to.

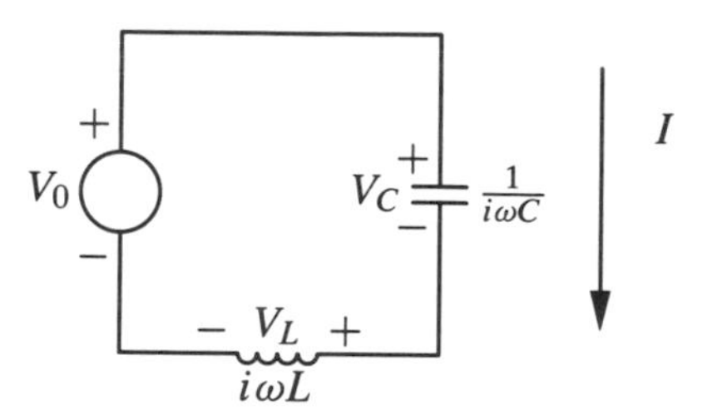

FIGURE 10.5. LC phasor circuit

10.3 Transient Circuits

The problems of incorrect assumptions in undetermined coefficients problems are avoided by using Laplace transforms to solve the problem. Because of this, it is not too surprising that there is a Laplace transform version of impedance analysis which actually sidesteps the hazards of AC analysis.

While AC impedance analysis is based on undetermined coefficients, transient impedance analysis can be thought of as computational way of solving the governing differential and algebraic equations by Laplace transforms.

The transient impedance relations follow simply by Laplace transforming the basic time domain circuit equations of Figure 2.3. For a simple resistor, the relations are

$$v(t) = Ri(t)$$

$$V(s) = RI(s).$$

The governing equation for a capacitor leads to

$$i(t) = C\frac{dv}{dt}$$

$$I(s) = C(sV(s) - v(0))$$

$$V(s) = \frac{1}{sC}I(s) + \frac{1}{s}\,v(0).$$

The differential equation for an inductor provides

$$v(t) = L\frac{di}{dt}$$

$$V(s) = sL\,I(s) - Li(0).$$

These equations are closely related to the AC impedance relations. The differences are a replacement of $i\omega$ in the AC case with s for the transient analysis, together with the presence of "initial condition terms" in the Laplace transform version of the equations.

The utility of impedance methods for circuit analysis lies in the possibility of going directly from the (time domain) circuit diagram to the impedance equivalent, without the intermediate step of writing out the governing differential equations. This idea is somewhat misleading, since the differential equations in question are implicit in the transformed versions written immediately above.

Be that as it may, there is a circuit diagram interpretation of the Laplace transformed circuit equations. The trick is to regard the initial condition terms in the formulas as (voltage) sources in the equivalent circuit. These circuit equivalent diagrams are in Figure 10.6.

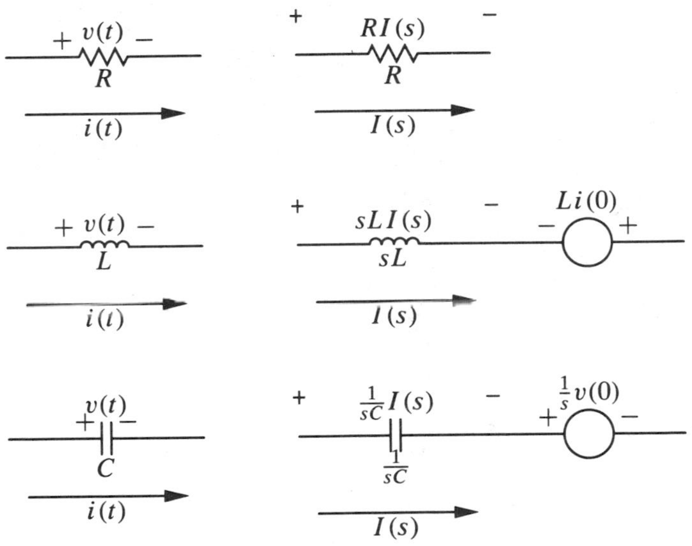

FIGURE 10.6. Transient impedance equivalents

Recall that an ideal voltage source delivers its appointed output voltage no matter what current the surrounding circuitry draws through the source. The $V(s)$ in the above represents the "terminal voltage" of the component. The voltage sources are "internal", and the internal node is inaccessible. The capacitor in effect has a series internal "battery" of value $v(0)$. The inductor series voltage source is of "opposite" polarity, corresponding to the "inductive kick" of an inductor.

10.4 Circuit Impedance Examples

Example
The simplest circuit which has some dynamical behavior is the RC circuit of Figure 6.2. The equivalent circuit in terms of impedances (that is, the Laplace transformed version of the original differential equations) is shown in Figure 10.7. This circuit

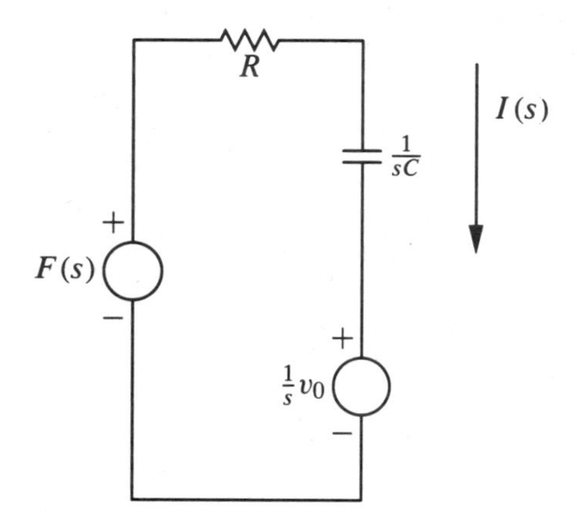

FIGURE 10.7. RC circuit impedances

can be solved as follows:

$$F(s) = RI(s) + \frac{1}{sC} I(s) + \frac{1}{s} v_0$$

$$F(s) - \frac{1}{s} v_0 = (R + \frac{1}{sC}) I(s)$$

$$I(s) = \frac{F(s) - \frac{1}{s} v_0}{R + \frac{1}{sC}}.$$

This expression represents the Laplace transform of the circuit "loop current". To find the Laplace transform of the capacitor voltage, recall from Figure 10.6 that the "terminal voltage" for the capacitor includes both the impedance drop $\frac{1}{sC} I(s)$ and the initial condition source contribution.

$$V_c(s) = \frac{1}{sC} I(s) + \frac{1}{s} v_0$$

$$= \frac{F(s) - \frac{1}{s} v_0}{RCs + 1} + \frac{1}{s} v_0$$

$$= \frac{F(s)}{RCs + 1} + \frac{RC v_0}{RCs + 1}.$$

Inverting this expression provides the expression

$$v_C(t) = e^{-\frac{t}{RC}} v_0 + \int_0^t e^{-\frac{t-\tau}{RC}} \frac{f(\tau)}{RC} \, d\tau.$$

Example
Another circuit that has a simple first order differential equation description is a source in series with a resistor and an inductor. Figure 10.8 has the impedance diagram of such a circuit.

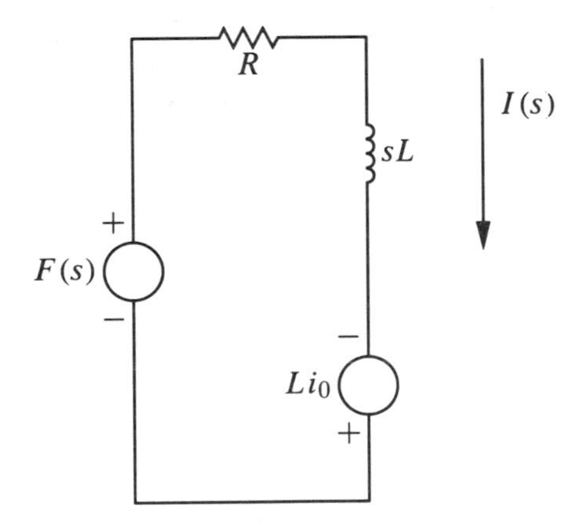

FIGURE 10.8. RL circuit impedance

$$\begin{aligned} F(s) &= RI(s) + sLI(s) - Li(0) \\ (R + sL)I(s) &= F(s) + Li(0) \\ I(s) &= \frac{F(s)}{R + sL} + \frac{i(0)}{s + \frac{R}{L}} \\ i(t) &= i(0)e^{-\frac{Rt}{L}} + \int_0^t e^{-\frac{R(t-\tau)}{L}} \frac{f(\tau)}{L} \, d\tau. \end{aligned}$$

Example
A circuit that contains both a capacitor and an inductor will correspond to a second order differential equation (more honestly, a two dimensional system of first order equations.) The "time domain" diagram of the series version of such a circuit is in Figure 2.18, and the impedance version (Laplace domain, or even frequency domain version) is shown in Figure 10.9 . The transform calculations for this

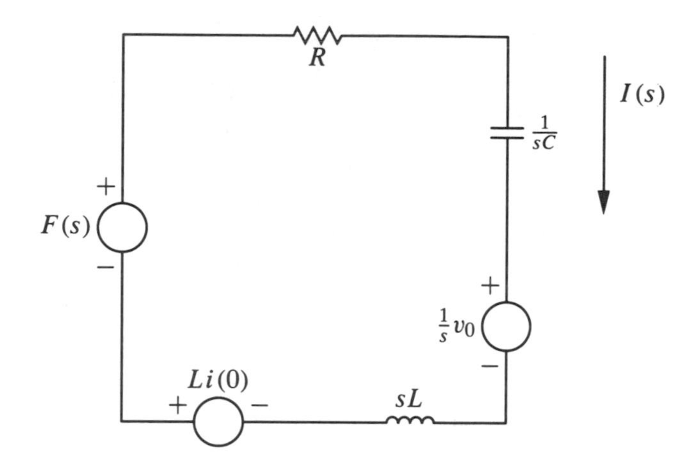

FIGURE 10.9. RLC Laplace domain circuit

example:

$$F(s) = RI(s) + sLI(s) - Li(0) + \frac{1}{sC}I(s) + \frac{1}{s}v(0)$$

$$(R + sL + \frac{1}{sC})I(s) = F(s) + Li(0) - \frac{1}{s}v(0)$$

$$I(s) = \frac{F(s)}{R + sL + \frac{1}{sC}} + \frac{Li(0)}{R + sL + \frac{1}{sC}} - \frac{v(0)}{Rs + s^2L + \frac{1}{C}}$$

$$= \frac{\frac{s}{L}F(s)}{s^2 + \frac{R}{L}s + \frac{1}{LC}} + \frac{si(0)}{s^2 + \frac{R}{L}s + \frac{1}{LC}} - \frac{\frac{v_0}{L}}{s^2 + \frac{R}{L}s + \frac{1}{LC}}.$$

This can be inverted in terms of the harmonic oscillator formulas, by identifying

$$\omega_n^2 = \frac{1}{LC}$$

$$\frac{R}{L} = 2\zeta\omega_n.$$

10.5 Loops and Nodes

At several points we have pointed out that capacitor voltages and inductor currents are the *natural* state variables when a circuit is described by differential equations.

The equations of motion follow by combining Kirchoff's law with the circuit element definitions.

A question that this description avoids is how to write down the *right* Kirchoff relations. This can be described in terms of the topology (geometric arrangement) of the circuit (graph), although this involves somewhat heavy machinery that is not of much aid in simple problems.

Approaches that have some merit are referred to as node voltages, and loop currents. These can be thought of as schemes which by construction eliminate the need for "half" of the Kirchoff equations for a given circuit. Assigning a voltage to each node in a circuit ensures that the sum of voltage drops around any loop vanishes, since the drops are signed differences of the node voltages. Using an independent set of circular loop currents as variables automatically means that the Kirchoff current law holds at any node, and again eliminates "half" of the equations from consideration.

As our interest here is in the relations of circuit problems to differential equation ideas and methods, we leave a catalog of loop and node examples to a circuits text.

10.6 Problems

1. Draw the AC impedance circuit for the diagram 10.10.

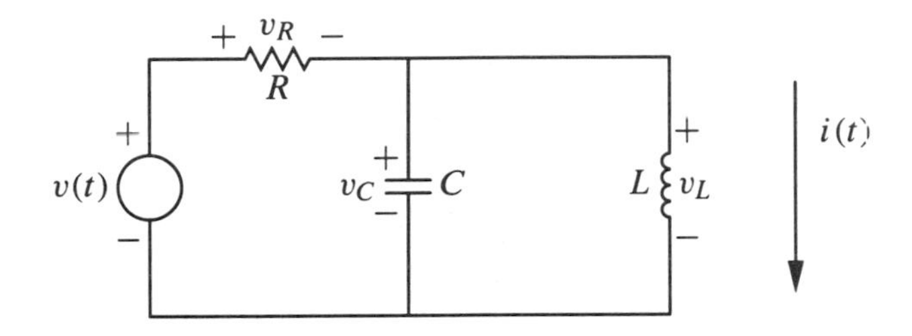

FIGURE 10.10. RLC parallel circuit

2. Draw the transient impedance version circuit for 10.10.

3. Find the phasor response for the circuit of Figure 10.10, given that $v(t) = e^{i\omega t}$.

4. Suppose that $v(t) = e^{i\omega t}$ in the circuit of Figure 10.10 . Using the transient Laplace equivalent circuit, compute the Laplace transform of the inductor current. Take the initial inductor current and capacitor voltage as arbitrary constants in the solution.

5. For the previous solution, use partial fractions methods to compute the part of the inductor current solution which is a multiple of $e^{i\omega t}$. Show that this is just the AC impedance solution calculated above in Problem 3.

6. Suppose the initial conditions (capacitor voltage and inductor current) vanish in Figure 10.10. Calculate the inductor current response for a forcing function of $f(t) = e^{i\omega t}$. Why is this different from the AC response calculation? What happens as $t \to \infty$ in that solution?

7. Find the general solution to the circuit of Figure 2.18 by inverting the Laplace impedance solution.

8. Write out the pair of differential equations (for capacitor voltage and inductor current) in the circuit of Figure 2.18. Solve the system by Laplace transforms, and verify that the solution Laplace transform equations duplicate the impedance method result.

9. For the RC circuit of Figure 2.20, find the phasor response for the capacitor voltages when the forcing voltage is $e^{i\omega t}$.

10. Find the general solution of the system of differential equations describing the circuit of Figure 2.20 by solving the transient response impedance equivalent circuit.

11. In addition to RLC components, transformers are commonly encountered. Transformers consist of independent (usually unconnected) windings which share a common core. The result of this is that rate of current change in one winding induces a voltage in the other winding (as well as itself.) The diagram of a transformer is in Figure 10.11. The equations relating currents

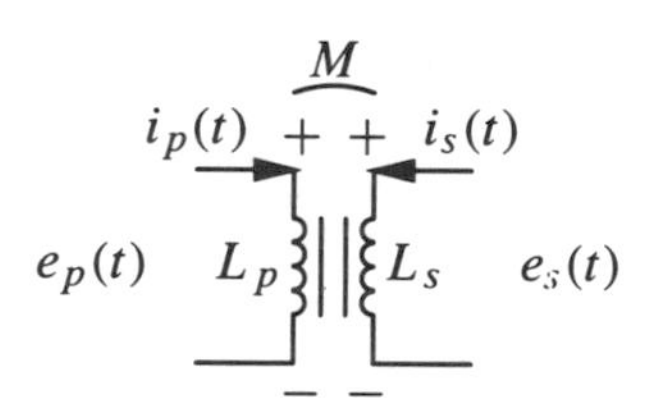

FIGURE 10.11. Basic transformer variables

and winding voltages for the transformer are

$$
\begin{aligned}
e_p(t) &= L_p \frac{di_p(t)}{dt} + M \frac{di_s(t)}{dt} \\
e_s(t) &= M \frac{di_p(t)}{dt} + L_s \frac{di_s(t)}{dt}.
\end{aligned}
$$

The L_p, L_s are the self inductances of the primary and secondary windings, and M is the mutual inductance of the transformer. Find the transient analysis impedance model for a transformer.

12. The transformer impedance model is discussed in Problem 11. Use it to find the general solution of the system of differential equations that describe the RC transformer circuit of Figure 10.12.

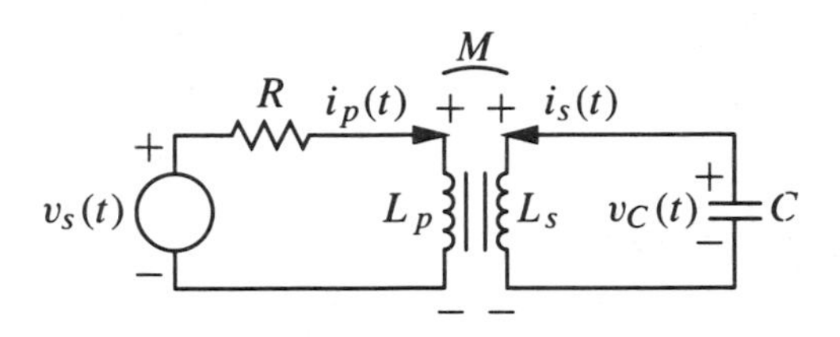

FIGURE 10.12. Transformer coupled RC circuit

11
Partial Differential Equations

11.1 Basic Problems

Many physical problems involve quantities that depend on more than one variable. The temperature within a "large" [1] solid body of conducting material varies with both time and location within the material. When such problems are modeled, what results is a differential equation involving partial derivatives, or a *partial differential equation.*.

Partial differential equations are broadly categorized according to the types of physical phenomena represented. The broad categories are heat or diffusion problems, wave phenomena, and equilibrium problems. In the following sections we derive models in these categories. The physical problems underlying the equations play an additional role in providing the *boundary conditions* that must be specified in order to pose and solve a partial differential equation.

11.1.1 Heat Equation

The general problem of heat conduction in a solid involves three spatial coordinates in addition to a time variable. For some problems, such as heat conduction through a wall, the transfer is primarily perpendicular to the wall, and so the distance through the wall is the only spatial coordinate that has to be considered.

---

[1]the notion of "large" is essentially that the spatial variations of temperature are the feature of interest.

The physical law governing heat conduction is Newton's law of cooling, paraphrased as the statement that heat flows downhill. The precise statement is that the *heat flux* (in units of, say, watts per unit area of cross section) is proportional to the spatial temperature gradient in the material. If the temperature in the material is denoted by $u(x, t)$, the flux at location x and time t is

$$-\alpha \frac{\partial u}{\partial x}(x, t).$$

The heat flux is illustrated in Figure 11.1 for a thin slice of heat conducting material (of unit cross sectional area in the Figure).

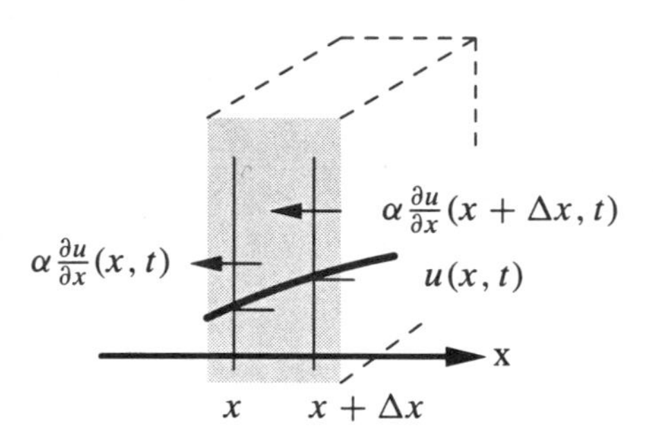

FIGURE 11.1. Heat conduction

The equation that governs this situation follows by conserving energy in the material slice. Assuming the material is of density ρ and has specific heat c, the heat energy stored in the slice is

$$\rho\, c\, u(x, t)\, \Delta x.$$

In the absence of any heat generation within the slice (from chemical, electrical, or nuclear processes, for example) the rate of change of the stored heat energy is entirely due to the heat flux across the edges of the slice. This observation leads to

$$\frac{\partial}{\partial t}(\rho\, c\, u(x, t)\, \Delta x) = \alpha \frac{\partial u}{\partial x}(x + \Delta x, t) - \alpha \frac{\partial u}{\partial x}(x, t).$$

Dividing by $\rho\, c\, \Delta x$ gives

$$\frac{\partial}{\partial t}(u(x, t)) = \frac{\alpha}{\rho c} \frac{\frac{\partial u}{\partial x}(x + \Delta x, t) - \alpha \frac{\partial u}{\partial x}(x, t)}{\Delta x},$$

and the differential equation results from calculating the limit as Δx goes to 0 in this expression for the energy rate. The result is the partial differential equation of heat conduction (heat equation)

$$\frac{\partial u}{\partial t} = \kappa \frac{\partial^2 u}{\partial x^2},$$

where κ is a shorthand for the combination of constants $\frac{\alpha}{\rho c}$.

This equation is first order with respect to time and involves a second spatial derivative. From this, and more particularly from the physical origin of the model, we expect that an initial temperature distribution

$$u(x, 0) = f(x)$$

will have to be provided to specify a solvable problem for that equation.

The same equation also governs diffusion problems in addition to heat conduction. That derivation is left to the problems below.

11.1.2 Wave Equation

Wave phenomena are governed, not surprisingly, by *wave equations*. These arise in various contexts, but the simplest case is that of the transverse vibrations of a string stretched between two supports. This is illustrated in Figure 11.2. The

FIGURE 11.2. A vibrating string

physical parameters of this model include the string length L, a mass per unit length ρ, and the tension T of the stretched string. In many ways the vibrating string is analogous to spring mass harmonic oscillator problems, and in particular, the governing equation is derived from force balance considerations.

The variable of interest in the vibrating string model is

$$u(x, t),$$

representing the transverse deflection of the string at position x and time t.

To derive the governing equation, consider the "free body diagram" of a short length (Δx) of the vibrating string. The string is fixed at the ends, and so is in equilibrium in the axial direction. The force balance problem for the transverse motion is shown in diagram 11.3. The string is assumed completely flexible, so that the tension force acts along the tangent of the string curve. For the transverse motion, we need to calculate the transverse component of the force exerted by the tension. This will be

$$T \sin(\theta),$$

where θ is the angle the moving string makes with respect to the horizontal x axis. The assumption is made that "small motions" are of interest, so that the angle is a very small quantity. Further, since

$$\sin(\theta) \approx \tan(\theta)$$

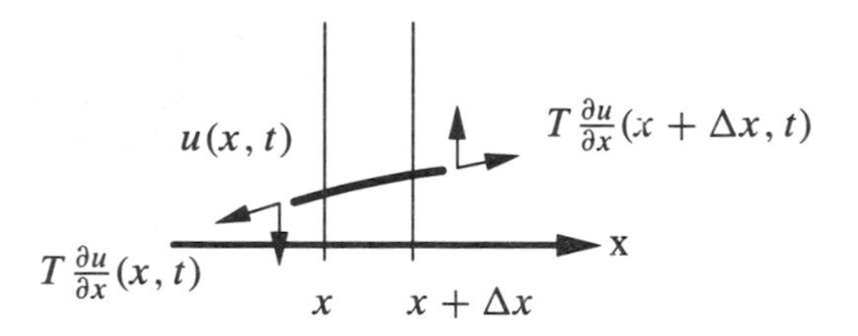

FIGURE 11.3. String force balance

for small angles, the transverse component of the tension force may be taken as

$$T\,\frac{\partial u}{\partial x}(x,t).$$

Equating the total force to the mass times acceleration for the string fragment leads to

$$\frac{\partial^2}{\partial t^2}(u(x,t)) = \frac{T}{\rho}\,\frac{\frac{\partial u}{\partial x}(x+\Delta x,t) - \alpha\frac{\partial u}{\partial x}(x,t)}{\Delta x}.$$

As $\Delta x \to 0$ this becomes the wave equation

$$\frac{1}{c^2}\frac{\partial^2 u}{\partial t^2} = \frac{\partial^2 u}{\partial x^2}.$$

The parameters of the problem have been collapsed into $\frac{1}{c^2} = \frac{\rho}{T}$.

The mechanical origin of this equation suggests that the problem requires specification of both an initial displacement function

$$u(x,0) = f(x),$$

and an initial velocity distribution

$$\frac{\partial u}{\partial t}(x,0) = g(x)$$

as part of the complete problem description.

11.1.3 Laplace's Equation

Laplace's equation generally models a condition of equilibrium. One example is steady state heat conduction (in a plane). The general heat equation for a two dimensional spatial region takes the form[2]

$$\frac{\partial u}{\partial t} = \kappa\left(\frac{\partial^2 u}{\partial x^2} + \frac{\partial^2 u}{\partial y^2}\right).$$

[2]This can be imagined as the effect of adding a second dimension to the original model, although the most natural derivations are based on the divergence theorem of vector calculus.

In the steady equilibrium state, the time derivative is zero, and so the governing equation becomes Laplace's equation,

$$\frac{\partial^2 u}{\partial x^2} + \frac{\partial^2 u}{\partial y^2} = 0.$$

Laplace's equation also arises in electromagnetic theory. In the case of two spatial dimensions, the problem can be described as a continuous limit of the resistor network illustrated in Figure 11.4. The physical situation is a region of the

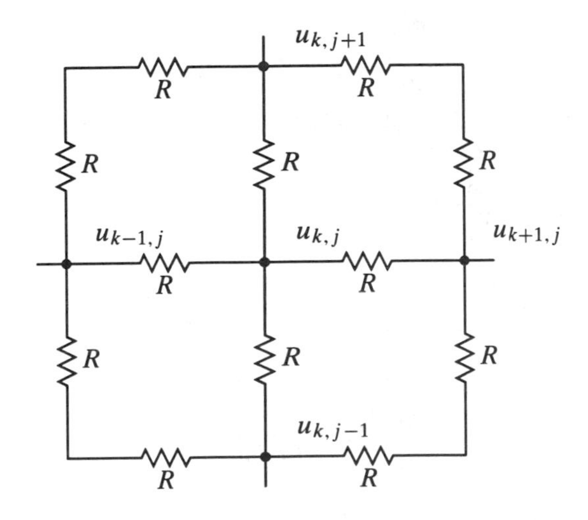

FIGURE 11.4. A grid of resistors

plane, made out of an electrically conducting material. By connecting sources to the boundaries of the region, an electric field

$$\mathbf{E}(x, y) = \begin{bmatrix} E_x(x, y) \\ E_y(x, y) \end{bmatrix}$$

will be set up in the region. In response to the field a steady current (determined by the material conductivity) will flow. Under this scenario the electric field (according to Maxwell's equations) will satisfy the condition that its divergence vanishes, so that the components must satisfy (the zero divergence condition)

$$\frac{\partial}{\partial x} E_x + \frac{\partial}{\partial y} E_y = 0.$$

There is a scalar potential function $\phi(x, y)$ associated with this electric field, defined as the work required to move a unit charge around in the field.[3] The field

[3]The work is independent of the path taken, and hence defines a potential.

is related to the potential by

$$\begin{bmatrix} E_x(x,y) \\ E_y(x,y) \end{bmatrix} = \begin{bmatrix} -\frac{\partial \phi}{\partial x} \\ -\frac{\partial \phi}{\partial y} \end{bmatrix},$$

because of the definition of ϕ in terms of work done along a path. Writing the zero divergence condition in terms of the potential function, we see that the electric potential ϕ must satisfy

$$\frac{\partial^2 \phi}{\partial x^2} + \frac{\partial^2 \phi}{\partial y^2} = 0,$$

which is again Laplace's equation.

Since Laplace's equation is an equilibrium condition, it is not really natural to speak of initial conditions for that equation. Common problems lead to boundary conditions involving potential values along the physical boundaries, and these are discussed along with with the other problem boundary conditions in Section 11.2 below.

11.1.4 Beam Vibrations

Deflection of beams was discussed earlier in Section 2.1.5. The issue there was the static deflection of beams under applied loads. As anyone who has ever shared a bridge with a heavy truck is aware, beams are also capable of vibration. The model that describes this is a partial differential equation, since both time and location with the span are variables of interest.

The problem unknown is the beam displacement

$$u(x,t),$$

oriented positive downward as in the earlier discussion in Section 2.1.5. The earlier notions of the beam shear and bending moment distributions also apply in beam vibrations. The only difference that there is now a time variable included in the function arguments, so that now we have

$$\begin{aligned} M &= M(x,t), \\ V &= V(x,t). \end{aligned}$$

What might at first seem surprising here is the fact that the relations between shear, bending moment, and the displacement are the same in this dynamical situation as they were in the static case of Section 2.1.5. This is the case because the vibrational motion is entirely transverse. The shear-moment-displacement relations are a consequence of the beam elements being in rotational equilibrium, and this is still the case in the dynamical model. As a result, the shear and moment are again related by

$$\frac{\partial M}{\partial x}(x,t) = V(x,t),$$

and as before the shear is connected to the displacement by

$$V(x,t) = -E\,I_{yy}\,\frac{\partial^3 u}{\partial x^3}(x,t).$$

The equation of motion for the vibrating beam follows from the "free body dia-

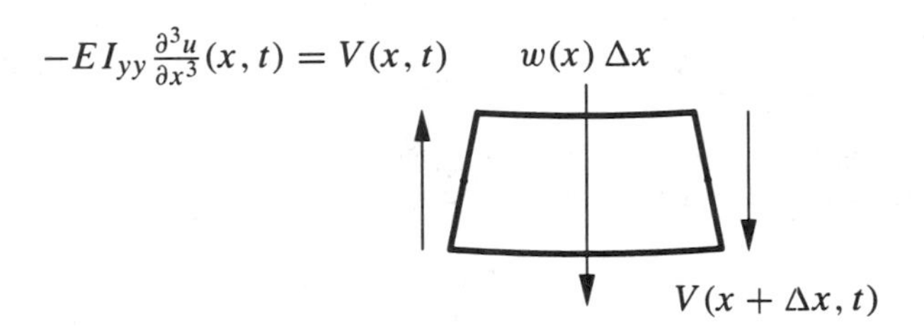

FIGURE 11.5. Beam vertical force balance

gram" of Figure 11.5. Proceeding as in the vibrating string derivation, (and bearing the beam coordinate orientation, positive downwards, in mind) the force balance is

$$\rho\,\frac{\partial^2 u}{\partial t^2}\,\Delta x = w(x)\,\Delta x + V(x+\Delta x,t) - V(x,t).$$

Dividing out Δx, and then letting $\Delta x \to 0$ we obtain

$$\begin{aligned}
\rho\,\frac{\partial^2 u}{\partial t^2} &= w(x) + \frac{\partial V}{\partial x}(x,t),\\
\rho\,\frac{\partial^2 u}{\partial t^2} &= w(x) - E\,I_{yy}\,\frac{\partial^4 u}{\partial x^4}.
\end{aligned}$$

Using the abbreviation $\frac{1}{c^2} = \frac{\rho}{E I_{yy}}$, the form of this is similar to the vibrating string model, except for the order of the spatial derivatives. The final form is just

$$\frac{1}{c^2}\,\frac{\partial^2 u}{\partial t^2} = -\frac{\partial^4 u}{\partial x^4} + \frac{w(x)}{E I_{yy}}.$$

Like the vibrating string, this is a model that is derived from Newton's second law, second order in time, and so we expect to specify both an initial displacement

$$u(x,0) = f(x),$$

and an initial velocity distribution

$$\frac{\partial u}{\partial t}(x,0) = g(x).$$

11.2 Boundary Conditions

Partial differential equations require the specification of conditions beyond the bare equation and initial conditions for equations of evolution. These boundary conditions originate in the physical problem being modeled, and must be carefully formulated lest one end up with an incorrectly specified problem.

11.2.1 Heat and Diffusion Problems

The simplest boundary conditions for a heat equation arise from imagining that the temperatures at the boundaries

$$x = 0, \; x = L$$

are held at the temperature of the surrounding environment. The conditions then would be

$$u(0, t) = T_0, \; u(L, t) = T_0.$$

In order to solve and analyze the partial differential equation, we need to have the problem set up in such a way that homogeneous boundary conditions apply. This can be done by redefining the temperature variable to measure deviation from the surroundings. The dependent variable is then

$$\tilde{u}(x, t) = u(x, t) - T_0,$$

and the equation and boundary conditions take the form

$$\begin{aligned}
\frac{\partial \tilde{u}}{\partial t} &= \kappa \frac{\partial^2 \tilde{u}}{\partial x^2}, \\
\tilde{u}(0, t) &= 0, \\
\tilde{u}(L, t) &= 0, \\
\tilde{u}(x, 0) &= f(x) - T_0.
\end{aligned}$$

In practice, the notation reverts to the original $u(x, t)$, with the implicit understanding that a variable redefinition that homogenizes the boundary conditions has taken place. The redefinition process is analogous to picking the variables in a spring mass model to measure deviation from the equilibrium position.

Another commonly encountered boundary condition in heat conduction problems arises from insulated boundaries. With perfect insulation, the heat flux through the material boundary must be zero. Since Newton's law of cooling stipulates that the heat flux is

$$-\alpha \frac{\partial u}{\partial x}(x, t),$$

this expression must vanish at a perfectly insulated boundary. The model for heat conduction in an insulated slab is then

$$
\begin{aligned}
\frac{\partial u}{\partial t} &= \kappa \frac{\partial^2 u}{\partial x^2}, \\
\frac{\partial u}{\partial x}(0, t) &= 0, \\
\frac{\partial u}{\partial x}(L, t) &= 0, \\
u(x, 0) &= f(x).
\end{aligned}
$$

11.2.2 Wave Equations

Vibrating strings have long been of interest[4] because of the use of vibrating strings in musical instruments. In these problems, the boundary condition is simply that the string is tied at the ends, so that the displacement is zero at the attachment points. The completely stated model them becomes

$$
\begin{aligned}
\frac{1}{c^2}\frac{\partial^2 u}{\partial t^2} &= \frac{\partial^2 u}{\partial x^2} \\
u(0, t) &= 0, \\
u(L, t) &= 0, \\
u(x, 0) &= f(x), \\
\frac{\partial u}{\partial t}(x, 0) &= g(x),
\end{aligned}
$$

including equation, boundary, and initial conditions.

11.2.3 Laplace's Equation

The most familiar instance of Laplace's equation is probably a problem of heat conduction in a rectangle, as illustrated in Figure 11.6. The Figure corresponds to a situation in which the temperature along the bottom edge of the rectangle is held at an ambient constant value. The lateral sides of the rectangle are taken as being perfectly insulated. With "the usual" variable redefinition, the boundary conditions becomes that the (new) temperature vanishes along the bottom edge. The fourth side condition along $y = b$ is that the temperature distribution along the top edge is fixed by some external agent well supplied with torches and ice, so that the temperature varies as a function of x along the top edge.

[4] The ancient Greeks knew the relationship between string tension and musical tones.

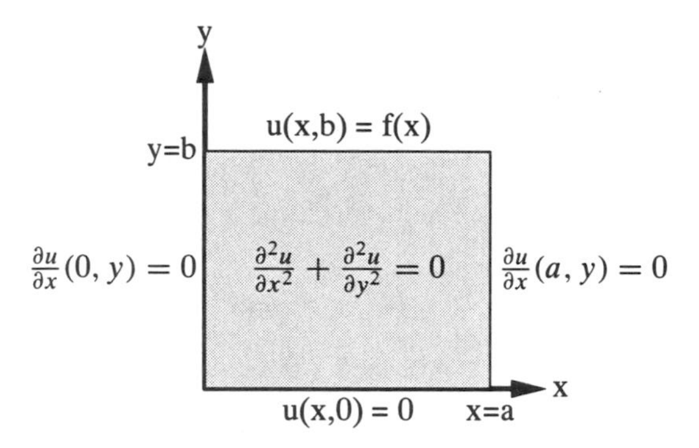

FIGURE 11.6. Heat conduction in a rectangle

11.3 Separation of Variables

Solution methods for partial differential equation models proceed by reducing the problem to a sequence of ordinary differential equation problems. Because the governing equations are linear and the boundary conditions homogeneous, the (separable) solutions obtained by ordinary differential equation methods can be superimposed to give a solution to the entire problem.

11.3.1 Cooling of a Slab

$$\begin{aligned}
\frac{\partial u}{\partial t} &= \kappa \frac{\partial^2 u}{\partial x^2}, \; 0 < x < L \\
u(0, t) &= u(L, t) = 0, \; t > 0 \\
u(x, 0) &= f(x), \; 0 < x < L.
\end{aligned}$$

To apply the separation of variables technique to the heat equation, we try a separable solution

$$u(x, t) = X(x)T(t),$$

where the variable dependence is confined to functions of a single argument. This is a trial solution, analogous to guessing an exponential for solution of a constant coefficient differential equation. If this trial solution works, we have

$$T^{'}(t)X(x) = \kappa T(t)X(x).$$

This can be algebraically rearranged to separate the independent variables x, t. Then

$$\frac{T^{'}(t)}{\kappa T(t)} = \frac{X^{''}(x)}{X(x)}.$$

Since the variables are independent, this relation can only hold in the case where both sides are equal to a common constant. That is,

$$\frac{T'(t)}{\kappa T(t)} = \frac{X''(x)}{X(x)} = -\lambda^2$$

for some *separation constant* called $-\lambda^2$.

In order that the separated solution should satisfy the required boundary condition

$$u(0,t) = u(L,t) = 0,$$

we need to have

$$T(t)X(0) = T(t)X(L) = 0,$$

for all t. To make this happen, the separated $X(\cdot)$ solution must vanish at $x = 0, L$. Thus we are led to consider a spatial ordinary differential equation with a pair of homogeneous boundary conditions:

$$\begin{aligned} X''(x) + \lambda^2 X(x) &= 0, \\ X(0) = X(L) &= 0. \end{aligned}$$

The general solution of the differential equation may be written as (for $\lambda^2 \neq 0$)[5]

$$X(x) = A\cos(\lambda x) + B\sin(\lambda x).$$

The requirement that $X(0) = 0$ forces the conclusion that $A = 0$. To satisfy the condition $X(L) = 0$ in a nontrivial fashion, without taking $B = 0$ and losing the whole solution, requires that the condition

$$\sin(\lambda L) = 0$$

must hold. From the periodic properties of the sine function it follows that

$$\begin{aligned} \lambda L &= n\pi,\ n = 1, 2, 3, 4, \ldots \\ \lambda &= \frac{n\pi}{L},\ n = 1, 2, 3, 4, \ldots \end{aligned}$$

This determines relevant possible separation constants. (Negative n in the above provides only a multiple of the solution already obtained, and $n = 0$ gives a zero solution.)

---

[5]Naming the separation constant $-\lambda^2$ leads to a convenient solution form, but other choices eventually lead to the same solutions.

Now that the allowable separation constants have been obtained, we can go back and solve the corresponding "T" equation. This equation takes the form

$$T'(t) = -\left(\frac{n\pi}{L}\right)^2 T(t).$$

This is a homogeneous, constant coefficient ordinary differential equation of first order, so we readily find by a variety of means that solutions are of the form

$$T_n(t) = b_n e^{-(\frac{n\pi}{L})^2 t}.$$

The b_n represents an initial condition for the time equation.

Combining this with the separated spatial solution, we conclude that the separable $u(x,t)$ solutions are of the form

$$u_n(x,t) = b_n e^{-\kappa(\frac{n\pi}{L})^2 t} \sin\left(\frac{n\pi x}{L}\right).$$

We must also consider (briefly) the possibility that perhaps $\lambda^2 = 0$. If this is the case, then the equation is simply

$$X''(x) = 0,$$

and the solution of the separated equation is not sinusoidal, but is of the form

$$X(x) = C + Dx.$$

Only the choice $C = D = 0$ provides a solution meeting the boundary conditions. Hence $\lambda^2 = 0$ in this problem provides no additional solutions, and the expression already obtained contains all nontrivial usable separated solutions.

The separated solutions all satisfy the heat equation as well as the spatial boundary conditions of the problem, but probably fail to satisfy the initial condition

$$u(x,0) = f(x)$$

specified with the problem.

Unless the initial temperature distribution happens to have exactly the shape of one of the sinusoidal separated solutions, the separable solutions individually will not satisfy the problem initial condition constraint. In order to meet the initial condition constraint, we form linear combinations of the solutions (superimpose the separated solutions), and try

$$u(x,t) = \sum_{n=1}^{\infty} b_n e^{-\kappa(\frac{n\pi}{L})^2 t} \sin(\frac{n\pi x}{L})$$

as a solution. To meet the initial condition at $t = 0$, we need

$$f(x) = u(x,0) = \sum_{n=1}^{\infty} b_n \sin(\frac{n\pi x}{L}).$$

This might be thought of as an infinite number of equations for the set of unknown coefficients $\{b_n\}_{n=1}^{\infty}$. Although this sounds imposing, the coefficients can be found from the orthogonality condition satisfied by the separated spatial solutions. These are the relations

$$\int_0^L \sin\left(n\pi\frac{x}{L}\right)\sin\left(m\pi\frac{x}{L}\right)dx = \begin{cases} \frac{L}{2} & \text{if } m = n, \\ 0 & \text{if } m \neq n. \end{cases}$$

Using these relations we can discover the expansion coefficients by a multiplication and integration. We have, in turn,

$$\begin{aligned} f(x)\sin\left(m\pi\frac{x}{L}\right) &= \sum_{n=1}^{\infty} b_n \sin\left(\frac{n\pi x}{L}\right)\sin\left(m\pi\frac{x}{L}\right) \\ \int_0^L f(x)\sin\left(m\pi\frac{x}{L}\right)dx &= \sum_{n=1}^{\infty} b_n \int_0^L \sin\left(\frac{n\pi x}{L}\right)\sin\left(m\pi\frac{x}{L}\right)dx, \\ \int_0^L f(x)\sin\left(m\pi\frac{x}{L}\right)dx &= b_m\frac{L}{2} \\ b_m &= \frac{2}{L}\int_0^L f(x)\sin\left(m\pi\frac{x}{L}\right)dx. \end{aligned}$$

11.3.2 Standing Wave Solutions

A standard boundary value problem for the wave equation is given by

$$\begin{aligned} \frac{1}{c^2}\frac{\partial^2 u}{\partial t^2} &= \frac{\partial^2 u}{\partial x^2} \\ u(x,0) &= f(x) \\ \frac{\partial u}{\partial t}(x,0) &= g(x) \\ u(0,t) &= 0 \\ u(L,t) &= 0. \end{aligned}$$

This may be interpreted as describing the transverse motion of a string with fixed ends and with given initial position and velocity.

We seek a separable solution

$$u(x,t) = X(x)T(t).$$

Substituting that form into the partial differential equation gives first

$$\frac{1}{c^2}T''(t)X(x) = T(t)X''(x),$$

and then the separated form

$$\frac{T''(t)}{c^2 T(t)} = \frac{X''}{X(x)}.$$

Since the variables are separated, the differential equations and initial conditions to be satisfied are

$$\begin{aligned} X'' + \lambda^2 X &= 0 \\ X(0) &= 0 \\ X(L) &= 0 \\ T'' + \lambda^2 c^2 T &= 0. \end{aligned}$$

The separation constant in the above is again $-\lambda^2$. For $\lambda^2 \neq 0$ (the only relevant case) the general X solution is

$$X(x) = A\cos(\lambda x) + B\sin(\lambda x).$$

Use of the boundary conditions leads to

$$\begin{aligned} A &= 0 \\ B\sin(\lambda L) &= 0, \end{aligned}$$

which are the same conditions encountered in the heat equation example. We find that[6]

$$\lambda_n = \frac{n\pi}{L},\ n = l, 2, \ldots$$

lists the values of the separation constant required. The spatial solutions are therefore

$$X_n(x) = \sin\left(n\pi\frac{x}{L}\right).$$

The corresponding time dependence is determined from

$$T_n'' + \left(\frac{n\pi}{L}\right)^2 c^2\, T_n = 0,$$

which in this wave equation case is a second order equation. The general solution of the time component is

$$T_n(t) = A_n\cos(n\omega_0 t) + B_n\sin(n\omega_0 t).$$

[6]This is the same argument as in the heat equation case immediately above. The separated equations have a tendency to make repeat appearances in a variety of partial differential equation problems.

Here $\omega_0 = \frac{\pi c}{L}$. The separated solutions are given by the product

$$u_n(x,t) = \sin\left(n\pi\frac{x}{L}\right)\left[A_n\cos(n\omega_0 t) + B_n\sin(n\omega_0 t)\right].$$

These separated solutions have a simple geometrical interpretation (and musical consequences as well.)

The solution may be rewritten (using trigonometric identities or phasors to combine the sinusoids) as

$$u_n(x,t) = D_n \sin\left(n\pi\frac{x}{L}\right)\cos(n\omega_0 t + \phi_n).$$

The interpretation of this solution is that it represents a standing wave. The shape of this standing wave is given by the spatial dependence

$$\sin\left(n\frac{\pi x}{L}\right),$$

and the frequencies of oscillation of the standing waves are all integer multiples of the fundamental frequency $\omega_0 = \frac{\pi c}{L}$.

11.3.3 Steady State Heat Conduction

One model problem is that of planar steady state heat conduction with lateral insulation and prescribed temperature on the top and bottom edges of the rectangle. The situations is illustrated in Figure 11.6.

$$\begin{aligned}
\frac{\partial^2 u}{\partial x^2} + \frac{\partial^2 u}{\partial y^2} &= 0,\ 0 < x < a,\ 0 < y < b \\
u(x,0) &= 0 \\
u(x,b) &= f(x) \\
\frac{\partial u}{\partial x}(0,y) &= 0 \\
\frac{\partial u}{\partial x}(a,y) &= 0.
\end{aligned}$$

Attempting separation of variables, try

$$u(x,y) = X(x)Y(y).$$

Proceeding as before we obtain the separated equations

$$\frac{X''}{X} = -\frac{Y''}{Y}.$$

The choice of a separation constant name is suggested by the realization that solutions of the separated equations will, because of the differing sign on the two sides of the equation, give trigonometric functions in one direction, and hyperbolic functions in the other. Since hyperbolic functions vanish at most once for real values of their argument, the boundary conditions suggest choosing the sign of the separation constant to obtain trigonometric functions in the x-variable.[7]

Taking the expedient way out, we choose the separation constant "λ^2" and consider first the problem

$$\begin{aligned} X'' + \lambda^2 X &= 0 \\ X'(0) = X'(a) &= 0. \end{aligned}$$

The general form of solution valid for $\lambda^2 \neq 0$ is

$$X(x) = A\cos(\lambda x) + B\sin(\lambda x).$$

Since finding allowable values for the separation constant is part of the solution process, we do not have any a-priori reason to think that the separation constant is non-zero. If the λ^2 happens to be zero, then the separated equation is actually

$$X''(x) = 0,$$

and the general solution of the equation in this case is a straight line

$$X(x) = A + B\,x.$$

Considering first the case where $\lambda^2 \neq 0$, applying the boundary conditions to the general solution leads to the conclusion that

$$\begin{aligned} \lambda_n^2 &= (\frac{n\pi}{b})^2 \\ X_n(x) &= a_n \cos\left(n\pi\frac{x}{a}\right), \quad n = 1, 2, 3, \ldots. \end{aligned}$$

If the condition $X'(0) = X'(a) = 0$ is applied to the straight line solution from the $\lambda^2 = 0$ case, we see that we get an additional solution

$$X_0(x) = a_0.$$

The corresponding homogeneous Y equation also has two different form solutions, depending on whether we are in the $\lambda^2 = 0$ case or not. For $\lambda^2 \neq 0$ the separated equation is

$$Y_n'' - (\frac{n\pi}{a})^2 Y_n = 0.$$

---

[7]Note that this is purely a cosmetic effect. If the "unfortunate" sign choice is made, the problem solution procedure will force the conclusion that the function arguments are purely imaginary, and the process will "self-correct".

This has a general solution that can be written as[8]

$$Y_n(y) = \alpha_n \sinh\left(n\pi \frac{y}{a}\right) + \beta_n \cosh\left(n\pi \frac{y}{a}\right).$$

To meet the boundary condition $u(x, 0) = 0$, we take $\beta_n = 0$ in the above.

For the $\lambda^2 = 0$ case, the separated Y equation is

$$Y''(y) = 0.$$

The solutions are straight lines

$$Y_0(y) = A + B\,y,$$

and the condition $Y_0(0) = 0$ means that $A = 0$, and so the $\lambda^2 = 0$ solution is linear in the y variable. Finally, combining the two cases, write the separable solutions in the form

$$\begin{aligned} u_n(x, y) &= a_n \cos\left(n\pi \frac{x}{a}\right) \sinh\left(n\pi \frac{y}{a}\right), \; n \neq 0 \\ u_0(x, y) &= a_0\, y. \end{aligned}$$

Since sums of such solutions still meet the homogeneous boundary conditions and satisfy the linear partial differential equation, we postulate

$$u(x, y) = a_0\, y + \sum_{n=1}^{\infty} a_n \cos\left(n\pi \frac{x}{a}\right) \sinh\left(n\pi \frac{y}{a}\right)$$

as the solution of the original problem. The $\{a_n\}$ are evaluated from the remaining boundary condition at $y = b$,

$$u(x, b) = a_0\, b + \sum_{n=1}^{\infty} a_n \cos\left(n\pi \frac{x}{a}\right) \sinh\left(n\pi \frac{b}{a}\right) = f(x).$$

The constants in the series expansion of the solution can be evaluated as above, using the orthogonality conditions satisfied by the separable solutions of this case. These are

$$\int_0^L \cos(n\pi \frac{x}{L}) \cos(m\pi \frac{x}{L})\, dx = \begin{cases} L & \text{if } m = n = 0, \\ \frac{L}{2} & \text{if } m = n \neq 0, \\ 0 & \text{if } m \neq n. \end{cases}$$

---

[8] The solution also could be written in terms of exponential functions, but the hyperbolic form is more convenient for handling the boundary condition.

The expansion constants then follow from the calculations

$$f(x)\cos\left(m\pi\frac{x}{a}\right)$$

$$= a_0\, b\cos\left(m\pi\frac{x}{a}\right) + \sum_{n=0}^{\infty} a_n \cos\left(n\pi\frac{x}{a}\right)\cos\left(m\pi\frac{x}{a}\right)\sinh\left(n\pi\frac{b}{a}\right)$$

$$\int_0^a f(x)\cos\left(m\pi\frac{x}{a}\right)\, dx$$

$$= \int_0^a \left[a_0\, b\cos\left(m\pi\frac{x}{a}\right) + \sum_{n=0}^{\infty} a_n \cos\left(n\pi\frac{x}{a}\right)\cos\left(m\pi\frac{x}{a}\right)\sinh\left(n\pi\frac{b}{a}\right)\right] dx$$

$$\int_0^a f(x)\cos\left(m\pi\frac{x}{a}\right)\, dx = a_m \frac{a}{2}\sinh\left(n\pi\frac{b}{a}\right)$$

$$a_m = \frac{2}{a}\frac{1}{\sinh\left(n\pi\frac{b}{a}\right)}$$

$$\int_0^a f(x)\, dx = \int_0^a \left[a_0\, b + \sum_{n=0}^{\infty} a_n \cos\left(n\pi\frac{x}{a}\right)\sinh\left(n\pi\frac{b}{a}\right)\right] dx$$

$$a_0 = \frac{1}{b}\frac{1}{a}\int_0^a f(x)\, dx.$$

11.4 Problems

1. Suppose that in a heat conduction problem there is an electric current running through the material causing heating of the material at a rate of W watts per unit length. Show that the governing equation becomes

$$\frac{\partial u}{\partial t} = \kappa\frac{\partial^2 u}{\partial x^2} + \frac{W}{\rho c}.$$

2. The equation for a vibrating string was derived assuming essentially that the motion was taking place in a horizontal plane, since no effect of gravity was included. What is the form of the governing equation if gravity is acting on the string in addition to the tension forces?

3. Fick's law of diffusion is that the flux of a dispersed material is proportional to the negative of the concentration gradient. Find the partial differential equation governing a diffusion process with one spatial variable.

4. What is the steady state (time invariant) string deflection for the Problem 2?

5. Show that the solution to the problem of Question 2 can be written in the form of a particular solution of Question 4 plus a solution of the homogeneous equation

$$\frac{1}{c^2}\frac{\partial^2 u}{\partial t^2} = \frac{\partial^2 u}{\partial x^2}.$$

6. Show that the beam deflection problems of Section 2.1.5 are actually the steady state (time invariant) solutions of the (inhomogeneous) vibrating beam equation

$$\rho \frac{\partial^2 u}{\partial t^2} = w(x) - E\, I_{yy} \frac{\partial^4 .u}{\partial x^4},$$

and so the solution of the (inhomogeneous) vibrating beam equation can be written in the form of a steady state deflection plus a solution of the homogeneous equation

$$\rho \frac{\partial^2 u}{\partial t^2} = -E\, I_{yy} \frac{\partial^4 .u}{\partial x^4}.$$

7. Find the solution to the heat equation for the case of a uniform initial temperature. That is,

$$\begin{aligned} \frac{\partial u}{\partial t} &= \kappa \frac{\partial^2 u}{\partial x^2},\ 0 < x < L \\ u(0,t) &= u(L,t) = 0,\ t > 0 \\ u(x,0) &= T_{initial},\ 0 < x < L. \end{aligned}$$

8. Show that the orthogonality condition

$$\int_0^L \sin\left(n\pi \frac{x}{L}\right) \sin\left(m\pi \frac{x}{L}\right)\, dx = \begin{cases} \frac{L}{2} & \text{if } m = n, \\ 0 & \text{if } m \neq n, \end{cases}$$

holds. Using Euler's formulas is easier than trigonometric identities.

9. What happens to the fundamental frequency of a vibrating string if the mass per unit length is doubled? A music interval of a fifth corresponds to a $\frac{3}{2}$ frequency ratio. Suppose that two strings are tuned a fifth apart, while the structural integrity of the instrument requires that the tension in the two strings be the same. What does the ratio of mass per unit length for the two strings have to be to accomplish this? If you own an accurate scale and a guitar, how close is the instrument to having a uniform tension acting on the bridge?

10. Find the solution of Laplace's equation problem

$$
\begin{aligned}
\frac{\partial^2 u}{\partial x^2} + \frac{\partial^2 u}{\partial y^2} &= 0, \ 0 < x < a, \ 0 < y < b \\
u(x, 0) &= 0 \\
u(x, b) &= T_b \\
\frac{\partial u}{\partial x}(0, y) &= 0 \\
\frac{\partial u}{\partial x}(a, y) &= 0.
\end{aligned}
$$

This corresponds to a uniform temperature at the far edge of a laterally insulated plate.

11. Solve Laplace's equation problem

$$
\begin{aligned}
\frac{\partial^2 u}{\partial x^2} + \frac{\partial^2 u}{\partial y^2} &= 0, \ 0 < x < a, \ O < y < b \\
u(x, 0) &= 0 \\
u(x, b) &= f(x) \\
u(0, y) &= 0 \\
u(a, y) &= f(y).
\end{aligned}
$$

12. Find the solution of the vibrating string problem for the case of a plucked string released from rest. Assume that the initial displacement is in the form of an equilateral triangle, and the initial velocity is zero.

Part III

Maple Application Topics

12
Introduction to Maple Applications

From the discussion of earlier chapters, it should be clear that Maple can be treated both as a high level symbolic calculator, and a program development environment with a particularly wide and powerful subroutine library. In this part we discuss applications of both aspects of Maple.

For many topics in differential equations, graphical display of solutions or approximations to solutions is useful. Maple has a generally useful set of plotting facilities, which in particular may be used to display differential equation solution curves. Greater use of the plotting facilities can be made if one understands how the plotting facilities are structured, since the appearance of the resulting plots can be controlled by the user.[1]

For linear differential equations, and more generally for coupled systems of such equations, explicit calculations are in principle possible. For substantial examples of such problems, the required calculations may well be infeasible for ordinary mortals. Maple has built in libraries for both Laplace transform calculations and linear algebra algorithms. This makes it possible to solve linear constant coefficient differential of "arbitrary" size, at least to the point where the scale of the resulting output becomes hard to comprehend.

In our earlier discussion of Runge–Kutta numerical methods, it was stated that in order to evaluate the truncation order of Runge–Kutta schemes in glorious generality, it was "only" necessary to calculate the terms in the Taylor series

[1]Since all Maple objects are effectively handled as symbolic expressions, it should not be a surprise that plots are actually generated by constructions of appropriate symbolic expressions. Appearances of plots are changed by substituting variants for the operands of these expressions.

expansion for the error associated with a single Runge–Kutta step. Of course, to make this of interest, it is required that the Runge–Kutta tableau parameters be dragged through the calculation, and that the number of stages of the method also be treated as a design parameter. When it is realized that the calculation also has to be done in a form that works for vector systems of equations (and not just single differential equations) it becomes clear that the project is easier described than carried out.

Maple code for carrying out these calculations is described in Chapter 16. This chapter illustrates the importance of the use of appropriate data structures for the problem at hand. It is not the case that unique appropriate representations of the problem data are immediately evident, and the "procedure writing" does not begin in earnest until the data structures are solidified.

Differential equations (and their discrete time relatives, difference equations) are the mathematical underpinning of control system theory. For "classical" control theory, the system models are linear time invariant ones, and so it should be no surprise that Laplace transform methods are used. For "modern" linear control theory, the models are coupled systems of linear differential equations, so that linear algebra methods are bound to make an appearance.

Since Maple supports both transforms and linear algebra calculations, it is a natural candidate for writing a control system design package. The term "package" is here used in the Maple technical sense, as a subroutine library that can be loaded by use of the Maple `with` command. A programming project that results in such a package is described below in Chapter 17. The fact that Maple is a symbolic calculation vehicle allows the discrete and continuous time versions of the problems to be handled at once, as the cases are conveniently distinguished by the conventional use of s as the variable in continuous time problems, with z holding sway for the discrete time case.

13
Plotting With Maple

Maple has a built in routine named `plot` which is intended as an all purpose plotting tool, handling functions and point sequences. There is actually a lower level to the Maple plotting facilities that can be used through the auspices of the `display` procedure. If this is used, it is possible to exercise more control on what gets plotted, when the default behavior of `plot` turns out to be not exactly what is wanted.

The first exploration to carry out discovers what is underneath the Maple plotting operations.

13.1 Maple Plotting Structures

Exercise
Try the following sequence of commands.

```
> with(plots);,
> F := plot(sin(x), x=-1..1);
```

Figure out in detail what the resulting `F` is, and what all its components are. When you are sufficiently mystified by the size of the `F` object, you can find an explanation in the help system.

```
> ?plot[structure]
```

Using an unevaluated function expression as a programming data structure is a valuable Maple programming technique. As this example shows, it is used internally by Maple to handle plotting. The Maple commands for extracting and substituting operands of expressions are used for manipulation of these data structures.

Exercise
Try

```
> display(F);
> G := plot(cos(x), x=-1..1);
> display({F,G});
```

What did this do anyhow? This simple exercise shows that plots can be made to order by manipulating the "plot structures" that are of the kind returned by the `plot` procedure, and then making the desired composite plot appear by an invocation of `display`. Displaying a set of plot structures effectively composites the separate images.

Exercise
The exercise above shows that generating a Maple expression that looks like a big function call with a particular sequence of arguments gives you something that can be plotted. Hence the secret to gaining ultimate power over the universe of Maple plotting relies only on being able to write a procedure that manufactures such an expression and returns it when called. To learn earn how to do this in stages start by writing a procedure that returns the expression `COW([1,2,3], GRASS)`.

Exercise
Write a procedure that takes an integer argument, and returns an expression consisting of the procedure name `COW` evaluated with three arguments: a list of integers from 1 to the one passed in, the integer passed in, and finally the cow's favorite food.

One of our interests in manipulating of Maple plot structures arises from the need to plot the output of numerical calculations, rather than Maple expressions. Further issues related to plotting the results of running a numerical method are encountered in the following exercises.

13.2 Remember Tables

Exercise
Try the following:

```
> mysequence := proc(n) option remember;
>  RETURN(n*h);
> end;
>
> h:= 1;
> seq(mysequence(n), n=1..10);
>
```

```
> h:= 2;
> seq(mysequence(n), n=1..10);
> seq(mysequence(n), n=1..12);
```

What is going on?

Exercise
Can you find the answer to the problem above somewhere in the Maple help system? Where do you start? The easy cure for the example above is to make the h a second argument to the procedure, but the example is actually a prototype for the parallel problem that arises when one wants to use the `remember` option to create the data storage painlessly when a numerical solution of a differential equation is done. The calculation will have potentially many parameters, with the step size among them. If some other parameter is introduced, all of the procedures will have to be changed unless one takes the cheap way out and uses the parameters as global variables in the Maple workspace. As soon as that option is followed, a means to clear the `remember` table is required.

At the risk of quiz show copyright infringement, if you do not remember something, you ... it ?

Exercise
The Maple plot routines (and many other supplied procedures) accept a sequence of "this=that" style arguments, which get exhaustively searched for keywords pulled apart. This might be argued to lead to long awkward argument lists. An alternative might be to put all the options in a table and just hand the table to the routine in question. A surrogate plot routine that does this is

```
> curvebuild := proc(ptseq, plotoptions)
>   local F,xrange,xxlabel,xylabel,xtitle,xrgb,argcount;
>
>   argcount := nargs;
>   xrange := 'DEFAULT';
>   xxlabel := 'x';
>   xylabel := 'y';
>   xtitle := 'graph title';
>   xrgb := [0,0,0];
>
>   if not nargs = 2 then
>     ERROR('need 2 arguments to build a plot structure from') fi;
>
>   if nargs >= 1 and not type(ptseq, 'list') then
>     ERROR('first argument must be a list of points') fi;
>
>   if not type(eval(plotoptions), table) then
>     ERROR('second argument must be a table of plot options') fi;
>
>   eqns := op(2, eval(plotoptions));
>
```

```
>   for i in eqns
>     do
>     if type(i,'=') and lhs(i) = 'xlabel' then xxlabel := rhs(i) fi;
>     if type(i,'=') and lhs(i) = 'ylabel' then xylabel := rhs(i) fi;
>     if type(i,'=') and lhs(i) = 'title' then xtitle := rhs(i) fi;
>     if type(i,'=') and lhs(i) = 'rgb' then xrgb := rhs(i) fi;
>     if type(i,'=') and lhs(i) = 'range' then xrange := rhs(i) fi;
>     od;
>
>   F := PLOT(CURVES(ptseq, COLOUR(RGB,op(xrgb))),
>                AXESLABELS(xxlabel, xylabel),TITLE(xtitle),
>                AXESTICKS(DEFAULT,DEFAULT),VIEW( xrange, DEFAULT));
>
>   RETURN(eval(F));
> end;
```

Build a table and extract its second operand, then read the code above. The Maple plot routine could stand improvement. If you plot a subrange of the x variable in a plot list, it gets the x range right, but the default y range picks out the extreme y values over the whole list, not just the subset of interest. This makes it awkward to zoom in on the interesting part of a graph. Modify the routine above so that it picks up the "right" range for the y variable, and makes that what gets plotted.

Could you modify things so that labels are put on the graph in the same color as the graph curve? Could you make the label appear at the end of the curve somehow? Note that the routine above is designed to plot one curve at a time – if you want several curves, you call the above several times and stuff what comes back in a call to `display`.

Exercise

Math background **Page:** 135

The amplitude response of a harmonic oscillator to a sinusoidal input turns out to be given by the formula

$$\frac{1}{\sqrt{(-\frac{\omega^2}{\omega_n^2})^2 + 4\zeta^2 \frac{\omega^2}{\omega_n^2}}},$$

which is a function of two variables. One is the ζ, the damping ratio of the oscillator, and the other is a ratio of the driving and resonant frequencies. The effect of the damping ratio on the resonance can be seen by plotting the above function of the frequency ratio for various values of the damping coefficient. To do this in Maple requires generating a set of PLOT structures to hand to the `display` routine.

Define the response function as a function of two variables.

```
> resp := (omega, zeta) -> (1/((1-omega^2)^2 + 4*zeta^2 * omega^2)^(1/2));
```

Generate nine plots with ζ ranging between .1 and .8.

```
> plotseq := seq( plot(resp(omega, .1*n), omega=0..3), n=1..8);
```

This should produce a relatively huge response looking like an expression sequence of plot structures:

```
PLOT(CURVES([[ 0, 1. ],
    [ .05978822437, 1.003515198581452],
    [ .1237745837, 1.015237730093892],
    [ .1881850862, 1.035925513545559 ],
    [ .2500000000, 1.065152858323015 ],
 ⋮
    [ .3117224143, 1.104997818410829 ],
    [ 2.750000000, .1265656438779994],
    [ 2.811659804, .1213366386056235 ],
    [ 2.872316305, .1164951165666322],
    [ 2.939667555, .1114436297716250], [ 3., .1071866157140680]]
    , COLOUR( RGB, 0, 0, 0 )), TITLE( ), %1, AXESLABELS( ω, ),
    VIEW( 0..3., DEFAULT ))
    %1 := AXESTICKS( DEFAULT, DEFAULT )
```

To actually use the `display` procedure, you must load it first.

```
> with (plots);
```

Then the resonance curves can be generated with the command

```
> display({plotseq});
```

13.3 Plotting Numerical Results

Exercise

Math background **Page:** 85

The exercise is to use Maple to numerically solve the differential equation

$$\frac{dy}{dx} = \sin(xy),$$

$$y(0) = a,$$

and to graphically display the resulting approximate solution curve. The numerical method to be used is Euler's method, so that the iteration takes the form

$$y_n = y_{n-1} + h\, \sin((n-1)\, h\, y_{n-1}),$$

started from the initial value $y_0 = a$.

Since this is a recursive algorithm (the next value is computed in terms of the last one) it is natural to use the Maple `remember` table facility to carry out the required iteration. The procedure `remember` table will then supply the storage for the computed data values, with no storage management programming effort required.

In order to generate a plot of the results, it is necessary to generate a list of $[x, y]$ pairs to feed to the Maple plot routine. This is also easily done by using the sequence generating library routine. The code for solving the above differential equation by Euler's method is given below.

```
> f := (y,x)->sin(x*y);
> a := 1;
> h := .001;
>
> iteration := proc(n) option remember;
>    if n = 0 then a
>    else
>      iteration(n-1) + h * f(iteration(n-1), (n-1)*h)
>    fi;
> end;
>
> ptslist := [seq([i*h, iteration(i)], i=0..1000)];
>
> plot(ptslist);
```

13.4 Plotting Vector Variables

Differential equations are often encountered as a system of first order equations or equivalently as a first order differential equation for a vector variable. This form is the natural one for many problems because the governing physical law is stated as a first derivative relation, and is also the form required for applying numerical methods to such problems.

Relations that are given in the form of a single higher order equation may be written in vector form by forming a vector whose successive components are identified as the corresponding derivatives of the original solution. An example is the harmonic oscillator problem, written in second order form as

$$\frac{d^2x}{dt^2} + \frac{K}{M}x = 0.$$

To convert this to vector form we write

$$\mathbf{x} = \begin{bmatrix} x(t) \\ \frac{dx}{dt} \end{bmatrix}.$$

Then

$$\frac{d\mathbf{x}}{dt} = \begin{bmatrix} \frac{dx}{dt} \\ \frac{d2x}{dt^2} \end{bmatrix} = \begin{bmatrix} \frac{dx}{dt} \\ -\frac{K}{M}x(t) \end{bmatrix}.$$

This last equation expresses the derivatives of the "state vector" **x** in terms of the state vector components, and so represents a vector differential equation equivalent to the original equation, but now in first order form.

If a vector differential equations is to be solved, then the answer must take the form of a vector-valued function of "time" (assuming that is the independent variable in the problem). When such a problem is given numerical treatment, then the result will take the form of a sequence of vectors, each term in the sequence corresponding to one time sample of the solution. If such calculations are done with Maple, then one must generate vector sequences and at the same time be able to "pull out" individual components of the vector for plotting purposes. To verify the feasibility of this approach, first try to generate a vector sequence whose components are the successive powers of 3, 5, and 7. With a view toward what will be required in a numerical differential equation solution, set up a subroutine F whose purpose is simply to multiply each component in the array fed to it by $3, 5$, or 7, as required. The memory storage for the sequence is provided by the sequential routine `iteration`; it both initializes the sequence to start appropriately and sets the "next" value to the result of F acting on the "last value". The space for the next value is allocated within `iteration`, and F is fed this as a modifiable copy of the previous output.

```
> F := proc(x)
>
>  x[1] := 3*x[1];
>  x[2] := 5*x[2];
>  x[3] := 7*x[3];
>
>  RETURN(NULL);
> end;
```

```
> iterate := proc(n) option remember;
>  local rv;
>  rv := array(1..3);
>  if n = 0 then
>   do
>    rv[1] := 1;
>    rv[2] := 1;
>    rv[3] := 1;
>    RETURN(eval(rv));
>   od;
>  else
>   do
>    rv[1] := iterate(n-1)[1];
```

```
>   rv[2] := iterate(n-1)[2];
>   rv[3] := iterate(n-1)[3];
>   F(rv);
>   RETURN(eval(rv));
>  od;
> fi;
> end;
```

```
> seq(iterate(n), n=0..3);
```

$$[\,1\,1\,1\,], [\,3\,5\,7\,], [\,9\,25\,49\,], [\,27\,125\,343\,]$$

```
> threepts := [seq([n,iterate(n)[1]], n=0..10)];
```

$threepts := [[\,0, 1\,], [\,1, 3\,], [\,2, 9\,], [\,3, 27\,], [\,4, 81\,], [\,5, 243\,], [\,6, 729\,], [\,7, 2187\,], [\,8, 6561\,], [\,9, 19683\,], [\,10, 59049\,]]$

```
> plot(threepts);
```

13.5 Further Plotting

Exercise
Although numerical methods get initially discussed as though the problems of interest are first order scalar differential equations, virtually all problem of interest are really first order vector differential equations. This means that the result of running a numerical method over such a problem will be to generate a sequence of vectors as the result. To get Maple to plot the results of such a calculation, one has to manufacture lists of pairs of points. The code examples below provide the background for doing such things.

```
>  iterate:= proc(n) option remember;
>  local value;
>
>  value := array(1..3);
>  value[1] := h*n;
>  value[2] := h*n*n;
>  value[3] := h*n*n*n;
>
>  RETURN(eval(value));
>  end;
>
>  h:= 1;
>  seq(iterate(n), n=1..10);
>  h:= 2;
```

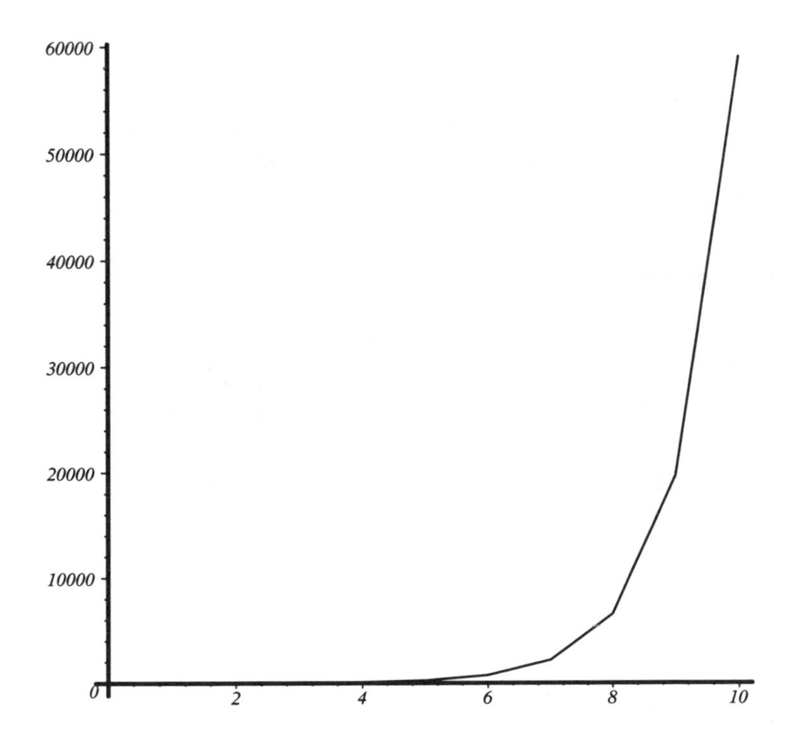

FIGURE 13.1. Powers of three

```
> seq(iterate(n), n=1..12);
>
> op(4, eval(iterate));
>
> reset := proc(f)
> f := subsop(4=NULL, eval(f));
> RETURN(NULL);
> end;
>
> reset(iterate);
> seq(iterate(n), n=1..12);
>
> op(4, eval(iterate));
>
> seq([n, iterate(n)[1]], n=1..10);
> seq([n, iterate(n)[3]], n=1..10);
>
```

```
> plot([%]);
```

Exercise

Maple has code provided for using some standard numerical methods as well as "hooks" that let users write and use their own numerical step routines. Naturally. given the huge size of the Maple documentation, it is hard to find out about that. The magic incantations that reveal these things are:

```
> ?dsolve/numeric
> ?dsolve/classical
```

Exercise

Math background **Page:** 95

To make up your own numerical method, you probably want to make it compatible with the methods Maple provides. Then it becomes easy to test your own routine by comparing the output against what a built in method produces. The output from the above in theory gives you that information, but seeing an example is more helpful.

```
> proc(n,hp,pt,F)
> local h,k1,k2,k3,k4,t,x,i,k,y,z;
> options 'Copyright 1993 by Waterloo Maple Software';
>   h := hp;
>   t := pt[1];
>   x := array(1 .. n);
>   for i to n do x[i] := pt[i+1] od;
>      k1 := array(1 .. n);
>      k2 := array(1 .. n);
>      y := array(1 .. n);
>      k3 := array(1 .. n);
>      k4 := array(1 .. n);
>      z := array(1 .. n);
>      F(k1,x,t);
>      for k to n do y[k] := x[k]+1/2*h*k1[k] od;
>      F(k2,y,t+1/2*h);
>      for k to n do z[k] := x[k]+1/2*h*k2[k] od;
>      F(k3,z,t+1/2*h);
>      for k to n do y[k] := x[k]+h*k3[k] od;
>      F(k4,y,t+h);
>      for i to n do
>        x[i] := x[i]+1/6*h*(k1[i]+2*k2[i]+2*k3[i]+k4[i]) od;
>      t := t+h
>   pt[1] := t;
>   for i to n do pt[i+1] := x[i] od
> end;
```

What is the above routine doing? What are all the arguments? What happens to all of the arguments involved? What do you have to provide in order to use this routine to see what it does on a particular problem?

We did the Maple code required to plot what Euler's method comes up with in an attempt to numerically solve $\frac{dx}{dt} = x(t), x(0) = 1$. Kutta's RK4 ought to do much better, and the above printout is Maple code for RK4 all ready and waiting.

Figure out how to plot what RK4 does on the same problem, by replacing the Euler's method hack with what is required to use `rk4`.

Exercise

Math background **Page:** 103

Stuff a copy of the above source code into an ascii editor, and turn it into `rk3`, as good a name as any for Heun's method with the table

$$\begin{array}{c|ccc} 0 & & & \\ 1/3 & 1/3 & & \\ 2/3 & 0 & 2/3 & \\ \hline & 1/4 & 0 & 3/4 \end{array}$$

Exercise

Math background **Page:** 100

Then figure out how you can write an adaptive step size routine using `rk4` and your new snazzy `rk3`. You want it to use arguments in the same way as `rk4` and `rk3` do, but you have to make an adjustment to account for the fact that the step size has to be varying. Given that you probably want to have a record of the step sizes used anyhow, the natural place to put the varying `h` is in an "extra" entry in the `pt[]` array.

Exercise

Math background **Page:** 124

One of the basic facts about higher order linear differential equations is the existence of a fundamental set of solutions of the homogeneous version of the equation. (see page 124) In the case where the equation has coefficients that depend on the independent variable of the problem, a fundamental set of solutions cannot be easily explicitly calculated. Then the argument for establishing existence of such a set relies on the fact that the Picard iteration procedure (see page 77) actually generates a sequence of functions that converges to a solution of the differential equation.

The equation which it "iterated upon" is the integral equation version

$$\mathbf{x}(t) = \mathbf{x}(0) + \int_0^t \mathbf{A}(t)\,\mathbf{x}(t)\,dt,$$

derived from the original time varying differential equation

$$\frac{d}{dt}\mathbf{x}(t) = \mathbf{A}(t)\,\mathbf{x}(t)$$

by integrating both sides of the equation between 0 and t.

The Picard iteration then takes the form

$$\mathbf{x}_n(t) = \mathbf{x}(0) + \int_0^t \mathbf{A}(t)\,\mathbf{x}_{n-1}(t)\,dt.$$

This result can be visualized (and calculated to essentially any desired accuracy) by using Maple to carry out the iterative procedure. To make this work with a minimum of fuss, one ought to choose an example in which Maple can easily evaluate the integrals that turn up in the generated sequence.

An example can be constructed from a constant coefficient problem. If the differential equation whose fundamental set is sought is

$$\frac{d^y}{dt^2} + \frac{dy}{dt} + 2\,y = 0,$$

then the equivalent first order vector equation is

$$\frac{d}{dt}\begin{bmatrix} x_1(t) \\ x_2(t) \end{bmatrix} = \begin{bmatrix} 0 & 1 \\ -2 & -1 \end{bmatrix}\begin{bmatrix} x_1(t) \\ x_2(t) \end{bmatrix}.$$

This is obtained from the original second order scalar equation by taking the vector components as

$$\begin{bmatrix} x_1(t) \\ x_2(t) \end{bmatrix} = \begin{bmatrix} y(t) \\ \frac{dy}{dt}(t) \end{bmatrix},$$

where y is the solution of the scalar problem.

```
> with(linalg):
> Warning: new definition for  norm
> Warning: new definition for  trace
>
> iterate := proc(n) option remember;
> global x0, A;
> local x, y, i;
>
> x := array(1..2);
> y := array(1..2);
>
> if n = 0 then
>   for i from 1 to 2 do
>     x[i] := x0[i];
>   od;
```

```
>   RETURN(eval(x));
> fi;
>
> y := evalm(A &* iterate(n-1));
> for i from 1 to 2 do
>   x [i]:= x0[i] + int( y[i], t=0..b);
> od;
>
> for i  from 1 to 2 do
>   x[i] := subs( b=t, x[i]);
> od;
>
> RETURN(eval(x));
> end;
```

To try the above Picard method code, we have to first define the global variables for the initial condition and the system coefficient matrix. The commands below then run the Picard iteration for two different initial conditions and generate plots of a basis for the homogeneous solution space.

```
> x0 := array(1..2, [1,0]);
> A:= matrix(2,2, [[0, 1],[-2, -1]]);
>
> y1 := iterate(20)[1];
>
> readlib(forget);
>
> forget(iterate);
>
> x0 := array(1..2, [0,1]);
>
> y2 := iterate(20)[1];
>
> plot({y1, y2}, t=0..2);
>
> plot({seq(iterate(n)[1], n=5..20)}, t=0..2);
```

The differential equation “solved” above by means of Picard iteration is a constant coefficient equation, so closed form expressions for the exact solution can be calculated. In order to get the solution corresponding to the numerical calculation, it is necessary to determine the values of the parameters in the general solution that meet the initial conditions of the numerical case.

Exercise

Math background **Page: 126**

Calculate explicitly the general solutions of

$$\frac{d^y}{dt^2} + \frac{dy}{dt} + 2\,y = 0,$$

either by using the characteristic equation, or by using the Maple `dsolve` routine. Note that the characteristic equation roots are the time constants in exponential solutions.

If these "obvious" solutions are called $y_3(t) = e^{\lambda_1 t}$ and $y_4(t) = e^{\lambda_2 t}$, use Maple to find the constants $\{c_1, c_2, c_3, c_4\}$ such that

$$y_1(t) = c_1 y_3(t) + c_2 y_4(t),$$

$$y_2(t) = c_3 y_3(t) + c_4 y_4(t),$$

are the solutions calculated by the Picard iteration routine above. This means we want solutions such that $y_1(0) = 1$, $\frac{dy_1}{dt}(0) = 0$, $y_2(0) = 0$, while $\frac{dy_2}{dt}(0) = 1$.

What you are doing is really finding the coefficient set that changes from one fundamental set (solution space basis) to another.

```
> ?solve
```

ought to help. It should solve both the characteristic equation and the subsystem of equations that has to be solved to find the c_j.

Maple can be tricked into giving you the solutions you are looking for, by forcing it to use Laplace transform methods of solution. (The special basis is the "natural" one produced by Laplace transform methods.)

```
> eqn := diff(y(t), t$2) + diff(y(t), t) + 2* y(t) = 0;
> dsolve({eqn},y(t), method=laplace);
```

14

Maple and Laplace Transforms

Math discussion **Page:** 153

Maple contains routines for computing both Laplace and inverse Laplace transforms. The built-in differential equation solver `dsolve` attempts to determine the type of proffered differential equation and to apply an appropriate solution method on the basis of type recognition. For constant coefficient problems, more recognizable solutions seem to emerge if Maple is forced to carry out the solution by Laplace transforms. This can be forced by using the `method` option of `dsolve`. A discussion of the `dsolve` method option choices may be found in the help facilities.

```
> dsolve(diff(y(t),t$2) + y(t) = sin(t), y(t), method=laplace); # $=order
```

Try this example both with and without the option, and compare the results.

As far as learning about Laplace transforms is concerned, watching Maple (presumably) solve problems by that method is not very instructive. On the other hand, if all you want is the answer . . . . It can be used to compute various ready examples of transforms and inverse transforms, verifying the manipulation rules (if there was any doubt). The general purpose algebraic manipulation routines are useful for carrying out Laplace transform calculations for purposes beyond solving particular differential equations.

The syntax of the `laplace` routine requires an expression, the original variable name, and the transform variable as arguments.

```
> laplace(sin(t), t,s);
```

$$\frac{1}{s^2+1}$$

14.1 Walking Maple Through Problems

As an example of transform manipulations, we propose to compute a set of our favorite fundamental solutions for a third order differential equation using Maple. The first problem is to pick a third order constant coefficient differential equation, and we do this by selecting small integer roots for the characteristic polynomial.

```
> p := (s+1)*(s+2)*(s+3);
```

$$p := (s+1)(s+2)(s+3)$$

```
> expand(p);
```

$$s^3 + 6s^2 + 11s + 6$$

Maple no doubt could be persuaded to substitute the derivatives into the above expression, but it is easier to enter what we want.

```
> de := diff(y(t),t$3) + 6*diff(y(t), t$2) + 11*diff(y(t),t) +6*y(t);
```

$$de := \left(\frac{\partial^3}{\partial t^3}y(t)\right) + 6\left(\frac{\partial^2}{\partial t^2}y(t)\right) + 11\left(\frac{\partial}{\partial t}y(t)\right) + 6\,y(t)$$

In spite of calling the variable **de**, it really is only the right hand side of the homogeneous differential equation we are investigating. Laplace transforming the above expression involves enough arithmetic so that Maple use is advised.

```
> lde := laplace(de, t, s);
```

$$\begin{aligned} lde := {} & ((\mathrm{laplace}(y(t), t, s)s - y(0))\,s - \mathrm{D}(y)(0))\,s - D^{(2)}(y)(0) \\ & + 6\,(\mathrm{laplace}(y(t), t, s)\,s - y(0))\,s - 6\,\mathrm{D}(y)(0) + 11\,\mathrm{laplace}(y(t), t, s)\,s \\ & - 11\,y(0) + 6\,\mathrm{laplace}(y(t), t, s) \end{aligned}$$

The conventional name of the transformed solution would be `Y`, so that is substituted for the expression Maple used. Perhaps `Y(s)` would work, but ...

```
> lde := subs(laplace(y(t),t,s) = Y, lde);
```

$$\begin{aligned} lde := {} & ((Y\,s - y(0))\,s - \mathrm{D}(y)(0))\,s - D^{(2)}(y)(0) + 6\,(Y\,s - y(0))\,s \\ & - 6\,\mathrm{D}(y)(0) + 11\,Y\,s - 11\,y(0) + 6\,Y \end{aligned}$$

We want to solve for `Y`, and the Maple `solve` routine only accepts equations and what to solve for. Make an equation:

```
> eqn := lde=0;
```

$$eqn := ((Y\,s - y(0))\,s - \mathrm{D}(\,y\,)(0))\,s - D^{(2)}(\,y\,)(0) + 6\,(Y\,s - y(0))\,s - 6\,\mathrm{D}(\,y\,)(0) + 11\,Y\,s - 11\,y(0) + 6\,Y = 0$$

Solving for `Y` will give the Laplace transformed version of the general solution to the homogeneous equation

$$\frac{d^3y}{dt^3} + 6\,\frac{d^2y}{dt^2} + 11\,\frac{dy}{dt} + 6y(t) = 0.$$

```
> solve(eqn, Y);
```

$$-\frac{-s^2\,y(0) - s\,\mathrm{D}(\,y\,)(0) - D^{(2)}(\,y\,)(0) - 6\,s\,y(0) - 6\,\mathrm{D}(\,y\,)(0) - 11\,y(0)}{s^3 + 6\,s^2 + 11\,s + 6}$$

To get this answer back in the time domain, compute the inverse Laplace transform. This routine takes the same arguments as the forward transform, again with the "output" variable last.

```
> invlaplace(%, s,t);
```

$$3\,y(0)\,e^{(-t)} + \frac{5}{2}\,\mathrm{D}(\,y\,)(0)\,e^{(-t)} + \frac{1}{2}\,D^{(2)}(\,y\,)(0)\,e^{(-t)} - 4\,\mathrm{D}(\,y\,)(0)\,e^{(-2\,t)} - 3\,y(0)\,e^{(-2\,t)} - D^{(2)}(\,y\,)(0)\,e^{(-2\,t)} + \frac{3}{2}\,\mathrm{D}(\,y\,)(0)\,e^{(-3\,t)} + y(0)\,e^{(-3\,t)} + \frac{1}{2}\,D^{(2)}(\,y\,)(0)\,e^{(-3\,t)}$$

Math discussion **Page: 160**

This expression represents a fairly messy function of the initial function value and derivatives. It is that expression which must be manipulated if we are to extract the desired fundamental set of solutions. Start the process by collection terms multiplying the initial condition values.

```
> y1 := collect(% , y(0));
```

$$y1 := \left(\frac{1}{2}\,e^{(-3\,t)} + \frac{1}{2}\,e^{(-t)} - e^{(-2\,t)}\right)\,D^{(2)}(\,y\,)(0) + \left(\frac{5}{2}\,e^{(-t)} - 4\,e^{(-2\,t)} + \frac{3}{2}\,e^{(-3\,t)}\right)\,\mathrm{D}(\,y\,)(0) + (\,3\,e^{(-t)} + e^{(-3\,t)} - 3\,e^{(-2\,t)}\,)\,y(0).$$

This separates the initial condition terms. Recall that a fundamental set for this problem can be obtained by picking out solutions satisfying the following conditions:

y_1 : At $t = 0$, we have $y_1(0) = 1$, while $\frac{dy_1}{dt}(0) = \frac{d^2y_1}{dt^2}(0) = 0$.

y_2 : At $t = 0$ we have $y_2(0) = 0$, with $\frac{dy_2}{dt}(0) = 1$, but $\frac{d^2y_2}{dt^2}(0) = 0$.

y_3 : At $t = 0$ we have $y_3(0) = \frac{dy_3}{dt}(0) = 0$, $\frac{d^2y_3}{dt^2}(0) = 1$.

The solutions satisfying these conditions can be read directly from the expressions collected above. y_1 is simply the coefficient function of $y(0)$ in the above so that

$$y_1(t) = (3\,e^{(-t)} + e^{(-3\,t)} - 3\,e^{(-2\,t)}).$$

The second basis vector (second member of the fundamental set) arises from taking the first derivative at 0 to be 1, and the other initial conditions to be 0, so that it amounts to the coefficient of $\frac{dy}{dt}(0)$ in the solution expression

$$y_2(t) = \left(\frac{5}{2}\,e^{(-t)} - 4\,e^{(-2\,t)} + \frac{3}{2}\,e^{(-3\,t)}\right).$$

The third fundamental solution in the set is the coefficient of the initial second derivative

$$y_3(t) = \left(\frac{1}{2}\,e^{(-3\,t)} + \frac{1}{2}\,e^{(-t)} - e^{(-2\,t)}\right).$$

These fundamental solutions naturally appear when an initial value problem is solved for a differential equation. This can be verified in a particular example by letting `dsolve` calculate the general solution to the forced version of the differential equation. The calculation proceeds as follows.

```
> dsolve(de=f(t), y(t), method=laplace);
```

$$\begin{aligned} y(t) = {} & y(0)\,e^{(-3\,t)} - 3\,y(0)\,e^{(-2\,t)} + 3\,y(0)\,e^{(-t)} \\ & + \frac{3}{2}\,\mathrm{D}(y)(0)\,e^{(-3\,t)} - 4\,\mathrm{D}(y)(0)\,e^{(-2\,t)} + \frac{5}{2}\,\mathrm{D}(y)(0)\,e^{(-t)} \\ & + \frac{1}{2}\,D^{(2)}(y)(0)\,e^{(-3\,t)} - D^{(2)}(y)(0)\,e^{(-2\,t)} + \frac{1}{2}\,D^{(2)}(y)(0)\,e^{(-t)} \\ & + \int_0^t \left(\frac{1}{2}\,e^{(-3\,\_U)} + \frac{1}{2}\,e^{(-\_U)} - e^{(-2\,\_U)}\right) \mathrm{f}(t - \_U)\,d\_U. \end{aligned}$$

```
> collect(%, {y(0), D(y)(0), D(D(y))(0)});
```

$$y(t) = (3\,e^{(-t)} + e^{(-3\,t)} - 3\,e^{(-2\,t)})\,y(0) + \left(\frac{5}{2}\,e^{(-t)} - 4\,e^{(-2\,t)} + \frac{3}{2}\,e^{(-3\,t)}\right)\mathrm{D}(y)(0) + \left(\frac{1}{2}\,e^{(-3\,t)} + \frac{1}{2}\,e^{(-t)} - e^{(-2\,t)}\right)D^{(2)}(y)(0) + \int_0^t \left(\frac{1}{2}\,e^{(-3\_U)} + \frac{1}{2}\,e^{(-\_U)} - e^{(-2\_U)}\right)\mathrm{f}(t - \_U)\,d\_U$$

14.2 Coding a Beam Problem Solver

Math background **Page: 43**

The differential equation for the displacement of a loaded Euler–Bernoulli beam is a simple constant coefficient, linear, fourth order one

$$E\,I_{yy}\frac{d^4}{dx^4}y(x) = w(x).$$

The general solution of such an equation will involve four coefficients for a fundamental set of homogeneous solutions, and these might as well be taken as the successive derivatives of the solution at the point $x = 0$. This choice is particularly useful if Laplace transforms are used to solve the differential equation.[1]

A prescription of a procedure for solving a conventionally posed beam problem is the following:

- Solve the governing differential equation, leaving the initial derivatives as parameters.
- Replace the initial conditions which happen to vanish by virtue of the specified left end beam attachment with zeroes.
- Calculate derivatives of the solution and evaluate these at the right hand end of the beam.
- Write out the equations that result from the imposed right end attachment conditions, and solve for the initial values that meet them.
- Plot the resulting shear, bending moment, and displacement curves.

[1]Using the Laplace methods allows representing localized beam loads with step functions as well as conceptual benefits.

Given the predictability of the solution process, it looks like a good candidate for a Maple implementation. On the other hand, it is a good example of a problem that can lead to "ugly code", if some thought is not taken.

The problem is that there are nine combinations of boundary conditions that must be handled. Structuring the code so that it tests which combination is desired, and then solves the problem "accordingly" is probably not a fruitful approach.

A better tack is to use a *data for code* tradeoff. The idea is to encode the operations that must take place in some data structure, and use the data structure to "drive" the procedure code.[2]

To implement this idea in Maple, we should first consider what *form* of data will be useful for the code in the procedure. Based on the procedure description above, consult the Maple *solve* and *subs* commands.

```
> ?solve
>
> ?subs
```

The relevant information is that the commands take Maple objects *sets* as arguments. In the case of *solve* it is a set of variables to solve for, while *subs* takes a set of equations defining the values to be substituted for subexpressions.

On this basis, including the substitution and equations sets in the data structure seems appropriate. The beam problems have numerical parameters as well as boundary conditions, so they can be included in the data. The user input (procedure arguments) should specify numerical values and boundary conditions, but for easy experimentation it might be useful to have default values for problem parameters.

Given the different types of data that must be handled, a Maple table is a natural candidate for a data structure. Since table data is accessed through "symbolic subscripts", readability of the code can be enhanced this way.

Our data table naturally breaks into two distinct sections. The variable data part reflects the numerical parameters and weight function for the beam problem. The fixed data portion consists of an entry for each type of end condition. For each end condition, there are three associated sets. These are the initial conditions that vanish when this condition is imposed at $x = 0$, the initial conditions remaining to be solved for in that case, and finally the terminal conditions applied when the condition is applied at the opposite end of the beam.

The user specifies the problem with an argument list consisting of a sequence of equations assigning table entries. The table subscripts that correspond to user settable variables are held in an auxiliary local variable, and the user input is checked for sanity by comparing the equation format to the assignable subscript list.

```
> beam_solve := proc(arguments)
>
```

[2]One extreme of this idea is the machine code (data) executed sequentially by a central processing unit.

```
> local y0L, y1L, y2L, y3L, shear, moment, displacement,
>     gen_soln, part_sol, beam_table, subscripts, i, ics,
>     EI, equations, ans, final_sol;
>
> beam_table := table([w = 160.0, E = 1000000, L = 6.0,
>   Iyy = .1e-1,    left = pin, right = pin,
>   pin = [{y00 = 0, y20 = 0}, {y10, y30}, {y0L = 0, y2L = 0}],
>   builtin = [{y00 = 0, y10 = 0}, {y20, y30}, {y1L = 0, y0L = 0}],
>   free = [{y20 = 0, y30 = 0}, {y00, y10}, {y2L = 0, y3L = 0}]]);
>
> subscripts := [w, E, L, Iyy, left, right];
>
> # ensure that arguments are table assignments
>
> for i from 1 to nargs
> do
>   if not type(args[i], '=') then
>     print(args[i]);
>     ERROR ('Invalid argument: not an equality')
>   fi;
> od;
>
> for i from 1 to nargs
> do
>   if has(subscripts, lhs(args[i])) then
>     beam_table[lhs(args[i])] := rhs(args[i]);
>   else
>     print(args[i]);
>     ERROR('invalid variable on left side of an equality');
>   fi;
> od;
>
> # catch suspicious beam load functions
>
> if not has(beam_table[w], 'x') then
>   print('Warning: beam load function does not depend on x');
> fi;
>
> EI := beam_table[E]*beam_table[Iyy];
>
> # The $ notation determines derivative order below
>
> gen_soln :=
>   dsolve(EI*diff(y(x), x$4) = beam_table[w], y(x), method=laplace);
>
> gen_soln := rhs(gen_soln);
>
> part_sol :=
>   subs({'@@'(D,3)(y)(0)=y30,'@@'(D,2)(y)(0)=y20,
```

```
>        `@@`(D,1)(y)(0)=y10,y(0)=y00}, gen_soln);
>
> # x = 0 corresponds to the left hand beam end
>
> ics := beam_table[beam_table[left]][1];
>
> part_sol := subs(ics, part_sol);
>
> # temporary assignments
>
> y0L := part_sol;
> y1L := diff(part_sol,x);
> y2L := diff(y1L, x);
> y3L := diff(y2L,x);
>
> # substitute end point into the derivative formulas
> # this assigns the local variables used in the end equations
>
> y0L := subs(x=beam_table[L], y0L);
> y1L := subs(x=beam_table[L], y1L);
> y2L := subs(x=beam_table[L], y2L);
> y3L := subs(x=beam_table[L], y3L);
>
> # use the correct set of equations for the wanted right hand end
>
> equations := eval(beam_table[beam_table[right]][3]);
>
> # compute the initial conditions required to solve equations
>
> ans := solve(equations, beam_table[beam_table[left]][2]);
>
> # put in the missing initial conditions, makes solution numerical
>
> final_sol := subs(ans, part_sol);
>
>
> # print(plot( ..)) is needed to get multiple plots from subroutines
>
> print(plot(final_sol, x=0..beam_table[L],
>      font=[TIMES,ITALIC,10], title=`displacement`));
>
> print(plot(-EI*diff(final_sol, x$2), x=0..beam_table[L],
>      font=[TIMES,ITALIC,10], title=`bending moment`));
>
> print(plot(-EI*diff(final_sol, x$3), x=0..beam_table[L],
>      font=[TIMES,ITALIC,10], title=`shear stress`));
>
> end;
```

An invocation of the procedure looks like:

```
> beam_solve(left=builtin, right=pin, w=200);
```

The bending moment plot that results is in Figure 14.1. The zero moment condition of the pin joint is evident, and the reaction at the built-in end $x = 0$ can also be determined from the plot.

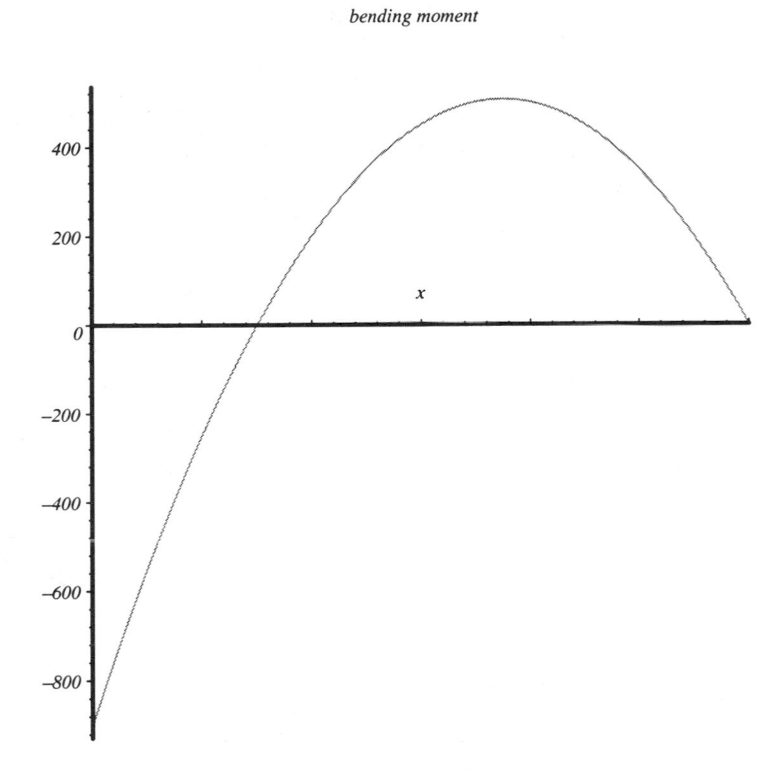

FIGURE 14.1. Beam_solve output plot.

15
Maple Linear Algebra Applications

15.1 Canonical Forms

Math background **Page:** 181

Maple has an extensive set of routines for linear algebraic manipulation; these are readily usable in problems involving systems of linear differential equations. The routines are accessed by loading the `linalg` package.

```
> with(linalg);
```

Descriptions of the use of the routines in the `linalg` package is available through the Maple Help facility.

15.1.1 Eigenvalues and Eigenvectors

Exercise

Math background **Page:** 187

Matrices are entered in a row-wise format.

```
> A := matrix([[0 ,1 ,0], [0 ,0 ,1], [-6 ,-11 ,-6]]);
```

$$A := \begin{bmatrix} 0 & 1 & 0 \\ 0 & 0 & 1 \\ -6 & -11 & -6 \end{bmatrix}$$

The eigenvalues are output as an expression sequence.

```
> eigenvals(A);
```

$$-3, -2, -1$$

The characteristic polynomial call must specify the variable to use.

```
> p := charpoly(A, s);
```

Rather than entering all matrix elements, one can construct a *companion matrix* from a prescribed characteristic polynomial.

$$p := s^3 + 6\,s^2 + 11\,s + 6$$

```
> ?companion
>
> companion(p,s);
```

$$\begin{bmatrix} 0 & 0 & -6 \\ 1 & 0 & -11 \\ 0 & 1 & -6 \end{bmatrix}$$

The standard form this text uses for representing high order scalar differential equations in vector form corresponds to the transpose of the output of the Maple `companion` routine. Both formats are called companion matrices in the literature, and the choice of format depends on whether one is using "row vectors" or "column vectors" for coordinate arrays.

```
> transpose(%);
```

$$\begin{bmatrix} 0 & 1 & 0 \\ 0 & 0 & 1 \\ -6 & -11 & -6 \end{bmatrix}$$

This can be hidden under a procedure definition.

```
> Companion := proc(p,s) RETURN(transpose(companion(p,s))); end;
```

Test it out:

```
> Companion(p,s);
```

$$\begin{bmatrix} 0 & 1 & 0 \\ 0 & 0 & 1 \\ -6 & -11 & -6 \end{bmatrix}$$

Maple can produce eigenvectors in various formats, depending on which Maple routine is used to compute them.

```
> evseq := eigenvects(A);
```

$$evseq := [-3, 1, \{[1\; -3\; 9]\}], [-2, 1, \{[1\; -2\; 4]\}],$$
$$[-1, 1, \{[1\; -1\; 1]\}]$$

The above output actually is in the form of an expression sequence of lists.

```
> whattype(");
```

$$exprseq$$

The vectors themselves are the third list elements. The sequence must be converted into a list in order to map an extraction function onto it (map will not "eat" sequences.)

```
> evlist := map( proc(x) op(op(3,x)) end, [evseq]);
```

$$evlist := [[1\ -3\ 9], [1\ -2\ 4], [1\ -1\ 1]]$$

The above has been built in the format of an argument `matrix` expects. Verify that we have the eigenvectors.

```
> EVS := matrix(evlist);
```

$$EVS := \begin{bmatrix} 1 & -3 & 9 \\ 1 & -2 & 4 \\ 1 & -1 & 1 \end{bmatrix}$$

This expression has the eigenvectors as the rows of the constructed matrix, so it must be transposed to end up with a matrix with the eigenvectors as columns.

```
> EVS := transpose(EVS);
```

$$EVS := \begin{bmatrix} 1 & 1 & 1 \\ -3 & -2 & -1 \\ 9 & 4 & 1 \end{bmatrix}$$

Check that the columns are eigenvectors:

```
> evalm(A &* EVS);
```

$$\begin{bmatrix} -3 & -2 & -1 \\ 9 & 4 & 1 \\ -27 & -8 & -1 \end{bmatrix}$$

Now try to diagonalize the matrix using the eigenvectors as a basis. Assign the change of basis matrix to the conventional name.

```
> P := EVS;
```

and calculate the inverse.

```
> Pinv := evalm(P^(-1));
```

$$Pinv := \begin{bmatrix} 1 & \frac{3}{2} & \frac{1}{2} \\ -3 & -4 & -1 \\ 3 & \frac{5}{2} & \frac{1}{2} \end{bmatrix}$$

Check that the computed matrices diagonalize the original matrix.

```
> evalm( Pinv &* A &* P);
```

$$\begin{bmatrix} -3 & 0 & 0 \\ 0 & -2 & 0 \\ 0 & 0 & -1 \end{bmatrix}$$

15.1.2 Jordan Form Calculations

Although the theory of linear algebra provides the result that every complex matrix can be put into Jordan form by a suitable change of basis, the required calculations are enormously burdensome. There is no reason to do the calculations by hand, as Maple has built in routines for this purpose.

Exercise

Math background **Page:** 184

Maple has a Jordan block calculation routine whose description is available through the help system.

```
> ?jordan
```

There are also Jordan block construction routines.

```
> ?JordanBlock
>
> J :=JordanBlock(lambda, 4);
```

$$J := \begin{bmatrix} \lambda & 1 & 0 & 0 \\ 0 & \lambda & 1 & 0 \\ 0 & 0 & \lambda & 1 \\ 0 & 0 & 0 & \lambda \end{bmatrix}$$

The cheap source of an identity matrix is the identity operation for matrix multiplication.

```
> alias(Id=&*());
```

The Jordan form consists of a block diagonal matrix with Jordan blocks along the main diagonal. If one desires to calculate the form of solution for a system of differential equations in Jordan form, it is sufficient to do the calculation corresponding to the case of a single Jordan block. The calculation in this case can be done by means of Laplace transforms. If

$$\frac{d}{dt}\mathbf{x} = \mathbf{J}\,\mathbf{x},$$

then solving by Laplace methods leads to

$$(\mathbf{I}\,s - \mathbf{J})\,\mathcal{L}\{\mathbf{x}\} = \mathbf{x}_0.$$

```
> evalm(s*Id - J);
```

$$\begin{bmatrix} -\lambda+s & -1 & 0 & 0 \\ 0 & -\lambda+s & -1 & 0 \\ 0 & 0 & -\lambda+s & -1 \\ 0 & 0 & 0 & -\lambda+s \end{bmatrix}$$

```
> resolvent :=evalm((s*Id -J)^(-1));
```

$$resolvent := \begin{bmatrix} -\frac{1}{\lambda-s} & \frac{1}{(\lambda-s)^2} & -\frac{1}{(\lambda-s)^3} & \frac{1}{(\lambda-s)^4} \\ 0 & -\frac{1}{\lambda-s} & \frac{1}{(\lambda-s)^2} & -\frac{1}{(\lambda-s)^3} \\ 0 & 0 & -\frac{1}{\lambda-s} & \frac{1}{(\lambda-s)^2} \\ 0 & 0 & 0 & -\frac{1}{\lambda-s} \end{bmatrix}$$

The Maple Laplace transform routines are not present by default, and must be loaded before use. The `inttrans` package loads integral transforms (including Laplace).

```
> with(inttrans);
```

In order to apply element-wise Laplace and inverse Laplace transforms with Maple, one must make use of the `map` operation, and this requires that a function of a single argument be available to be applied to each element. The basic Maple Laplace operations take more than one argument, so that convenience routines that build in the standard variable names must be built.

```
> lapinv := proc(expr)
>
>  RETURN(eval(invlaplace(expr, s, t)));
>
> end;
>
> evalm(map(lapinv, resolvent));
```

$$\begin{bmatrix} e^{(\lambda t)} & t\,e^{(\lambda t)} & \frac{1}{2}t^2 e^{(\lambda t)} & \frac{1}{6}t^3 e^{(\lambda t)} \\ 0 & e^{(\lambda t)} & t\,e^{(\lambda t)} & \frac{1}{2}t^2 e^{(\lambda t)} \\ 0 & 0 & e^{(\lambda t)} & t\,e^{(\lambda t)} \\ 0 & 0 & 0 & e^{(\lambda t)} \end{bmatrix}$$

Exercise

Math background **Page:** 180

The Jordan calculator will return the change of basis matrix, if a name is provided as the second argument. The form of the call is:

```
> jordan(J, 'Q');
```

For matrices with distinct eigenvalues, the Jordan form is the diagonalized form, and the call will return the eigenvector change of basis matrix. This actually is much easier than the process of extracting the change of basis matrix from a call to `eigenvectors`. A simple matrix with distinct eigenvalues was used in an earlier exercise.

```
> A := matrix([[0 ,1 ,0], [0 ,0 ,1], [-6 ,-11 ,-6]]);
```

$$A := \begin{bmatrix} 0 & 1 & 0 \\ 0 & 0 & 1 \\ -6 & -11 & -6 \end{bmatrix}$$

The `jordan` procedure returns the Jordan form, and optionally stores the change of basis matrix in a second argument.

```
> jordan(A, 'Q');
```

$$\begin{bmatrix} -1 & 0 & 0 \\ 0 & -2 & 0 \\ 0 & 0 & -3 \end{bmatrix}$$

Since `Q` is a matrix, evaluation must be used to force it to display the entries.

```
> eval(Q);
```

$$\begin{bmatrix} 1 & \frac{5}{6} & \frac{1}{6} \\ 1 & \frac{4}{3} & \frac{1}{3} \\ 1 & \frac{3}{2} & \frac{1}{2} \end{bmatrix}$$

15.2 Matrix Exponentials

The Maple `linalg` package actually contains a routine that will compute the exponential function of a matrix. The name of the routine is `exponential`. The user should be wary however, as the also included procedure named `exp` returns the result of applying the exponential function element wise, which is not the same thing at all. Maple can also be used to experiment with calculation of matrix exponentials by Laplace transforms, and this is examined below.

```
> with(linalg):
> alias(Id=&*());
```

Exercise

Math background **Page: 191**

Do the calculation for a general harmonic oscillator problem as an example.

```
> A := matrix(2,2, [[0,1],[-omega_n^2, -2*zeta*omega_n]]);
```

$$A := \begin{bmatrix} 0 & 1 \\ -omega\_n^2 & -2\,\zeta\; omega\_n \end{bmatrix}$$

```
> s*Id - A;
```

$$s\,Id - A$$

Matrices are passed to the Maple output routine as names (see the discussion in section) and hence are not evaluated. To force this, evaluate the expression as a matrix.

```
> evalm(%);
```

$$\begin{bmatrix} s & -1 \\ omega\_n^2 & 2\,\zeta\; omega\_n + s \end{bmatrix}$$

```
> resolvent := evalm( %^(-1));
```

$$resolvent := \begin{bmatrix} \dfrac{2\,\zeta\; omega\_n + s}{s^2 + 2\,\zeta\; omega\_n\, s + omega\_n^2} & \dfrac{1}{s^2 + 2\,\zeta\; omega\_n\, s + omega\_n^2} \\ -\dfrac{omega\_n^2}{s^2 + 2\,\zeta\; omega\_n\, s + omega\_n^2} & \dfrac{s}{s^2 + 2\,\zeta\; omega\_n\, s + omega\_n^2} \end{bmatrix}$$

```
> interface(labelling=false);
>
> lapinv := proc(expr)
>   RETURN(eval(invlaplace(expr, s,t)));
> end;
>
> eat := evalm(map(lapinv, resolvent));
```

$$\begin{bmatrix} \dfrac{\zeta e^{(-\zeta\omega_n t)}\sin\left(\sqrt{1-\zeta^2}\omega_n t\right)}{\sqrt{1-\zeta^2}} + e^{(-\zeta\omega_n t)}\cos\left(\sqrt{1-\zeta^2}\omega_n t\right), & \dfrac{e^{(-\zeta\omega_n t)}\sin\left(\sqrt{1-\zeta^2}\omega_n t\right)}{\sqrt{1-\zeta^2}\omega_n} \\ -\dfrac{\omega_n e^{(-\zeta\omega_n t)}\sin\left(\sqrt{1-\zeta^2}\omega_n t\right)}{\sqrt{1-\zeta^2}}, & -\dfrac{\zeta e^{(-\zeta\omega_n t)}\sin\left(\sqrt{1-\zeta^2}\omega_n t\right)}{\sqrt{1-\zeta^2}} + e^{(-\zeta\omega_n t)}\cos\left(\sqrt{1-\zeta^2}\omega_n t\right) \end{bmatrix}$$

Exercise
An example with obviously distinct eigenvalues follows.

```
> p(s) := expand((s-1)*(s-2)*(s-3));evalm
```

$$p(s) := s^3 - 6s^2 + 11s - 6$$

```
> B := matrix(3,3, [[0,1,0],[0,0,1],[6, -11, 6]]);
```

$$B := \begin{bmatrix} 0 & 1 & 0 \\ 0 & 0 & 1 \\ 6 & -11 & 6 \end{bmatrix}$$

```
> resolvent_B :=evalm( (s*Id - B)^(-1));
```

resolvent_B :=

$$\begin{bmatrix} \frac{-6s+s^2+11}{s^3-6s^2+11s-6} & \frac{-6+s}{s^3-6s^2+11s-6} & \frac{1}{s^3-6s^2+11s-6} \\ 6\frac{1}{s^3-6s^2+11s-6} & \frac{s(-6+s)}{s^3-6s^2+11s-6} & \frac{s}{s^3-6s^2+11s-6} \\ 6\frac{s}{s^3-6s^2+11s-6} & -\frac{11s-6}{s^3-6s^2+11s-6} & \frac{s^2}{s^3-6s^2+11s-6} \end{bmatrix}$$

```
> ebt :=map(lapinv, resolvent_B);
```

ebt :=

$$\begin{bmatrix} e^t - 3e^{(2t)} + e^{(3t)} & -\frac{5}{2}e^t + 4e^{(2t)} - \frac{3}{2}e^{(3t)} & \frac{1}{2}e^t - e^{(2t)} + \frac{1}{2}e^{(3t)} \\ 3e^t - 6e^{(2t)} + 3e^{(3t)} & -\frac{5}{2}e^t + 8e^{(2t)} - \frac{9}{2}e^{(3t)} & \frac{1}{2}e^t - 2e^{(2t)} + \frac{3}{2}e^{(3t)} \\ 3e^t - 12e^{(2t)} + 9e^{(3t)} & -\frac{5}{2}e^t + 16e^{(2t)} - \frac{27}{2}e^{(3t)} & \frac{1}{2}e^t - 4e^{(2t)} + \frac{9}{2}e^{(3t)} \end{bmatrix}$$

Exercise
Verify that a matrix exponential has the identity matrix as the value at $t = 0$.

```
> initialvalue := proc(expr)
>   RETURN(eval(subs(t=0, expr)));
> end;
>
> evalm(map(initialvalue, ebt));
```

$$\begin{bmatrix} 1 & 0 & 0 \\ 0 & 1 & 0 \\ 0 & 0 & 1 \end{bmatrix}$$

```
> evalm(map(initialvalue, eat));
```

$$\begin{bmatrix} 1 & 0 \\ 0 & 1 \end{bmatrix}$$

15.3 Stability Examples

The classification of stability diagrams into spirals, stable nodes, and unstable nodes is based on the phase plots of the harmonic oscillator model. As the damping coefficient of the model is varied, the phase plots change through the various categories. The standard stability pictures can be generated using maple by using a matrix exponential solution expression to generate a set of Maple parametric plots.

Exercise

Math background **Page:** 199

Generate the matrix exponential corresponding to a harmonic oscillator.

```
> A := matrix(2,2, [[0,1],[-omega_n^2, -2* zeta*omega_n]]);
```

$$A := \begin{bmatrix} 0 & 1 \\ -\omega_n{}^2 & -2\,\zeta\,\omega_n \end{bmatrix}$$

Prevent output abbreviations by using the `interface` command. Without this, Maple will use "percent variables" as subexpression abbreviations.

```
> interface(labelling=false);
```

Create an alias for the identity matrix as the identity operation for matrix multiplication.

```
> alias(Id=&*());
```

Invert the Laplace transform matrix. This is the calculation required when solving the problem by means of Laplace transforms.

```
> resolvent := evalm((s*Id - A)^(-1));
```

$$resolvent := \begin{bmatrix} \dfrac{2\,\zeta\,\omega_n + s}{2\,s\,\zeta\,\omega_n + s^2 + \omega_n{}^2} & \dfrac{1}{2\,s\,\zeta\,\omega_n + s^2 + \omega_n{}^2} \\ -\dfrac{\omega_n{}^2}{2\,s\,\zeta\,\omega_n + s^2 + \omega_n{}^2} & \dfrac{s}{2\,s\,\zeta\,\omega_n + s^2 + \omega_n{}^2} \end{bmatrix}$$

Term-by-term Laplace inversion requires "mapping" the inversion operation onto each matrix element.

```
> eAt := map(expr -> invlaplace(expr, s,t), evalm((s*Id - A)^(-1)));
```

$$\begin{bmatrix} \dfrac{\zeta e^{(-\zeta\omega_n t)}\sin\left(\sqrt{1-\zeta^2}\omega_n t\right)}{\sqrt{1-\zeta^2}} + e^{(-\zeta\omega_n t)}\cos\left(\sqrt{1-\zeta^2}\omega_n t\right), & \dfrac{e^{(-\zeta\omega_n t)}\sin\left(\sqrt{1-\zeta^2}\omega_n t\right)}{\sqrt{1-\zeta^2}\omega_n} \\ -\dfrac{\omega_n e^{(-\zeta\omega_n t)}\sin\left(\sqrt{1-\zeta^2}\omega_n t\right)}{\sqrt{1-\zeta^2}}, & -\dfrac{\zeta e^{(-\zeta\omega_n t)}\sin\left(\sqrt{1-\zeta^2}\omega_n t\right)}{\sqrt{1-\zeta^2}} + e^{(-\zeta\omega_n t)}\cos\left(\sqrt{1-\zeta^2}\omega_n t\right) \end{bmatrix}$$

To generate plots of the solution trajectories, numerically valued (rather than symbolic) solutions are required. Otherwise, the dreaded "empty Iris plot" message appears.

```
> eAtn:=map(expr -> subs(zeta=.99, omega_n=2, expr), eAt);
```

To pick out a particular solution, multiply an initial condition vector by the matrix exponential. This gives a two dimensional vector function of time (actually a Maple expression with "t" as parameter.) representing the solution starting at the initial condition $\begin{bmatrix} 1 \\ 1 \end{bmatrix}$.

```
> state := evalm(eAtn&*[1,1]);
```

$$state :=$$
$$[\; 10.56232996\, e^{(-1.98\,t)} \sin(\,.2821347196\, t\,) + e^{(-1.98\,t)} \cos(\,.2821347196\, t)$$
$$-\, 21.19554803\, e^{(-1.98\,t)} \sin(.2821347196\, t) + e^{(-1.98\,t)} \cos(.2821347196\, t)\;]$$

The position component of the solution is the first component of the solution vector.

```
> state[1];
```

$$10.56232996\, e^{(-1.98\,t)} \sin(.2821347196\, t) + e^{(-1.98\,t)} \cos(.2821347196\, t)$$

The Maple syntax for parametric plots is different from the one used to plot sets of functions. Plot a circle to see the syntax for a parametric plot of a vector function in the plane.

```
> plot({[cos(t), sin(t), t=0..6.28]});
```

Plot two solution trajectories, starting from opposite sides of the origin, using the curves generated above.

```
> plot({[state[1], state[2], t=0..15], [-state[1], -state[2], t=0..15]});
```

To get a representation of the "phase trajectories" or "phase plane plot" for the system, generate a sequence starting from initial conditions located around the unit circle in the plane. The model is the plot generated above, except for the positioning of the initial points along a unit circle.

```
> curves := seq( [evalm(eAtn &* [cos(n*.2), sin(n*.2)])[1],
>
> evalm(eAtn &* [cos(n*.2), sin(n*.2)])[2], t=0..15], n=0..25);
```

$$
\begin{aligned}
curves := [&7.017923930\, e^{(-1.98\, t)} \sin(.2821347196\, t) \\
&+ e^{(-1.98\, t)} \cos(.2821347196\, t), \\
&-14.17762410\, e^{(-1.98\, t)} \sin(.2821347196\, t), t = 0..15], [\\
&7.582197462\, e^{(-1.98\, t)} \sin(.2821347196\, t) \\
&+ .9800665778\, e^{(-1.98\, t)} \cos(.2821347196\, t), \\
&-15.28926178\, e^{(-1.98\, t)} \sin(.2821347196\, t) \\
\vdots \\
&-1.408097337\, e^{(-1.98\, t)} \sin(.2821347196\, t) \\
&+ .2836621855\, e^{(-1.98\, t)} \cos(.2821347196\, t), \\
&2.708001777\, e^{(-1.98\, t)} \sin(.2821347196\, t) \\
&- .9589242747\, e^{(-1.98\, t)} \cos(.2821347196\, t), t = 0..15]
\end{aligned}
$$

As generated, the `curves` variable is a sequence of parametric plot expressions, so we convert it to a set in order to generate a plot.

```
> plot({curves});
```

Other curves can be generated by repeating the above steps with different values for the damping and fundamental frequency in the model equation. When the parameter values are made to correspond to the conventional standard cases, the plots shown in Figures 8.1, 8.2 and 8.3 are the result.

15.4 Periodic Solutions with Maple

Historically, many periodic solutions for linear differential equations were computed using Fourier series approximations, as these were analytically tractable using paper and pencil. Time domain calculations have always been theoretically possible, but often so imposing that a hand calculation is not really feasible. The availability of Maple to do the calculations changes the situation, and makes it possible to generate and plot exact periodic solution curves for many problems.

Exercise

Math background Page: 230

This section carries out the calculations for the response of a harmonic oscillator (with general parameters) with sawtooth shape periodic forcing input. The matrix exponential calculations are presumed done as above, and the follow-on manipulations for periodic solution calculation are described. The `A` is the coefficient matrix for the harmonic oscillator equation. The Maple variable `eAt` is the matrix exponential solution of the harmonic oscillator equations.

The periodic solution process requires the matrix exponential, evaluated at the end of the first oscillation period.

```
> eAT := map(expr -> subs(t=T, expr), eAt);
```

$$\left[\begin{array}{cc} \frac{\zeta e^{(-\zeta\omega_n T)}\sin\left(\sqrt{1-\zeta^2}\omega_n T\right)}{\sqrt{1-\zeta^2}} + e^{(-\zeta\omega_n T)}\cos\left(\sqrt{1-\zeta^2}\omega_n T\right), & \frac{e^{(-\zeta\omega_n T)}\sin\left(\sqrt{1-\zeta^2}\omega_n T\right)}{\sqrt{1-\zeta^2}\omega_n} \\ -\frac{\omega_n e^{(-\zeta\omega_n T)}\sin\left(\sqrt{1-\zeta^2}\omega_n T\right)}{\sqrt{1-\zeta^2}}, & -\frac{\zeta e^{(-\zeta\omega_n T)}\sin\left(\sqrt{1-\zeta^2}\omega_n T\right)}{\sqrt{1-\zeta^2}} + e^{(-\zeta\omega_n T)}\cos\left(\sqrt{1-\zeta^2}\omega_n T\right) \end{array}\right]$$

The coefficient matrix in the equation

$$\left(\mathbf{I} - e^{\mathbf{A}T}\right)\mathbf{x}(0) = \mathbf{G}(T)$$

for the periodic initial condition is constructed next. Calculate the matrix inverse required to solve for the periodic initial condition directly.

```
> Coeffinv := evalm((Id -eAT)^(-1));
```

Next, construct the right hand side of the system of equations. The following is the Laplace transform corresponding to a sawtooth periodic forcing function response.

```
> G := evalm( resolvent &* [0, 1/(s^2)]);
```

$$G := \left[\begin{array}{c} \frac{1}{(2s\zeta\omega_n + s^2 + \omega_n{}^2)s^2} \\ \frac{1}{s(2s\zeta\omega_n + s^2 + \omega_n{}^2)} \end{array}\right]$$

To get the time domain version, invert the transform term-by-term.

```
> Gt := map( expr -> invlaplace(expr,s, t), G);
```

$$Gt := \left[\begin{array}{c} \frac{e^{(-\zeta\omega_n t)}\sin\left(\sqrt{1-\zeta^2}\,\omega_n t\right)}{\omega_n{}^3\sqrt{1-\zeta^2}} + 2\frac{\zeta^2 e^{(-\zeta\omega_n t)}\sin\left(\sqrt{1-\zeta^2}\,\omega_n t\right)}{\omega_n{}^3\sqrt{1-\zeta^2}} \\ +2\frac{\zeta e^{(-\zeta\omega_n t)}\cos\left(\sqrt{1-\zeta^2}\,\omega_n t\right)}{\omega_n{}^3} - 2\frac{\zeta}{\omega_n{}^3} + \frac{t}{\omega_n{}^2} \\ \frac{1}{\omega_n{}^2} - \frac{\zeta e^{(-\zeta\omega_n t)}\sin\left(\sqrt{1-\zeta^2}\,\omega_n t\right)}{\omega_n{}^2\sqrt{1-\zeta^2}} - \frac{e^{(-\zeta\omega_n t)}\cos\left(\sqrt{1-\zeta^2}\,\omega_n t\right)}{\omega_n{}^2} \end{array}\right]$$

The right hand side of the system of equations is the value of this expression at the final time. This could be computed by substituting `T` in the above, but it is simpler to paste and edit the previous expression and recompute it.

```
> GT := map( expr -> invlaplace(expr,s, T), G);
```

Since the coefficient matrix has already been inverted, the initial condition required for a periodic solution is computed with a matrix multiplication. The answer is rather large, but this is the cost of dragging all the problem parameters through the calculation.

```
> x0 := evalm( Coeffinv &* GT);
```

$$x0 := \Big[\Big(-2\,\zeta\,(e^{(-\zeta\,\omega_n\,T)})^2 \sinh\left(\omega_n\sqrt{-1+\zeta^2}\,T\right)\sin\left(\sqrt{1-\zeta^2}\,\omega_n\,T\right)$$

$$+2\,\zeta^3\,(e^{(-\zeta\,\omega_n\,T)})^2 \sinh\left(\omega_n\sqrt{-1+\zeta^2}\,T\right)\sin\left(\sqrt{1-\zeta^2}\,\omega_n\,T\right)+2$$

$$\zeta^2\,(e^{(-\zeta\,\omega_n\,T)})^2 \sinh\left(\omega_n\sqrt{-1+\zeta^2}\,T\right)\cos\left(\sqrt{1-\zeta^2}\,\omega_n\,T\right)\sqrt{1-\zeta^2}$$

$$-2\,\zeta^2\,e^{(-\zeta\,\omega_n\,T)} \sinh\left(\omega_n\sqrt{-1+\zeta^2}\,T\right)\sqrt{1-\zeta^2}$$

$$+\zeta\,e^{(-\zeta\,\omega_n\,T)} \sinh\left(\omega_n\sqrt{-1+\zeta^2}\,T\right)\sqrt{1-\zeta^2}\,\omega_n\,T$$

$$+(e^{(-\zeta\,\omega_n\,T)})^2 \cosh\left(\omega_n\sqrt{-1+\zeta^2}\,T\right)\sqrt{-1+\zeta^2}\sin\left(\sqrt{1-\zeta^2}\,\omega_n\,T\right)$$

$$-2$$

⋮

⋮

⋮

$$-(e^{(-\zeta\,\omega_n\,T)})^2 \cosh\left(\omega_n\sqrt{-1+\zeta^2}\,T\right)\sqrt{-1+\zeta^2}\cos\left(\sqrt{1-\zeta^2}\,\omega_n\,T\right)$$

$$\sqrt{1-\zeta^2}-\sqrt{-1+\zeta^2}\sqrt{1-\zeta^2}$$

$$+\sqrt{-1+\zeta^2}\,\zeta\,e^{(-\zeta\,\omega_n\,T)}\sin\left(\sqrt{1-\zeta^2}\,\omega_n\,T\right)$$

$$+\sqrt{-1+\zeta^2}\,e^{(-\zeta\,\omega_n\,T)}\cos\left(\sqrt{1-\zeta^2}\,\omega_n\,T\right)\sqrt{1-\zeta^2}\Big)\Big/\Big({\omega_n}^2$$

$$\sqrt{-1+\zeta^2}\Big(1-2\,e^{(-\zeta\,\omega_n\,T)}\cosh\left(\omega_n\sqrt{-1+\zeta^2}\,T\right)$$

$$+(e^{(-\zeta\,\omega_n\,T)})^2 \cosh\left(\omega_n\sqrt{-1+\zeta^2}\,T\right)^2$$

$$-(e^{(-\zeta\,\omega_n\,T)})^2 \sinh\left(\omega_n\sqrt{-1+\zeta^2}\,T\right)^2\Big)\sqrt{1-\zeta^2}\Big)\Big]$$

The periodic solution trajectories also can be explicitly calculated, once the periodic initial condition is known. This is a huge expression, because of the free parameters in all of the terms.

```
> persoln := evalm(evalm(eAt &* x0) + Gt);
```

$$persoln := \Big[-e^{(-\zeta\,\omega_n\,t)} \Big(-\sinh\Big(\omega_n \sqrt{-1+\zeta^2\,t}\Big) \Big(-1+\zeta^2\Big)^{3/2} \sqrt{1-\zeta^2}$$

$$+ 2\cosh\Big(\omega_n \sqrt{-1+\zeta^2\,t}\Big) \zeta^5 \sqrt{1-\zeta^2} - \cosh\Big(\omega_n \sqrt{-1+\zeta^2\,t}\Big)$$

⋮

⋮

$$- 2\,e^{(-\zeta\,\omega_n\,T)} \cosh\Big(\omega_n \sqrt{-1+\zeta^2\,T}\Big)$$

$$+ (e^{(-\zeta\,\omega_n\,T)})^2 \cosh\Big(\omega_n \sqrt{-1+\zeta^2\,T}\Big)^2$$

$$- (e^{(-\zeta\,\omega_n\,T)})^2 \sinh\Big(\omega_n \sqrt{-1+\zeta^2\,T}\Big)^2\Big) \sqrt{1-\zeta^2}\Big) + \frac{1}{{\omega_n}^2}$$

$$- \frac{\zeta\,e^{(-\zeta\,\omega_n\,t)} \sin\Big(\sqrt{1-\zeta^2}\,\omega_n\,t\Big)}{{\omega_n}^2\sqrt{1-\zeta^2}} - \frac{e^{(-\zeta\,\omega_n\,t)} \cos\Big(\sqrt{1-\zeta^2}\,\omega_n\,t\Big)}{{\omega_n}^2}\Big]$$

To generate plots of the periodic solution, name the expressions that are the vector components. The first component of the harmonic oscillator vector is the position and the second is the velocity.[1]

```
> x := persoln[1]:
>
> xdot := persoln[2]:
```

Plotting requires numerical values for all of the parameters except the time variable. Leaving the parameters in the solutions allows one to check what happens as the forcing frequency approaches the natural frequency of the system.

```
> x1 := subs({omega_n=2, zeta=.1, T=4}, x):
>
> x1dot := subs({omega_n=2, zeta=.1, T=4}, xdot):
```

First produce a conventional plot of the position and velocity curves.

```
> plot( {x1, x1dot}, t=0..4);
```

The periodicity of the trajectory is evident from that plot. However, this example is a two dimensional system, and hence the periodic solution should show up as a closed curve when plotted in the phase plane. In Maple terms, this takes only a parametric plot with time as the parameter.

[1]The terminology is from the mechanical model problem and probably should be varied when one has an RLC circuit in mind.

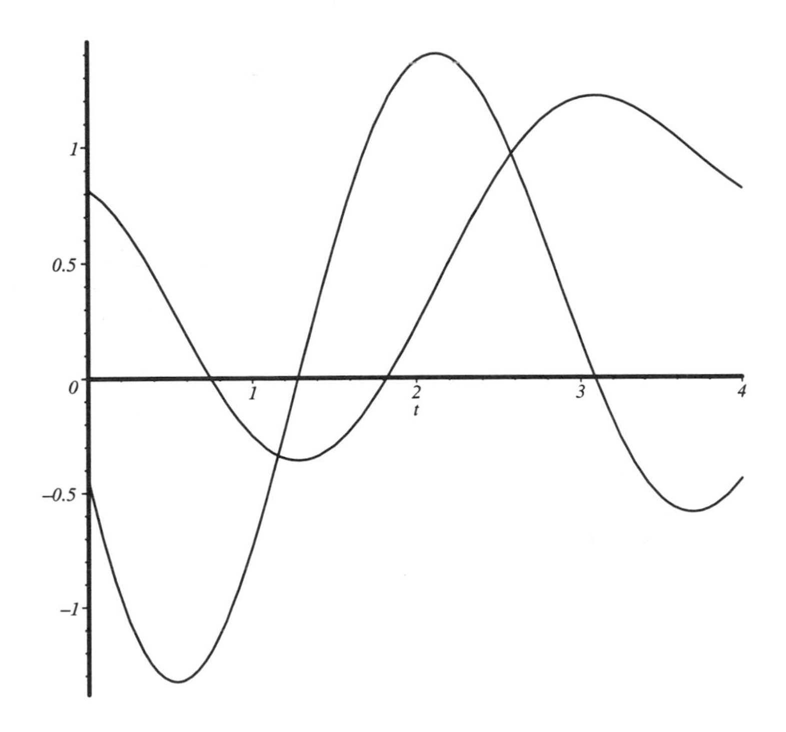

FIGURE 15.1. Periodic solution components

```
> plot([x1, x1dot, t=0..4]);
```

Math discussion **Page:** 234

A combination of the two views can be generated using the Maple `tubeplot` routine. This will plot a curve in three dimensional space, and at the same time surround the locus with a "tube", making it more visible. For this example, we can use the phase variables for two of the coordinates, and time as the third axis.

```
> tubeplot([x1, t, x1dot], t=0..4, radius=.05, axes=BOXED);
```

The ordering of the coordinates plots the velocity vertically, position horizontally, and time toward the viewer. The other plots can be thought of as projections of the information appearing in Figure 15.3. The phase plot of figure 15.2 is a projection onto the phase plane formed by the position and velocity axes. The solution curves of Figure 15.1 are projections onto the time-position and time-velocity planes. Figure 15.3 graphically illustrates how the solution vector returns to its original state at the end of the time period.

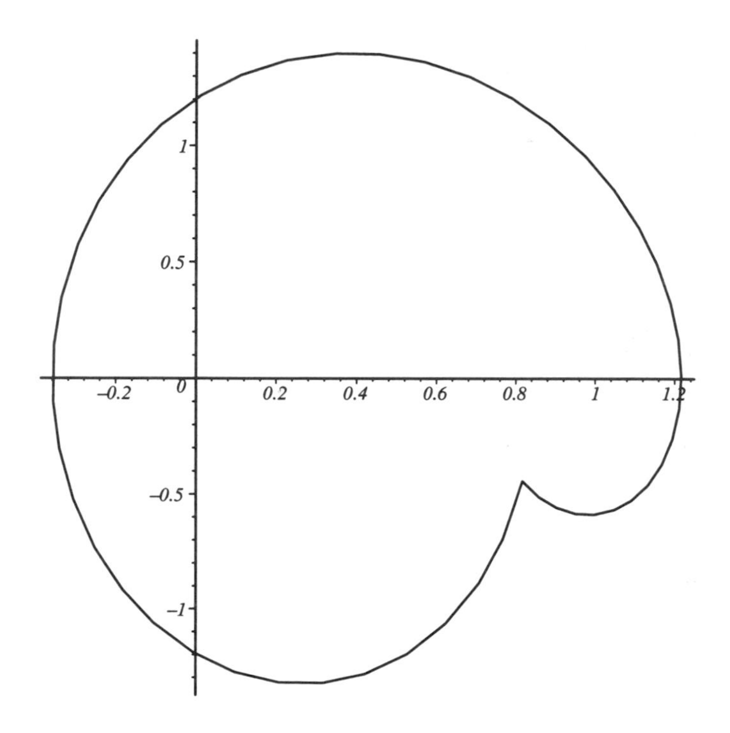

FIGURE 15.2. Periodic solution phase plot

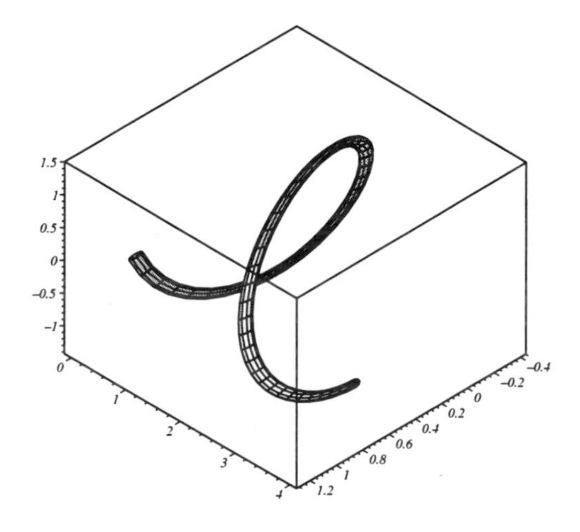

FIGURE 15.3. Periodic solution phase time trajectory

16

Runge–Kutta Designs

16.1 Orientation and Overview

Most of the other Maple examples in this text have made use of Maple built-in facilities, and have required the user to write at most one or two "wrapper" routines to accomplish the desired end. When the problem at hand is to derive a set of order (truncation accuracy) condition for Runge–Kutta schemes, more serious Maple programming is required. Rather than "high-level" Maple routines being available, it is largely the "low level" expression manipulation operations that are used.

Also, attempting to do a calculation of this sort is not just a problem of getting the correct formulas embedded in Maple procedures. The calculations are sufficiently imposing (tedious, long, complicated, ...) that it takes an effort to get Maple to do them at all, while it is at the same time clear that doing the calculations by hand (they were originally done this way, [1]) is a very difficult prospect. In short, the problem puts up a good fight.

There are a lot of programming details embedded in the resulting code, and mentioning them all is probably not useful. Some of the methods employed are useful in contexts other than Runge–Kutta order condition calculations. Almost all attempts to deal with nonlinear systems of differential equations in any fashion involve what are at the base Taylor series calculations; the subroutines below are adaptable to such purposes.

16.2 Notational Issues

In the numerical analysis literature it is common to use x as the independent variable, with y as the notation for the solution function. In the case of a vector system of differential equations, the problem can be taken as

$$\begin{aligned} \frac{d}{dx}\mathbf{y} &= \mathbf{f}(\mathbf{y}, x), \\ \mathbf{y}(0) &= \mathbf{y}_0. \end{aligned}$$

If Maple is to be applied to such problems, the first thought is that the Maple capacity for vectors or arrays might be used, but this is mistaken. We have to make calculations that are valid independent of the dimension of the state vector $\mathbf{y}$, while the Maple linear algebra routines support vector objects whose dimension is known at run time. If one attempts to use Maple with symbolic variables thought of as vector valued, one soon discovers that Maple routines treat undefined symbols as scalars, and commute expressions accordingly.

In order to carry out noncommutative array manipulations, it is more useful to utilize a subscript notation for vector components. These notations are often associated with the subject area of tensor analysis, where the compactness of the notation is noticeable. The compactness of the notation is also a major issue here, since one of the crucial issues is to keep the "size" of the symbolic calculations under control. The "tensor" notation is very helpful for this.

In terms of subscript (component) notation the system of differential equations whose numerical solution is under discussion might be written as

$$\frac{d}{dx} y_i = f_i(\mathbf{y}, x)$$

where i ranges over the dimensions of the vector variable. Subscripted variables can be used to do the calculations, and noncommutative routines for multiplication and differentiation can be written to work in terms of the subscript notation.

Numerical method truncation error is analyzed on the basis of a single step of length h. A more or less arbitrary numerical method propagating a step of length h from the initial value $(\mathbf{y}, x)$ can be represented symbolically as

$$\hat{\mathbf{y}} = \mathbf{M}(\mathbf{y}, x, h).$$

The true solution at $x + h$ is

$$\mathbf{y}(x + h),$$

so the error between the true solution and the numerical approximation is

$$\mathbf{error}(h) = \mathbf{y}(x + h) - \mathbf{M}(\mathbf{y}, x, h).$$

This is a vector-valued function, and it must be differentiated with respect to h in order to compute the order conditions. To derive the order conditions, what is required is to differentiate the above expression with respect to h (an "arbitrary" number of times) and then write down the expression equating the successive Taylor series coefficients of the error components to zero. The interpretation of the resulting equations is that they represent the conditions on the numerical method $\mathbf{M}$ which make the method accurate to the given order. To do this calculation, formulas for the derivatives must be calculated in an efficient notation, both for the true solution (the first term in the error equation) and the proposed numerical method.

For the case of Runge–Kutta algorithm numerical methods, it turns out that these calculations can be done in terms of the entries in the coefficient matrix which serves to define the particular method. The result is a set of "order conditions" which must be satisfied by the method coefficients in order to guarantee that a given truncation accuracy is obtained.

16.3 True Solution Derivatives

To introduce the approach to the problem, we first discuss the calculation of successive derivatives of the true solution in terms of subscript (component) notation.

The first derivative of the true solution is just the right hand side of the differential equation

$$\begin{aligned} \frac{d}{dh} y_i(x+h) &= f_i, \\ &= f_i(\mathbf{y}(x+h), x+h). \end{aligned}$$

Higher derivatives follow by applying the chain rule to the above to get in the case of the second derivative

$$\frac{d^2}{dh^2} y_i(x+h) = \Sigma_{j=1}^{n} \frac{\partial}{\partial y_j} f_i \, f_j + \frac{\partial}{\partial x} f_i.$$

A useful convention (from tensor analysis) to use for such calculations is the "summation convention". It arises from the observation that the summation signs in such expressions are associated with repeated subscript index pairs. The convention is that repeated index pairs indicate an implicit summation over the repeated indices. The summation sign may thus be omitted with no danger of any ambiguity in the expressions. With this notation, the above expression looks like

$$\frac{d^2}{dh^2} y_i(x+h) = \frac{\partial}{\partial y_j} f_i \, f_j + \frac{\partial}{\partial x} f_i.$$

The partial derivatives involved may also be expressed in compact notation by using the subscript notation for partial derivatives. In vector notation the second derivative may be indicated as

$$\frac{d^2}{dh^2} \mathbf{y}(x+h) = \mathbf{f}_{\mathbf{y}} \, \mathbf{f} + \mathbf{f}_x.$$

The term $\mathbf{f_y}$ really ought to be thought of as a matrix, multiplying the vector velocity to produce a vector. In other terminology, a "second order tensor" is closer to the mark. The required derivatives can be calculated in the vector notation, as long as care is taken not to inadvertently commute vector and matrix (tensor, really) quantities whose composition order must be preserved.

In order to carry out this type of calculation in Maple, the form of expressions which must be handled has to be identified. We observe (from long experience with chain rule calculations) that the most general form of expression that is involved in any of the contemplated calculations is in the form of a sum of terms, each of which may be a product of numbers, symbolic constants, and functions of the pair (y, x). As is clear from the second derivative calculation, the multiplications must be treated as noncommutative, since otherwise Maple will take it upon itself to rearrange terms to its own liking, not "knowing" that the terms do not commute. The way to prevent the unwanted commutation is to carry the products around as the arguments of (symbolic) functions. It is actually useful to use two symbolic function expressions for this. One is for the functions of (y, x) which must be differentiated using the chain rule. These are carried in symbolic MPY expressions. It is also necessary to retain order in products of the Runge–Kutta routine coefficient arrays so that common terms can be appropriately collected in symbolic mpy expressions.

Since function argument order matters, Maple will not commute variables in such a context. For the case of true solution derivatives, only use of MPY is required. The fact that mpy is also required should not be regarded as obvious, and only becomes evident when sufficiently high order Runge–Kutta methods are tried. Since the true solution derivatives have to be subtracted from the Runge–Kutta formula derivatives, the true solution differentiator must operate in these terms to keep uniform data formats in play even though the Runge–Kutta constants are not involved in the true solution calculations.

As long as this problem is managed, the required differentiations can be done just by manipulation of the argument lists inside the MPY expressions. These represent noncommutative products, and so must be differentiated with the appropriate application of the chain rule. The solution differentiation in these terms can be done with the Maple code

```
# Differentiate SUM of n*mpy()*MPY(..,g(y,x),..)
d := proc(expr)
 local name, parx, pary, nsummands, nprods, i, j, subscripts,
     returnvalue, lexpr, summand, localfunlist, locallexpr, localfunderlist,
     funlist, coefflist, funderlist, constlist;
 lexpr := eval(expr);
 if op(0, lexpr) = '+' then
   funlist := [op(lexpr)];
 else
   funlist := [lexpr];
 fi ;
```

```
coefflist := map( elt → op(nops(elt)−1, elt), funlist);
constlist :=map(elt →if nops(elt)=2 then 1
        else convert_mpy(op(1..nops(elt)−2, elt)) fi , funlist);
funlist := map( elt → op(nops(elt), elt), funlist);
for i from 1 to nops(funlist)
do
  lexpr := op(i, funlist);
# product terms : keep values expanded into a sum of products at all stages
# must have an MPY in hand, but test for safety
  if evalb( op(0, lexpr) = 'MPY') then
   nprods := nops(lexpr);
   localfunlist := [op(lexpr)];
   for j from 1 to nprods
   do
   locallexpr := localfunlist[j];
# functions of two arguments have to have y,x arguments
   if type(locallexpr, anyfunc(anything, anything)) then
     name := eval(op(0, locallexpr));
     subscripts := op(1..nops(name), name);
     pary := subscripts,y;
     parx := x,subscripts;
     returnvalue :=
      mpy()*MPY(f[pary](y,x), f[](y,x))+ mpy()*MPY(f[parx](y,x));
     localfunderlist[j] := returnvalue;
   fi ;
  od;
# plug derivatives in place of functions and multiply out
  summand := 0;
  localfunlist := map(elt→mpy()*MPY(elt), localfunlist);
  for j from 1 to nprods
  do
   summand := summand + MULT(op(subsop(j=localfunderlist[j], localfunlist)));
  od;
  funderlist[i] := summand;
 else
  ERROR('unexpected argument: not expected MPY, so fell through');
 fi ;
od;
summand := 0;
for j from 1 to nops(funlist)
 do
  summand := summand + MULT(constlist[j]*coefflist[j]*MPY(), funderlist[j]);
 od;
```

```
 RETURN(summand);
end;
```

For this to work, the procedure $MULT$ must be written in such a way that MPY-ication obeys the appropriate noncommunicative law, while reducing expressions to the appropriate sum of formal triple product terms. This routine will be the workhorse of the whole project and so has to be squeaky tight, as far as this is possible with an interpreted language like Maple. The routine operates by taking each term of one factor, and placing the factor multiplication arguments inside the multiplications of the other factor on the appropriate side.

```
# A noncommutative multiplication routine, as far as functions
# are concerned. Associates MULT calls, and handles mpy() for keeping
# constants in sync with MPY arguments.
SUB :=  SUB ;
# VEEERRRRY dangerous hack: SUB has to be anonymous symbol ?????
MULT :=proc()
 local ARGLIST,i, k, summand_list, MPYS, mpys, ns,
     ACC_MPYS, ACC_mpys, ACC_ns, lexpr, summand ; # SUB;
   ARGLIST := [args];
   if member(0,ARGLIST) then
     RETURN(0)
   fi ;
   if nops(ARGLIST) = 1 then RETURN(ARGLIST[1]) fi ;
   if ARGLIST = [] then RETURN(1) fi ;
   lexpr := op(1, ARGLIST);
   if type(lexpr, '+') then
     summand_list := [op(lexpr)];
   else
     summand_list := [lexpr];
   fi ;
# allow for no number in front, but require mpy, MPY (maybe empty arguments)
   ACC_MPYS := map(elt→op(nops(elt), elt), summand_list);
   ACC_mpys := map(elt→op(nops(elt)−1, elt), summand_list);
   ACC_ns := map(elt→if nops(elt)=2 then 1
           else convert_mpy(op(1..nops(elt)−2, elt)) fi , summand_list);
   for i from 2 to nops(ARGLIST)
   do
    lexpr := op(i, ARGLIST);
    if type(lexpr, '+') then
     summand_list := [op(lexpr)];
    else
     summand_list := [lexpr];
```

```
    fi ;
    MPYS := map(elt→op(nops(elt), elt), summand_list);
    mpys := map(elt→op(nops(elt)−1, elt), summand_list);
    ns := map(elt→if nops(elt)=2 then 1
            else convert_mpy(op(1..nops(elt)−2, elt)) fi , summand_list);
    for k from 1 to nops(ACC_mpys)
    do
     ACC_mpys := subs(SUB=op(op(k, ACC_mpys)),
                subsop(k=map(elt→mpy(SUB , op(elt)), mpys), ACC_mpys));
     ACC_MPYS := subs(SUB=op(op(k, ACC_MPYS)),
                subsop(k=map(elt→MPY(SUB , op(elt)), MPYS), ACC_MPYS));
     ACC_ns := subs(SUB=op(k, ACC_ns),
                subsop(k=map(elt→SUB * op(elt), ns), ACC_ns));
    od;
    ACC_mpys := map(op, ACC_mpys);
    ACC_MPYS := map(op, ACC_MPYS);
    ACC_ns := map(op, ACC_ns);
   od;
   summand := 0;
   for k from 1 to nops(ACC_MPYS)
   do
    summand := summand + op(k, ACC_ns)*op(k, ACC_mpys)*op(k, ACC_MPYS);
   od;
   RETURN(summand);
end;
```

16.4 Runge–Kutta Derivatives

The Runge–Kutta methods operate by constructing several "guesses" at the appropriate "slope" to use in taking a single step. These guesses take the particular form involving a linear combination of the current slope and the previous guesses

$$\mathbf{k}_j = \mathbf{f}(\mathbf{y} + h * \Sigma_{l=1}^{j-1} a_{j,l}\mathbf{k}_l,\ x + h * c_j).$$

Note that this involves the "guessed slopes" on both sides of the equation, although for so-called explicit Runge–Kutta methods the fact that the sum only extends to $j-1$ makes the equations explicit rather than implicit as far as calculation of the slope value $\mathbf{k}_j$ is concerned. There are implicit Runge–Kutta routines (useful for "stiff problems") which require that the system of equations be solved implicitly for the slopes in order to carry out a solution step. The term inside the slope evaluation essentially involves a matrix multiplication with the array **a** which is part of the definition of the method.

After the guessed slopes have been calculated, the numerical step is taken as

$$\hat{\mathbf{y}} = \mathbf{y} + h * \Sigma_{j=1}^{s} b_j \mathbf{k}_j.$$

This expression represents the numerical method expression $M(\mathbf{y}, x, h)$ for the case of a Runge–Kutta method.

As part of the calculation of the order conditions, we must evaluate the derivatives of the above expression with respect to h, at the point $h = 0$. From the form of these equations it is evident that the quantities

$$\frac{d^m}{dh^m}\mathbf{k}_j$$

are what is required to evaluate the Runge–Kutta step derivatives. The successive derivatives can be explicitly calculated (using the tensor notation convention) as

$$\begin{aligned}
\frac{d}{dh}\hat{\mathbf{y}} &= b_j\,\mathbf{k}_j + h\,b_j\,\frac{d}{dh}\mathbf{k}_j \\
\frac{d^2}{dh^2}\hat{\mathbf{y}} &= 2\,b_j\,\frac{d}{dh}\mathbf{k}_j + h\,b_j\,\frac{d^2}{dh^2}\mathbf{k}_j \\
\ldots &= \ldots \\
\frac{d^m}{dh^m}\hat{\mathbf{y}} &= m\,b_j\,\frac{d^{m-1}}{dh^{m-1}}\mathbf{k}_j + h\,b_j\,\frac{d^m}{dh^m}\mathbf{k}_j.
\end{aligned}$$

At $h = 0$ the last term drops out, so that only derivatives $\mathbf{k}_j$ of one less order than the truncation order are required.

The calculation of the slope derivatives can be carried out by differentiating the equation governing the slope guess iteration implicitly. The stage updates are (in summation convention notation)

$$\mathbf{k}_j = \mathbf{f}(\mathbf{y} + h\,a_{j,l}\mathbf{k}_l,\ x + h * c_j).$$

The derivative of this expression with respect to h takes the form

$$\begin{aligned}
\frac{d}{dh}\mathbf{k}_j &= \mathbf{f}_{\mathbf{y}}\frac{d}{dh}(h\,a_{j,l}\mathbf{k}_l) + c_j\,\mathbf{f}_x \\
&= a_{j,l}\,\mathbf{f}_{\mathbf{y}}\mathbf{k}_l + h\,a_{j,l}\,\mathbf{f}_{\mathbf{y}}\frac{d}{dh}\mathbf{k}_l + c_j\,\mathbf{f}_x,
\end{aligned}$$

resulting from the chain rule calculations.

Notice that the above expression involves h derivatives of the stage slopes on both sides of the equation, which might seem to cause a problem. However, the right hand side equation occurrence has a factor of h and disappears when the expression is evaluated at zero, as required. The only remaining problem is how to organize the Maple calculation in this fashion.

The Maple code handles this by treating the Maple expression $slope(j, m, h)$ as the mth derivative of slope component j. The routines to calculate Runge–Kutta method derivatives consist of recursive procedures, combined with a chain-rule differentiator specific to the form of the expressions encountered in differentiating Runge–Kutta slopes.

Needless to say, the differentiator is the result of a certain amount of "code enhancement" after an initial draft. The peculiar code for $n < 0$ is a (groan) Maple hack which allows the remember table of `slope_der` to act as a data structure for unique dummy subscripts.

```
# Recursive evaluation of the derivatives of a Runge--Kutta slope
# formula for an arbitrary slope stage number j. It calculates
# it implicitly, to keep the expressions manageable while the
# calculations are made. The implicit loop is closed below when
# the h=0 is set.
slope_der := proc(n, extra) option remember;
 if n < 0 then RETURN(extra) fi ;
 if n = 0 then
   RETURN(mpy()*MPY(f[](y + h* a[j, l[0]] * slope(l[0], 0, h), x + h*c[j])));
 else
  RETURN(dh(slope_der(n-1)));
 fi ;
end;


# Differentiate a Runge--Kutta formula with respect to h
# answer can be eval(subs(h=0, answer)) to get the Taylor derivatives
dh := proc(expr)
 local name, parx, pary, nsummands, nprods, i, j, subscripts,
     returnvalue, lexpr, summand,localfunlist, locallexpr, localfunderlist,
     funlist, next_index, stage_index, coefflist, constderlist, funderlist,
     constlist ;
 lexpr := eval(expr);
 if op(0, lexpr) = '+' then
   funlist := [op(lexpr)];
 else
   funlist := [lexpr];
 fi ;
 coefflist := map( elt -> op(nops(elt)-1, elt), funlist);
 constlist :=map(elt ->if nops(elt)=2 then 1
    else convert_mpy(op(1..nops(elt)-2, elt)) fi , funlist);
 funlist := map( elt -> op(nops(elt), elt), funlist);
 constderlist := map( elt -> diff(elt, h), constlist);
 for i from 1 to nops(funlist)
```

```
 do
  lexpr := op(i, funlist);
# product terms : keep values expanded into a sum of products at all stages
# must have an MPY in hand, but test for safety
  if evalb( op(0, lexpr) = 'MPY') then
   nprods := nops(lexpr);
   localfunlist := [op(lexpr)];
   for j from 1 to nprods
   do
   locallexpr := localfunlist[j];
# if hit a slope, direct treatment
   if evalb(op(0, op(2,locallexpr)) = 'slope') then
    localfunderlist[j] :=
        mpy()*MPY(subsop(2=subsop(2= op(2, op(2,locallexpr))+1,
                op(2,locallexpr)), locallexpr));
# any other functions, chain rule it
# functions of two arguments have to have RK argument format
   elif type(locallexpr, anyfunc(anything, anything)) then
    name := eval(op(0, locallexpr));
    subscripts := op(1..nops(name), name);
    pary := subscripts,y;
    parx := x,subscripts;
    next_index := get_free_index();
    stage_index := op(1, op(2, op(2, op(2, locallexpr))));
    returnvalue :=
          mpy()*MPY(f[pary](op(1..2, locallexpr)),
         mpy(a[stage_index, next_index])* slope(next_index, 0, h));
    next_index := get_free_index();
    returnvalue := returnvalue +
       h*mpy()*MPY(f[pary](op(1..2, locallexpr)),
         mpy( a[stage_index, next_index])* slope(next_index, 1, h))
     + mpy(diff(op(2, locallexpr),h)) *MPY(f[parx](op(1..2, locallexpr)));
    localfunderlist[j] := expand(eval(returnvalue));
   fi ;
  od;
# plug derivatives in place of functions and multiply: reformat localfunlist
  summand := 0;
  localfunlist := map(elt->mpy()*MPY(elt), localfunlist);
  for j from 1 to nprods
  do
   summand := summand + MULT(op(subsop(j=localfunderlist[j], localfunlist)));
  od;
```

```
  funderlist[i] := summand;
 else
  ERROR('unexpected argument: not expected MPY, so fell through');
 fi ;
od;
summand := 0;
for j from 1 to nops(funlist)
  do
   summand := summand + MULT(constlist[j]*coefflist[j]*MPY(), funderlist[j])
            + MULT(constderlist[j]*coefflist[j]*MPY(),mpy()*funlist[j]);
  od;
RETURN(summand);
end;
```

16.5 Traps and Tricky Bits

The code is written to use the "summation convention", while the calculation uses recursion heavily in order to calculate derivatives. One aspect of using the summation convention is the danger of trying to use the same "dummy subscript" as already appears somewhere in the expression. The code keeps track of what has been used and inserts "fresh" subscripts as required. This solves one problem (wrong answers) at the expense of another.

The "other" problem generally is one of recognizing when two expressions are "the same" so that Maple's built in term collection agency will operate. At a late stage of the calculation the dummy subscripts are replaced with "low" subscripts to normalize such expressions.

Another form of normalization is required because (assuming continuity of partial derivatives of the $f(y, x)$) of equality of cross-partials, so that, for example,

$$\frac{\partial^3}{\partial y_j x y_l} f_k = \frac{\partial^3}{\partial y_l y_j x} f_k.$$

In the Maple code, this means that (write out the subscript equivalents of the "vector notation" Maple code)

```
MPY(f[yxy] f f[y] f )
```

is actually the same as

```
MPY(f[yyx] f[y] f f)
```

so that such expressions have to be reduced to some sort of standard form in order for identical terms to be collected. The code that does this has to cope with the standard form sums that the data is carried about in, as well as (this is the tricky bit) the fact that the terms inside the `mpy` expressions are coupled to those in the `MPY` expressions. The `mpy` expressions contain Runge–Kutta routine coefficient combinations that arose out of chain rule calculations of the Runge–Kutta formula derivatives. They have to be "shifted" in parallel to the function derivative terms in order to end up with expressions that collapse to the simplest forms for the final answer. Getting that code to work is the hardest part of the project. Another trick that is used to speed up the calculation in several places is dummy procedure name substitution. This can hold off a Maple evaluation loop call until an expression is entirely built, when otherwise a huge recursive evaluation call will take place. This is not only a matter of code speed enhancement, as there are even places where this hold off is required in order for the code to execute correctly.

The code has a lot of use of Maple remember tables in the procedures. The same expressions appear repeatedly in a run, and the remember tables speed it up enormously. This is an absolutely classic space-time tradeoff in the code. Take the memory to store the answers rather than redoing the calculation. A remember table is also used to govern the normal form reduction for the "cross partials". Effectively the first expression encountered becomes the normal form representative for its class.

16.6 A Sample Run

This represents a stress test of the code. It really is a lot of calculation, and took 437 seconds (cold) on an AMD DX4-100 running the Linux release of Maple V-3. Upgrading to an AMD Athlon-550 brought the time down to 22.7 seconds, running under Maple V-5.1.

```
> read(‘rkcond.txt‘);
>
> RK_equations(6);
```

$$\left(\sum_{j=1}^{s} b_j\right) - 1 = 0$$

$$\left(\sum_{j=1}^{s} (2\, b_j\, c_j)\right) - 1 = 0$$

$$\left(\sum_{j=1}^{s} \left(3\, b_j\, {c_j}^2\right)\right) - 1 = 0$$

$$\left(\sum_{l_0=1}^{s}\left(\sum_{j=1}^{s}(6\,b_j\,a_{j,l_0}\,c_{l_0})\right)\right)-1=0$$

$$\left(\sum_{l_0=1}^{s}\left(\sum_{j=1}^{s}(24\,b_j\,c_j\,a_{j,l_0}\,c_{l_0})\right)\right)-3=0$$

$$-1+\left(\sum_{l_1=1}^{s}\left(\sum_{l_0=1}^{s}\left(\sum_{j=1}^{s}(24\,b_j\,a_{j,l_0}\,a_{l_0,l_1}\,c_{l_1})\right)\right)\right)=0$$

$$\left(\sum_{l_0=1}^{s}\left(\sum_{j=1}^{s}\left(12\,b_j\,a_{j,l_0}\,{c_{l_0}}^2\right)\right)\right)-1=0$$

$$-1+\left(\sum_{j=1}^{s}\left(4\,b_j\,{c_j}^3\right)\right)=0$$

$$-4+\left(\sum_{l_0=1}^{s}\left(\sum_{j=1}^{s}\left(60\,b_j\,c_j\,a_{j,l_0}\,{c_{l_0}}^2\right)\right)\right)=0$$

$$-3+\left(\sum_{l_1=1}^{s}\left(\sum_{l_0=1}^{s}\left(\sum_{j=1}^{s}(60\,b_j\,a_{j,l_0}\,c_{l_0}\,a_{j,l_1}\,c_{l_1})\right)\right)\right)=0$$

$$\left(\sum_{l_0=1}^{s}\left(\sum_{j=1}^{s}\left(60\,b_j\,{c_j}^2\,a_{j,l_0}\,c_{l_0}\right)\right)\right)-6=0$$

$$-1+\left(\sum_{l_2=1}^{s}\left(\sum_{l_1=1}^{s}\left(\sum_{l_0=1}^{s}\left(\sum_{j=1}^{s}(120\,b_j\,a_{j,l_0}\,a_{l_0,l_1}\,a_{l_1,l_2}\,c_{l_2})\right)\right)\right)\right)=0$$

$$\left(\sum_{l_1=1}^{s}\left(\sum_{l_0=1}^{s}\left(\sum_{j=1}^{s}(120\,b_j\,a_{j,l_0}\,c_{l_0}\,a_{l_0,l_1}\,c_{l_1})\right)\right)\right)-3=0$$

$$\left(\sum_{l_1=1}^{s}\left(\sum_{l_0=1}^{s}\left(\sum_{j=1}^{s}\left(60\,b_j\,a_{j,l_0}\,a_{l_0,l_1}\,{c_{l_1}}^2\right)\right)\right)\right)-1=0$$

$$-1+\left(\sum_{j=1}^{s}\left(5\,b_j\,{c_j}^4\right)\right)=0$$

$$\left(\sum_{l_1=1}^{s}\left(\sum_{l_0=1}^{s}\left(\sum_{j=1}^{s}\left(120\,b_j\,c_j\,a_{j,l_0}\,a_{l_0,l_1}\,c_{l_1}\right)\right)\right)\right)-4=0$$

$$\left(\sum_{l_0=1}^{s}\left(\sum_{j=1}^{s}\left(20\,b_j\,a_{j,l_0}\,{c_{l_0}}^3\right)\right)\right)-1=0$$

$$\left(\sum_{l_2=1}^{s}\left(\sum_{l_1=1}^{s}\left(\sum_{l_0=1}^{s}\left(\sum_{j=1}^{s}\left(360\,b_j\,a_{j,l_0}\,a_{l_0,l_1}\,c_{l_1}\,a_{l_0,l_2}\,c_{l_2}\right)\right)\right)\right)\right)-3=0$$

$$\left(\sum_{l_1=1}^{s}\left(\sum_{l_0=1}^{s}\left(\sum_{j=1}^{s}\left(360\,b_j\,a_{j,l_0}\,c_{l_0}\,a_{l_0,l_1}\,{c_{l_1}}^2\right)\right)\right)\right)-4=0$$

$$\left(\sum_{l_1=1}^{s}\left(\sum_{l_0=1}^{s}\left(\sum_{j=1}^{s}\left(360\,b_j\,c_j\,a_{j,l_0}\,a_{l_0,l_1}\,{c_{l_1}}^2\right)\right)\right)\right)-5=0$$

$$\left(\sum_{l_2=1}^{s}\left(\sum_{l_1=1}^{s}\left(\sum_{l_0=1}^{s}\left(\sum_{j=1}^{s}\left(720\,b_j\,c_j\,a_{j,l_0}\,a_{l_0,l_1}\,a_{l_1,l_2}\,c_{l_2}\right)\right)\right)\right)\right)-5=0$$

$$\left(\sum_{l_1=1}^{s}\left(\sum_{l_0=1}^{s}\left(\sum_{j=1}^{s}\left(720\,b_j\,c_j\,a_{j,l_0}\,c_{l_0}\,a_{l_0,l_1}\,c_{l_1}\right)\right)\right)\right)-15=0$$

$$\left(\sum_{l_0=1}^{s}\left(\sum_{j=1}^{s}\left(120\,b_j\,c_j\,a_{j,l_0}\,{c_{l_0}}^3\right)\right)\right)-5=0$$

$$\left(\sum_{l_2=1}^{s}\left(\sum_{l_1=1}^{s}\left(\sum_{l_0=1}^{s}\left(\sum_{j=1}^{s}\left(720\, b_j\, a_{j,l_0}\, c_{l_0}\, a_{j,l_1}\, a_{l_1,l_2}\, c_{l_2}\right)\right)\right)\right)\right) - 10 = 0$$

$$-3 + \left(\sum_{l_2=1}^{s}\left(\sum_{l_1=1}^{s}\left(\sum_{l_0=1}^{s}\left(\sum_{j=1}^{s}\left(720\, b_j\, a_{j,l_0}\, a_{l_0,l_1}\, c_{l_1}\, a_{l_1,l_2}\, c_{l_2}\right)\right)\right)\right)\right) = 0$$

$$-1 + \left(\sum_{l_1=1}^{s}\left(\sum_{l_0=1}^{s}\left(\sum_{j=1}^{s}\left(120\, b_j\, a_{j,l_0}\, a_{l_0,l_1}\, {c_{l_1}}^3\right)\right)\right)\right) = 0$$

$$\left(\sum_{l_0=1}^{s}\left(\sum_{j=1}^{s}\left(120\, b_j\, {c_j}^3\, a_{j,l_0}\, c_{l_0}\right)\right)\right) - 10 = 0$$

$$\left(\sum_{l_2=1}^{s}\left(\sum_{l_1=1}^{s}\left(\sum_{l_0=1}^{s}\left(\sum_{j=1}^{s}\left(720\, b_j\, a_{j,l_0}\, c_{l_0}\, a_{l_0,l_1}\, a_{l_1,l_2}\, c_{l_2}\right)\right)\right)\right)\right) - 4 = 0$$

$$\left(\sum_{l_3=1}^{s}\left(\sum_{l_2=1}^{s}\left(\sum_{l_1=1}^{s}\left(\sum_{l_0=1}^{s}\left(\sum_{j=1}^{s}\left(720\, b_j\, a_{j,l_0}\, a_{l_0,l_1}\, a_{l_1,l_2}\, a_{l_2,l_3}\, c_{l_3}\right)\right)\right)\right)\right)\right) - 1 = 0$$

$$\left(\sum_{l_2=1}^{s}\left(\sum_{l_1=1}^{s}\left(\sum_{l_0=1}^{s}\left(\sum_{j=1}^{s}\left(360\, b_j\, a_{j,l_0}\, a_{l_0,l_1}\, a_{l_1,l_2}\, {c_{l_2}}^2\right)\right)\right)\right)\right) - 1 = 0$$

$$\left(\sum_{l_0=1}^{s}\left(\sum_{j=1}^{s}\left(30\, b_j\, a_{j,l_0}\, {c_{l_0}}^4\right)\right)\right) - 1 = 0$$

$$-15 + \left(\sum_{l_1=1}^{s}\left(\sum_{l_0=1}^{s}\left(\sum_{j=1}^{s}\left(360\, b_j\, c_j\, a_{j,l_0}\, c_{l_0}\, a_{j,l_1}\, c_{l_1}\right)\right)\right)\right) = 0$$

$$-10 + \left(\sum_{l_1=1}^{s}\left(\sum_{l_0=1}^{s}\left(\sum_{j=1}^{s}\left(360\, b_j\, a_{j,l_0}\, c_{l_0}\, a_{j,l_1}\, {c_{l_1}}^2\right)\right)\right)\right) = 0$$

$$\left(\sum_{l_1=1}^{s}\left(\sum_{l_0=1}^{s}\left(\sum_{j=1}^{s}\left(360\, b_j\, a_{j,l_0}\, c_{l_0}{}^2\, a_{l_0,l_1}\, c_{l_1}\right)\right)\right)\right) - 6 = 0$$

$$-10 + \left(\sum_{l_1=1}^{s}\left(\sum_{l_0=1}^{s}\left(\sum_{j=1}^{s}\left(360\, b_j\, c_j{}^2\, a_{j,l_0}\, a_{l_0,l_1}\, c_{l_1}\right)\right)\right)\right) = 0$$

$$-1 + \left(\sum_{j=1}^{s}\left(6\, b_j\, c_j{}^5\right)\right) = 0$$

$$-10 + \left(\sum_{l_0=1}^{s}\left(\sum_{j=1}^{s}\left(180\, b_j\, c_j{}^2\, a_{j,l_0}\, c_{l_0}{}^2\right)\right)\right) = 0$$

The form of the constraint equations can be elaborated upon, and every set of parameters meeting the order equations leads to a usable numerical method of the Runge–Kutta type. The subject of Runge–Kutta methods has been developed to the point where solutions can be described in some detail. The text [1] can be consulted for results of this sort.

16.7 Order Condition Code

This is a code listing for the whole package. It includes some comments that at least hint at what is going on, while the session record above has the code listing stripped out.

```
# Package for generating Runge--Kutta order conditions

last_index_counter := 0;
readlib('forget');
interface(labelling=false);

# Recursive evaluation of the derivatives of a Runge--Kutta slope
# formula for an arbitrary slope stage number j. It calculates
# it implicitly, to keep the expressions manageable while the
# calculations are made.  The implicit loop is closed below when
# the h=0 is set.
slope_der := proc(n, extra) option remember;
  if n < 0 then RETURN(extra) fi ;
  if n = 0 then
```

```
  RETURN(mpy()*MPY(f[](y + h* a[j, l[0]] * slope(l[0], 0, h), x + h*c[j])));
 else
  RETURN(dh(slope_der(n-1)));
 fi ;
end;

# Evaluate derivatives of the RK slopes at h=0, and back substitute
# for the lower order derivatives.
SLOPE_der := proc(n) option remember;
 local i, k, m, lexpr, last_index, var, free_index, subs_seq, summand,
     funlist, coefflist, subsfunlist, locallexpr, origcoefflist,
     subscripts, subs_term, ind, jj, constlist, origconstlist, temp;
 lexpr := eval(subs(h=0, slope_der(n)));
 if n = 0 then RETURN(lexpr) fi ;
 if op(0,lexpr) = '+' then
  funlist := [op(lexpr)];
 else
  funlist := [lexpr];
 fi ;
 coefflist := map(elt -> op(nops(elt)-1, elt),funlist);
 constlist := map(elt->if nops(elt)=2 then 1 else
                  convert_mpy(op(1..nops(elt)-2, elt)) fi , funlist);
 funlist := map(elt -> op(nops(elt),elt),funlist);
 subsfunlist:= funlist;
 origcoefflist := coefflist;
 origconstlist := constlist;
 summand := 0;
 for m from 1 to nops(funlist)
 do
  locallexpr := funlist[m];
  if op(0, locallexpr) = 'MPY' then
   subs_seq := [op(locallexpr)];
  else
   ERROR('hit a funlist without an MPY');
  fi ;
  subsfunlist := subs_seq;
  for ind from 2 to nops(subs_seq)
  do
   i := op(2, op(2, subs_seq[ind]));
   subs_term := SLOPE_der(i);
   subscripts := [NULL];
   for jj from 1 to get_last_index()
   do
```

```
      if hastype(subsfunlist, l[jj]) then
          subscripts := [op(subscripts), l[jj]];
      fi ;
    od;
    if nops(subscripts) > 0 then
      for jj from 1 to nops(subscripts)
      do
        if hastype(subs_term, subscripts[jj]) then
            subs_term := subs(subscripts[jj]=get_free_index(), subs_term);
        fi ;
      od;
    fi ;
    subs_term :=  subs(j=op(1, op(2,subs_seq[ind])), subs_term);
    subs_term := MULT(op(1, subs_seq[ind])*MPY(), subs_term);
    subsfunlist := subsop(ind=subs_term, subsfunlist);
 od;
# put expanded term on outgoing sum
# first re-format the first (unsubstituted) term
  subsfunlist:=subsop(1=constlist[m]*coefflist[m]*MPY(op(1,subsfunlist)),
              subsfunlist);
  lexpr := MULT(op(subsfunlist));
  summand := summand + lexpr;
  od;
  lexpr := summand;
  if op(0,lexpr) = '+' then
      funlist := [op(lexpr)];
  else
      funlist := [lexpr];
  fi ;
  coefflist := map(elt → op(nops(elt)-1, elt),funlist);
  constlist := map(elt→ if nops(elt)=2 then 1 else
                    convert_mpy(op(1..nops(elt)-2, elt)) fi , funlist);
  funlist := map(elt → op(nops(elt),elt),funlist);

  i := nops(funlist);
  for i from 1 to nops(funlist)
   do
    temp := normalize_MPY(eval([ op(i, coefflist), op(i, funlist)]));
    coefflist := subsop(i=op(1, temp), coefflist);
    funlist := subsop(i=op(2, temp), funlist);
   od;
  summand := 0;
  for i from 1 to nops(funlist)
```

```
 do
  summand := summand + constlist[i]*coefflist[i]*funlist[i];
 od;
 RETURN(summand);
end;

# Constructs the term occurring in the RK expression derivative
RK_der := proc(n) option remember;
 local lexpr;
 lexpr := SLOPE_der(n-1);
 RETURN(expand(MULT(n* mpy(b[j])*MPY(), lexpr)));
end;

# Increments the dummy variable counter and returns a free
# I subscript
get_free_index := proc()
 local i;
 global last_index_counter;
  last_index_counter := last_index_counter+1;
  i := last_index_counter;
# install the new symbol in the remember table
  slope_der(-i, I[i]);
  RETURN(I[i]);
end;

# Left over from earlier code that figured it out;
# It is faster to have a global counter instead.
get_last_index := proc()
 global last_index_counter;
  RETURN(last_index_counter);
end;

# Utility to convert a symbolic multiplication entity
# into a real product of the arguments
convert_mpy := proc(mpylist)
 local factorlist, answer, i;
  factorlist := [args];
  answer := 1;
  if nops(factorlist) = 0 then
   RETURN(answer);
  else
   for i from 1 to nops(factorlist)
   do
```

```
      answer := answer * factorlist[i];
    od;
    RETURN(answer);
  fi ;
end;

# A noncommutative multiplication routine, as far as functions
# are concerned. Associates MULT calls, and handles mpy() for keeping
# constants in sync with MPY arguments.
SUB :=  SUB ;
# VEEERRRRY dangerous hack: SUB has to be anonymous symbol ?????
MULT :=proc()
 local ARGLIST,i, k, summand_list, MPYS, mpys, ns,
     ACC_MPYS, ACC_mpys, ACC_ns, lexpr, summand ; # SUB;
   ARGLIST := [args];
   if member(0,ARGLIST) then
     RETURN(0)
   fi ;
   if nops(ARGLIST) = 1 then RETURN(ARGLIST[1]) fi ;
   if ARGLIST = [] then RETURN(1) fi ;
   lexpr := op(1, ARGLIST);
   if type(lexpr, '+') then
      summand_list := [op(lexpr)];
   else
      summand_list := [lexpr];
   fi ;
# allow for no number in front, but require mpy, MPY (maybe empty arguments)
   ACC_MPYS := map(elt->op(nops(elt), elt), summand_list);
   ACC_mpys := map(elt->op(nops(elt)-1, elt), summand_list);
   ACC_ns := map(elt->if nops(elt)=2 then 1
            else convert_mpy(op(1..nops(elt)-2, elt)) fi , summand_list);
   for i from 2 to nops(ARGLIST)
   do
    lexpr := op(i, ARGLIST);
    if type(lexpr, '+') then
     summand_list := [op(lexpr)];
    else
     summand_list := [lexpr];
    fi ;
    MPYS := map(elt->op(nops(elt), elt), summand_list);
    mpys := map(elt->op(nops(elt)-1, elt), summand_list);
    ns := map(elt->if nops(elt)=2 then 1
           else convert_mpy(op(1..nops(elt)-2, elt)) fi , summand_list);
```

```
   for k from 1 to nops(ACC_mpys)
   do
    ACC_mpys := subs(SUB=op(op(k, ACC_mpys)),
              subsop(k=map(elt→mpy(SUB , op(elt)), mpys), ACC_mpys));
    ACC_MPYS := subs(SUB=op(op(k, ACC_MPYS)),
              subsop(k=map(elt→MPY(SUB , op(elt)), MPYS), ACC_MPYS));
    ACC_ns := subs(SUB=op(k, ACC_ns),
              subsop(k=map(elt→SUB * op(elt), ns), ACC_ns));
   od;
   ACC_mpys := map(op, ACC_mpys);
   ACC_MPYS := map(op, ACC_MPYS);
   ACC_ns := map(op, ACC_ns);
  od;
  summand := 0;
  for k from 1 to nops(ACC_MPYS)
  do
   summand := summand + op(k, ACC_ns)*op(k, ACC_mpys)*op(k, ACC_MPYS);
  od;
  RETURN(summand);
end;

# Differentiate a Runge--Kutta formula with respect to h
# answer can be eval(subs(h=0, answer)) to get the Taylor derivatives
dh := proc(expr)
 local name, parx, pary, nsummands, nprods, i, j, subscripts,
     returnvalue, lexpr, summand,localfunlist, locallexpr, localfunderlist,
     funlist, next_index, stage_index, coefflist, constderlist, funderlist,
     constlist ;
 lexpr := eval(expr);
 if op(0, lexpr) = '+' then
   funlist := [op(lexpr)];
 else
   funlist := [lexpr];
 fi ;
 coefflist := map( elt → op(nops(elt)-1, elt), funlist);
 constlist :=map(elt →if nops(elt)=2 then 1
    else convert_mpy(op(1..nops(elt)-2, elt)) fi , funlist);
 funlist := map( elt → op(nops(elt), elt), funlist);
 constderlist := map( elt → diff(elt, h), constlist);
 for i from 1 to nops(funlist)
 do
   lexpr := op(i, funlist);
# product terms : keep values expanded into a sum of products at all stages
```

```
# must have an MPY in hand, but test for safety
  if evalb( op(0, lexpr) = 'MPY') then
    nprods := nops(lexpr);
    localfunlist := [op(lexpr)];
    for j from 1 to nprods
    do
    locallexpr := localfunlist[j];
# if hit a slope, direct treatment
    if evalb(op(0, op(2,locallexpr)) = 'slope') then
      localfunderlist[j] :=
           mpy()*MPY(subsop(2=subsop(2= op(2, op(2,locallexpr))+1,
                op(2,locallexpr)), locallexpr));
# any other functions, chain rule it
# functions of two arguments have to have RK argument format
    elif type(locallexpr, anyfunc(anything, anything)) then
      name := eval(op(0, locallexpr));
      subscripts := op(1..nops(name), name);
      pary := subscripts,y;
      parx := x,subscripts;
      next_index := get_free_index();
      stage_index := op(1, op(2, op(2, op(2, locallexpr))));
      returnvalue :=
             mpy()*MPY(f[pary](op(1..2, locallexpr)),
            mpy(a[stage_index, next_index])* slope(next_index, 0, h));
      next_index := get_free_index();
      returnvalue := returnvalue +
         h*mpy()*MPY(f[pary](op(1..2, locallexpr)),
            mpy( a[stage_index, next_index])* slope(next_index, 1, h))
       + mpy(diff(op(2, locallexpr),h)) *MPY(f[parx](op(1..2, locallexpr)));
      localfunderlist[j] := expand(eval(returnvalue));
   fi ;
  od;
# plug derivatives in place of functions and multiply: reformat localfunlist
  summand := 0;
  localfunlist := map(elt->mpy()*MPY(elt), localfunlist);
  for j from 1 to nprods
  do
   summand := summand + MULT(op(subsop(j=localfunderlist[j], localfunlist)));
  od;
  funderlist[i] := summand;
 else
  ERROR('unexpected argument: not expected MPY, so fell through');
```

```
 fi ;
od;
summand := 0;
for j from 1 to nops(funlist)
  do
    summand := summand + MULT(constlist[j]*coefflist[j]*MPY(), funderlist[j])
            + MULT(constderlist[j]*coefflist[j]*MPY(),mpy()*funlist[j]);
  od;
RETURN(summand);
end;

# Uses remember table to keep sorting order consistent --
compare_consistently := proc(temp1, temp2)  option remember;
 local temp;
 if evalb( temp1 = op(1, {temp1, temp2})) then
    compare_consistently(temp2, temp1) := temp1;
    RETURN(temp1)
 else
    compare_consistently(temp2, temp1) := temp2;
    RETURN(temp2)
 fi ;
end;

# A function that disappears when invoked
COLLAPSE := proc() RETURN(args) end;

# Ordering routine for sorting atomic partial derivative arguments
# --- mostly it is handling different formats for the atoms in a
# consistent way. The actual comparison is above.
Tree_Order := proc(arg1, arg2)
 local inlist, outlist, temp1, temp2;
      temp1 := arg1;
      temp1 := subs(ATOM=COLLAPSE, temp1);
      temp1 :=  COLLAPSE  (eval(temp1));
      temp2 := arg2;
      temp2 := subs(ATOM=COLLAPSE, temp2);
      temp2 :=  COLLAPSE  (eval(temp2));
      if type(temp1, function) then
        temp1 := map( elt->op(1, elt), [op(temp1)]);
      else
        temp1 := [op(1, temp1)];
      fi ;
      if type(temp2, function) then
```

```
        temp2 := map( elt→op(1, elt), [op(temp2)]);
      else
        temp2 := [op(1, temp2)];
      fi ;
      if temp1 = compare_consistently(temp1, temp2) then
       RETURN(true)
      else
       RETURN(false)
      fi ;
end;

ZERO_MULT := proc();
 if member(0, [args]) then
   RETURN(0);
 else
  RETURN(MPY(args));
 fi ;
end;

# rewrite products in a standard order, so equal partials collapse
normalize_MPY := proc(expr)
 local lexpr, subexpr, flist, coefflist, pairlist, xs, ys, atom_args,
     factor, stack, i, k, element, nprods, summand, last, nsubs;
 lexpr := eval(expr);
# print( lexpr);
# put the sequence in a list to allow access toelements
 flist := [op(op(2,lexpr))];
 coefflist := [op(op(1,lexpr))];
 subexpr := NULL;
 last := nops(coefflist);
 k := nops(flist);
 while last > 0
 do
   xs := number_of_xs(op(k, flist));
   ys := number_of_ys(op(k, flist));
   if k > 1 then
     subexpr := [op(k, flist), [op(last−xs...last, coefflist)]], subexpr;
   else
     subexpr := [op(k, flist), [op(1...last, coefflist)]], subexpr;
   fi ;
   last := last − xs −1;
   k := k −1;
 od;
 while k > 0
```

```
 do
  subexpr := [op(k, flist), [NULL]], subexpr;
  k := k -1;
 od;
 pairlist := [subexpr];
# put in a marker for stack underflow
 stack := ['STACKERROR'];
 i := nops(pairlist);
 while not i = 0
 do
   element := op(i, pairlist);
   ys := number_of_ys(op(1,element));
   if ys > 0 then
     atom_args := op(1..ys, stack);
     atom_args := sort([atom_args], Tree_Order);
     atom_args := op(1..nops(atom_args), atom_args);
     stack := [ATOM(element, atom_args), op((1+ys)..nops(stack), stack)];
   else
     stack := [ATOM(element), op(1..nops(stack), stack)];
   fi ;
   i := i-1;
 od;
 lexpr := COLLAPSE(op(1..(nops(stack)-1), stack));
 lexpr := subs(ATOM=COLLAPSE, lexpr);
 lexpr := 'COLLAPSE'(eval(lexpr));
 lexpr := [op(lexpr)];
 flist := map(elt->op(1, elt), lexpr);
 coefflist := map(elt->op(op(2, elt)), lexpr);
 coefflist := normalize_COEFFS(coefflist);
 lexpr := [mpy(op(coefflist)), MPY(op(flist))];
 RETURN(eval(lexpr));
end;

# Differentiate SUM of n*mpy()*MPY(..,g(y,x),..)
d := proc(expr)
 local name, parx, pary, nsummands, nprods, i, j, subscripts,
     returnvalue, lexpr, summand, localfunlist, locallexpr, localfunderlist,
     funlist, coefflist, funderlist, constlist;
 lexpr := eval(expr);
 if op(0, lexpr) = '+' then
   funlist := [op(lexpr)];
 else
   funlist := [lexpr];
```

```
fi ;
coefflist := map( elt → op(nops(elt)−1, elt), funlist);
constlist :=map(elt →if nops(elt)=2 then 1
        else convert_mpy(op(1..nops(elt)−2, elt)) fi , funlist);
funlist := map( elt → op(nops(elt), elt), funlist);
for i from 1 to nops(funlist)
do
  lexpr := op(i, funlist);
# product terms : keep values expanded into a sum of products at all stages
# must have an MPY in hand, but test for safety
  if evalb( op(0, lexpr) = 'MPY') then
   nprods := nops(lexpr);
   localfunlist := [op(lexpr)];
   for j from 1 to nprods
   do
   locallexpr := localfunlist[j];
# functions of two arguments have to have y,x arguments
   if type(locallexpr, anyfunc(anything, anything)) then
     name := eval(op(0, locallexpr));
     subscripts := op(1..nops(name), name);
     pary := subscripts,y;
     parx := x,subscripts;
     returnvalue :=
      mpy()*MPY(f[pary](y,x), f[](y,x))+ mpy()*MPY(f[parx](y,x));
     localfunderlist[j] := returnvalue;
   fi ;
  od;
# plug derivatives in place of functions and multiply out
  summand := 0;
  localfunlist := map(elt→mpy()*MPY(elt), localfunlist);
  for j from 1 to nprods
  do
   summand := summand + MULT(op(subsop(j=localfunderlist[j], localfunlist)));
  od;
  funderlist[i] := summand;
 else
  ERROR('unexpected argument: not expected MPY, so fell through');
 fi ;
od;
summand := 0;
for j from 1 to nops(funlist)
 do
  summand := summand + MULT(constlist[j]*coefflist[j]*MPY(), funderlist[j]);
```

```
  od;
 RETURN(summand);
end;

# count the number of y partials, to know how many arguments to
# pull off the normalization stack
number_of_ys := proc(func)
 local i, lexpr,count, subscripts, name;
  lexpr := eval(func);
  name := eval(op(0, lexpr));
  subscripts := [op(1..nops(name), name)];
  count := 0;
  for i from 1 to nops(subscripts)
  do
    if op(i, subscripts) = 'y' then count := count+1; fi ;
  od;
  RETURN(count);
end;

# Count x subscripts for coefficient grouping
number_of_xs := proc(func)
 local i, lexpr,count, subscripts, name;
  lexpr := eval(func);
  name := eval(op(0, lexpr));
  subscripts := [op(1..nops(name), name)];
  count := 0;
  for i from 1 to nops(subscripts)
  do
    if op(i, subscripts) = 'x' then count := count+1; fi ;
  od;
  RETURN(count);
end;

# make a Taylor series for the true solution
# records the Taylor sequence in the remember table of the procedure
SOLN_der := proc(n) option remember;
 local summand, i, constlist, coefflist, funlist, lexpr, ones;
  if n = 0 then RETURN(mpy()*MPY(y)) fi ;
  if n = 1 then RETURN(mpy()*MPY(f[](y,x))) fi ;
  lexpr := d(SOLN_der(n-1));
  if op(0,lexpr) = '+' then
      funlist := [op(lexpr)];
  else
```

```
    funlist := [lexpr];
  fi ;
  coefflist := map(elt → op(nops(elt)−1, elt),funlist);
  constlist := map(elt→ if nops(elt)=2 then 1 else
                        convert_mpy(op(1..nops(elt)−2, elt)) fi , funlist);
  funlist := map(elt → op(nops(elt),elt),funlist);
# dummy constants for the normalize_MPY call
  ones := mpy(eval(1$n));
   for i from 1 to nops(funlist)
   do
     funlist := subsop(i=op(2, normalize_MPY([ones, funlist[i]])), funlist);
   od;
# mpy(1) needed to handle convert_mpy substitution
  summand := 0;
  for i from 1 to nops(funlist)
  do
   summand := summand + constlist[i]∗mpy(1)∗funlist[i];
  od;
  RETURN(summand);
end;

# This pulls factors out of the sum of factors times MPY's format,
# and returns them in a list.
list_of_coeffs := proc(expr)
 local lexpr, locallexpr, coeff_seq, j, coeff_seq_tail;
# if it is a sum, we have had to collect terms first
  lexpr := eval(expr);
  if op(0, lexpr) = '+' then
   coeff_seq := NULL;
   for j from 1 to nops(lexpr)
   do
    locallexpr := op(j, lexpr);
    if op(0, locallexpr) = '∗' then
      coeff_seq_tail :=
        subs(op(nops(locallexpr), locallexpr) =1, locallexpr);
      coeff_seq := coeff_seq, coeff_seq_tail;
    else
     print( expected_a_product );
    fi ;
   od;
   RETURN([coeff_seq]);
  else
   RETURN([subs(op(nops(lexpr), lexpr) =1, lexpr)]);
```

```
  fi ;
end;

normalize_COEFFS := proc(expr)
 local coeff_seq, subs_list, j, tracer, flag, m, temp,
      a_list, b_list, c_list, array_list;
# everything should be coefficients
     array_list := eval(expr);
     subs_list := map(op, array_list);
# get the subscripts involved
     tracer := NULL;
     for j from 1 to nops(subs_list)
       do
        if not has([tracer], subs_list[j]) then
          tracer := tracer, subs_list[j];
        fi ;
       od;
       subs_list := [tracer];
       m := 0;
     for j from 1 to nops(subs_list)
       do
         if type(subs_list[j], I[integer]) then
            array_list := subs(subs_list[j]=k[m], array_list);
            m := m+1;
         fi ;
       od;
# put back I on the way out
     RETURN(subs(k=I,array_list));
end;

# Produces a list of order conditions obtained by setting the coefficients
# of the error Taylor series to zero, for derivative orders from 1 through
# n_order, using a RK scheme with a general number of stages
order_conditions := proc(n_order) option remember;
 local eqn_seq, i, coeff_list, j,
      lexpr, fun_seq, n, current, fun_list;
  eqn_seq := NULL;
  fun_seq := NULL;
#n=1 is a special simple form
  if n_order = 1 then
     RETURN([b[j]-1 = 0]);
  fi ;
# recall the earlier calculations
```

```
# and throw away the list brackets
    eqn_seq := op(order_conditions(n_order-1));
    lexpr := eval(RK_der(n_order) - SOLN_der(n_order));
    lexpr := subs(mpy=convert_mpy,lexpr);
    lexpr := subs({f[](y,x)=0, MPY=ZERO_MULT}, lexpr);
    lexpr := eval(lexpr);
    if op(0, lexpr) = '+' then
     n := nops(lexpr);
     for j from 1 to n
     do
       current := op(nops(op(j, lexpr)), op(j, lexpr));
       if not has([fun_seq], current) then
        fun_seq := fun_seq, current;
       fi ;
     od;
     n := nops([fun_seq]);
     fun_list := [fun_seq];
     for j from 1 to n
     do
       lexpr := collect(lexpr, fun_list[j]);
     od;
    fi ;
    coeff_list := list_of_coeffs(lexpr);
    for j from 1 to nops(coeff_list)
    do
      eqn_seq := eqn_seq, op(j, coeff_list) = 0;
    od;
  RETURN(eval([eqn_seq]));
end;

# Routine to pretty print the order conditions
# puts in summation signs for the dummy indices
RK_equations := proc(order)
 local temp, i;
 temp := order_conditions(order);
 for i from 1 to nops(temp)
 do
  write_sum(temp[i], i);
 od;
end;

# The pretty-print work horse.
# --- convert and write out one equation at a time
```

```
write_sum := proc(equation, position)
 local summand_list, temp, lexpr, summand,
      i, m, n, indices, subs_list, sigma_list;
  lexpr := lhs(equation);
# special treatment for the first equation
  if position = 1 then
    temp := Sum(b[j], j=1..S) -1 =0;
  else
    if  op(0,lexpr) = '+' then
      summand_list := [op(lexpr)];
    else
      summand_list := [ lexpr ];  # this really is an error ???
    fi ;
    sigma_list := summand_list;
# make a list of subscripts involved
    indices := map(op, summand_list);
    indices := map(op, indices);
    subs_list := NULL;
    for m from 1 to nops(indices)
    do
      temp := indices[m];
      if not has([subs_list], indices[m]) and
            type(indices[m], {identical(j), l[integer]}) then
         subs_list := subs_list, indices[m];
      fi ;
    od;
    subs_list := [subs_list];
# replace subscripts with summed out subscripts
    for i from 1 to nops(summand_list)
    do
      for m from 1 to nops(subs_list)
       do
        if hastype(summand_list[i], identical(subs_list[m])) then
          sigma_list :=
            subsop(i=Sum(sigma_list[i], subs_list[m]=1..S), sigma_list);
        fi ;
       od;
# construct left hands of equations with sums
      summand := 0;
      for m from 1 to nops(sigma_list)
      do
        summand := summand + sigma_list[m];
      od;
```

```
    temp := subsop(1=summand, equation);
   od;
 fi ;
  print(temp);
end;

# Clear out various remember tables
# --- of use mostly for debugging
reset := proc()
 forget(slope_der);
 forget(SLOPE_der);
 forget(order_conditions);
 forget(RK_der);
 forget(SOLN_der);
 forget(d);
 forget(dh);
end;
```

17
Maple Packages

17.1 Introduction

For trying out a simple Maple procedure (of several lines, say) typing the procedure definition directly into the Maple interpreter is a tolerable approach.

For a more substantial project (like the Runge–Kutta routine analysis codes of Chapter 16) there is a collection of interrelated procedures whose development takes a certain amount of backing, filling, redesign, and recoding during development. The only feasible way to handle this is to type the procedure definitions into what is in effect a source code text file. The text file can be loaded with a command like

```
> read('rkcond.txt');
```

which amounts to the Maple equivalent of compiling the source code file.

What is actually happening is not the equivalent of a conventional compiler, but an operation in which the Maple interpreter treats the contents of the text file *as though* you had interactively typed the sequence of characters in the file directly into the Maple interpreter window. The contents of the file must contain valid expressions, and the interpreter behaves (printing procedure bodies according to the current `verboseproc` value, for example) just as it would if the text had been typed.

For Maple programming projects that result in generally useful code, a better approach than reading a text file is the construction of a Maple package. The use of packages is familiar (`with(linalg);`), and the construction of packages is actually surprisingly simple.

17.2 Constructing Packages

A package consists of a collection of related procedures and variables, and computing lore would suggest that the "natural" way to organize such data (in the Maple context) would be in the form of a table. In fact, a Maple package is basically a table of procedures (which after all, are just Maple variables), saved in a file that is read by the `with()` command.

As one might expect, there are file naming conventions and details that must be attended to in order that the `with` command can locate and load the saved file.

Since the publication of the original (and arguably still best) book [12] on C programming, a "hello, world" example seems to be required in any programming language discussion. We fulfill the requirement with a "hello, world" package.

First test the concept.

```
> print('hello, world');
```

hello, world

Next create a one element table, whose only entry is a procedure that prints the greeting.

```
> hello[world] := proc()
>
>   printf('hello, world');
>
> end;
```

```
hello[world] := proc() printf('hello, world') end
```

Test that the procedure works as expected (it ignores the spurious argument).

```
> hello[world](donkey);
```

```
hello, world
```

Save the table in a Maple binary format file. In order for the `with` command to work, the filename affix must match the name of the package table.

```
> save(hello, '/home/jon/maple/lib/hello.m');
```

To test the operation of the `with` command, clear the Maple workspace.

```
> restart;
```

Maple has a global variable called `libname`, which is a sequence of directories searched by Maple in response to `with` as well as `readlib` commands. In response to

```
> libname;
```

Maple returns the default search directories (mileage varies, values and format are operating system dependent).

$$libname := /usr/local/maple/lib/update,\\ /usr/local/maple/lib$$

In order that Maple locate the `hello.m` file, add the directory where it was saved to the `libname` search sequence.

```
> libname := `/home/jon/maple/lib`, libname;
```

$$libname := /home/jon/maple/lib, /usr/local/maple/lib/update,\\ /usr/local/maple/lib$$

The search path includes the saved location, so try to load the package using the `with` command.

```
> with (hello);
```

Maple's response indicates that it has loaded a single procedure by the name `world`.

$$[\,world\,]$$

Verify that it works.

```
> world();
```

```
hello, world
```

One can also (out of curiosity) check that the original saved table is now present in the workspace.

```
> hello;
```

```
table([
   world = proc() printf(`hello, world`) end
])
```

You can also check that the file `hello.m` is saved in the directory used in the `save` command.

For packages of enduring value, it is more usual to save packages in a library archive, rather than saving individual "dot m" files for each package. The considerations for doing this are discussed below.

17.3 A Control Design Package

Maple has facilities for algebraic manipulation as well as numerical linear algebraic calculation, numerical equation solution, and plotting. It is very easy to adapt these to the calculations required for control system design and analysis. Such a project serves as an example of a (useful) package.

Some of the procedures involve algorithms for making calculations associated with somewhat advanced design methods. Rather than try to explain all of the control theory involved, the commentary gives a more or less general explanation, and comments on useful Maple hacks employed. It goes without saying that many of these emerged by interactive experimentation after a first code pass went down in flames.

17.3.1 Root Locus

Math background **Page:** vi

The root locus problem arises from a single scalar (constant coefficient) differential equation, which happens to contain a single "variable" parameter. The form of the equation is

$$\frac{d^n x}{dt^n} + p_{n-1}\frac{d^{n-1}x}{dt^{n-1}} + \ldots + p_0 x + k\left(q_m \frac{d^m x}{dt^m} + q_{m-1}\frac{d^{m-1}x}{dt^{m-1}} + \ldots + q_0 x\right) = 0,$$

where k is the adjustable parameter, and the other $\{p_j\}$ and $\{q_k\}$ are assumed known and given numerical values for the problem duration. The character of the solutions is (see Section 5.5) completely determined by the solutions of the auxiliary polynomial

$$s^n + p_{n-1}s^{n-1} + \ldots + p_0 + k(q_m s^m + \ldots + q_0) = 0.$$

Assuming $n > m$ (this is enforced by the problem context) for each value of k we obtain a set of roots

$$\{\lambda_1(k), \lambda_2(k), \ldots, \lambda_n(k)\}$$

where some of the roots will generally turn out to be complex valued. The root locus problem is to produce a plot of the locations of the roots of the polynomial, when the parameter varies over some range.

Conceptually, one can accomplish this in Maple by generating a *set* of parametric equations, as in

```
> plot( { [sin(x), cos(x), x=0..6], [t, cos(2*t), t=-2..2]});
```

This form is not usable since there is no way to get an explicit formula for the roots as a function of the parameter k. However, Maple also supports plotting *functions* without explicit specification of an argument, as the following illustrates.

```
> plot( { [sin, cos, 0..2*Pi]});
```

If somehow a function representing the k dependence of the roots could be manufactured, then that syntax would be usable. This is a common enough problem that Maple has a built in facility for handling it.

The solution is to use `unapply` to manufacture a function expression with anonymous argument, which can be plotted with a bare range argument rather than a "variable = range" expression of the sort used above. The code also uses a phony procedure name `listofroots` to "hold off" evaluation until the `unapply` acts.

```
# Source for Maple control systems routines
# (c) Jon H. Davis
# NO WARRANTY: free for non-commercial use as long as this
# copyright notice is retained.

# classical control design routines

# returns list of zeroes of p+k*q
control[poles] := proc(p,q,k)
 local poly,answer;
  poly := p+k*q;
  poly := subs(z=s, poly);
  answer := fsolve(poly,s,complex);
  RETURN([answer]);
end;

# returns list of zeroes of p+k*q
control[Poles] := proc(G, k)
 local poly, answer, p, q;
  p := numer(G);
  q := denom(G);
  poly := p+k*q;
  poly := subs(z=s, poly);
  answer := fsolve(poly,s,complex);
  RETURN([answer]);
end;

# returns a plot function argument with a dummy name for "poles"
control[plotset] := proc(p, q, n, parmrange)
 local i,j, temp, item;
  temp := NULL;
  for j from 1 to n do
   item := subs(i=j, [Re(listofroots(p,q,k)[i]),
                Im(listofroots(p,q,k)[i]), parmrange]);
   item := subsop(1=unapply(op(1, item), k), 2=unapply(op(2, item),k), item);
   temp := temp,item;
```

```
  od;
 RETURN({temp});
end;

# displays a root locus for a given range of k
control[rootlocus] := proc(p, q, krange)
 local toplot, ran;
  if type(krange, relation) then
   ran := rhs(krange);
  else
   ran := krange;
  fi ;
 toplot := plotset(p, q, degree(p), ran);
 toplot := subs(listofroots=poles, toplot);
 plot(toplot);
end;

# a variant that takes a transfer function and a possible k=range argument
control[Rootlocus] := proc(trans_fun, k_range)
 local p, q;
  p := denom(trans_fun);
  q := numer(trans_fun);
  rootlocus( p, q, k_range);
end;
```

17.3.2 Nyquist Locus

The frequency response associated with the scalar differential equation

$$\frac{d^n x}{dt^n} + p_{n-1}\frac{d^{n-1}x}{dt^{n-1}} + \ldots p_0 x = q_m \frac{d^m u}{dt^m} + q_{m-1}\frac{d^{m-1}u}{dt^{m-1}} + \ldots q_0 u$$

can be defined as the (complex) amplitude of the particular solution resulting from the complex sinusoidal input

$$u(t) = e^{i\omega t}.$$

The usual undetermined coefficients calculation leads to

$$x(t) = \frac{q(i\omega)}{p(i\omega)}\, e^{i\omega t} = G(i\omega)\, e^{i\omega t},$$

so that the frequency response is

$$\frac{q(i\omega)}{p(i\omega)} = G(i\omega).$$

The Nyquist plot can be viewed as the locus of the frequency response, in principle as the frequency ranges from $]-\infty < \omega < \infty[$. The only complications occur because there are situations (resonance) in which an oscillatory solution does not exist because the homogeneous equation has a solution of the form $e^{i\omega_0 t}$. Since this means that $i\omega_0$ is a root of the characteristic polynomial, the formula for the frequency response will end up dividing by zero at the value $\omega = \omega_0$.

This possibility is accommodated by making the routine second argument be a list of frequency ranges. In the case of ranges with gaps, the routine fills in the locus with the image of a semi-circle about the singular point. This is in accord with the use of Nyquist plots in feedback stability problems.

Finally, there are discrete time versions of these problems. The code determines which case is wanted on the basis of the use of the Laplace transform variable s or the Z transform variable z in the transform expression.

```
# Nyquist plotting routine
control[Nyquist] := proc(trans_fun, freq_range_list)
 local real_part, imag_part, i, left_part, right_part,
     temp, plotlist, omega, center, radius, circle_list, singularity_list,
     alpha, theta_0, theta_1, delta_theta, radius_sq, E;
  E:=evalf(exp(1));
  singularity_list := NULL;
  if nops(freq_range_list) > 1 then
   for i from 2 to nops(freq_range_list) do
    left_part := op(i-1, freq_range_list);
    left_part := rhs(left_part);
    right_part := op(i, freq_range_list);
    right_part := lhs(right_part);
    singularity_list := singularity_list, left_part .. right_part;
   od;
   singularity_list := [singularity_list];
 fi;
 if hastype(trans_fun, identical(s)) then
   temp := subs(s=I*omega, trans_fun);
 else
   if hastype(trans_fun, identical(z)) then
    temp := subs(z= E^(I*omega), trans_fun);
   else
     ERROR(`need a transfer function for Nyquist plot`);
   fi;
 fi;
 real_part := Re(temp);
 imag_part :=Im(temp);
 real_part := evalc(real_part);
 imag_part := evalc(imag_part);
```

```
real_part := unapply(real_part, omega);
imag_part := unapply(imag_part, omega);
plotlist := freq_range_list;
for i from 1 to nops(freq_range_list)
do
  plotlist := subsop(i=[ real_part(omega), imag_part(omega),
                  omega=op(i, freq_range_list)], plotlist);
od;
circle_list := NULL;
if singularity_list <> NULL then
  circle_list := singularity_list;

  for i from 1 to nops(singularity_list)
  do
    temp := op(i, singularity_list);
    center := .5 *(lhs(temp) + rhs(temp));
    if hastype(trans_fun, identical(s)) then
      radius := .5 *(rhs(temp) - lhs(temp));
      theta_0 := -Pi/2;
      theta_1 := Pi/2;
    else
      delta_theta := .5 *(rhs(temp) - lhs(temp));
      alpha := Pi/2 + delta_theta/2;
      theta_0 := center - alpha;
      theta_1 := center + alpha;
      radius_sq := (cos(center) -cos(theta_0))^2
              +(sin(center) - sin(theta_0))^2;
      radius := sqrt(radius_sq);
    fi ;
    if hastype(trans_fun, identical(s)) then
      temp := subs(s=I*center+ radius*E^(I*omega), trans_fun);
    else
     if hastype(trans_fun, identical(z)) then
      temp := subs(z= cos(center) + I*sin(center) + radius *E^(I*omega),
              trans_fun);
     else
      ERROR(`need a transfer function for a Nyquist plot`);
     fi ;
    fi ;
    real_part := Re(temp);
    imag_part :=Im(temp);
    real_part := evalc(real_part);
    imag_part := evalc(imag_part);
    real_part := unapply(real_part, omega);
```

```
    imag_part := unapply(imag_part, omega);

    circle_list :=
    subsop(i=[ real_part(omega), imag_part(omega),
            omega= theta_0..theta_1], circle_list);
  od;
 circle_list := op(circle_list);
 fi ;
 temp := op(plotlist);
 plotlist := {temp, circle_list};
 plot(plotlist);
end;
```

17.3.3 Bode Plots

Bode plots convey exactly the same information as is contained in a Nyquist plot. The essence of the difference is that the Nyquist plot uses the rectangular form

$$G(i\omega) = x + i\, y,$$

for the complex frequency response, while the Bode plot uses the polar form

$$G(i\omega) = re^{i\theta},$$

and plots the amplitude (r) and phase (θ) information on separate graphs using a logarithmic frequency scale.

The Maple novelty in the code is the sequence of two `print` statements in the `Bode` procedure. The arguments are Maple $PLOT$ structures. The unexpected use of the `print` calls is what allows a single procedure to cause two plot windows to appear. It was sufficiently hard to find that out that it seems worth mentioning the required hack explicitly.

```
control[Magnitude] := proc(trans_fun, freq_range)
 local logminfreq, logmaxfreq, temp, log_omega, omega, E;
 E:=evalf(exp(1));
 if hastype(trans_fun, identical(s)) then
  logminfreq := (1/ln(10))*log(lhs(freq_range));
  logmaxfreq := (1/ln(10))* log(rhs(freq_range));
  temp := (20/ln(10.0)) * log( abs(subs(s=I*10^log_omega, trans_fun)));
  plot(temp, log_omega=logminfreq..logmaxfreq,
     title='log magnitude vs.  log frequency');
 else
  logminfreq := lhs(freq_range);
  logmaxfreq := rhs(freq_range);
  temp := (20/ln(10.0)) * log( abs(subs(z=E^(I*omega), trans_fun)));
```

```
  plot(temp, omega=logminfreq..logmaxfreq,
      title=`log magnitude vs. frequency`);
 fi ;
end;

control[Phase] := proc(trans_fun, freq_range)
 local logminfreq, logmaxfreq, temp, log_omega, omega, E;
 E:=evalf(exp(1));
 if hastype(trans_fun, identical(s)) then
   logminfreq := (1/ln(10)) *log(lhs(freq_range));
   logmaxfreq := (1/ln(10))* log(rhs(freq_range));
   temp := (180/Pi) *Im( log( (subs(s=I*10^log_omega, trans_fun))));

   plot(temp, log_omega=logminfreq..logmaxfreq,
      title=`phase vs.  log frequency`);
 else
   if hastype(trans_fun, identical(z)) then
     logminfreq := lhs(freq_range);
     logmaxfreq := rhs(freq_range);
     temp := (180/Pi) * Im( log(subs(z=E^(I*omega), trans_fun)));

     plot(temp, omega=logminfreq..logmaxfreq,
        title=`phase vs. frequency`);
   else
     ERROR(`first argument has to be a transfer function`);
   fi ;
 fi ;
end;

control[Bode] := proc(trans_fun, freq_range)
   print(Phase( trans_fun, freq_range));
   print(Magnitude(trans_fun, freq_range));
end;
```

17.3.4 Classical vs. State Space

The previous routines are those associated with classical control systems. The models are single-input single-output, linear, and time-invariant.

The routines that follow are associated with state-space models (multiple-input multiple-output) and time invariant least squares problems.

17.3.5 LQR Utility Code

Least squares optimal control problems involve solution of a quadratic matrix equation known as a Riccati equation. A linear algebraic method for solution known as "Potter's Method" is implemented below. The algorithm requires sorting the eigenvalues of a certain matrix, either according to real part, or according to complex magnitude. The routines below implement the sort comparison functions.

```
# eigenvalue ordering routines needed
# for are and dare routines
control[leftrightorder] := proc(a, b)
 local aval, bval;
  aval := convert(evalc(a[1]), float);
  bval := convert(evalc(b[1]), float);
  if Re(aval) < Re(bval) then
          true
  else
          false
  fi ;
end;

control[axialorder] := proc(a, b)
 local aval, bval;
  aval := convert(evalc(a[1]), float);
  bval := convert(evalc(b[1]), float);
  if abs(aval) < abs(bval) then
          true
  else
         false
  fi ;
end;

control[insertcol] := proc(a, n, c)
 local i;
 for  i to rowdim(a)
 do
  a[i,n] := c[i]
 od;
RETURN(eval(a))
end;
```

17.3.6 Least Squares Optimal Control

The algebraic Riccati equation

$$0 = \mathbf{A}^* \mathbf{K} + \mathbf{K}\mathbf{A} - \mathbf{K}\mathbf{B}\mathbf{R}^{-1}\mathbf{B}^*\mathbf{K} + \mathbf{Q}$$

can be explicitly solved in terms of the eigenvectors of an associated Hamiltonian matrix. A direct translation of Potter's method is below in the routine `are`.

The discrete time version of the algebraic Riccati equation is

$$\mathbf{K} = \mathbf{A}^*\mathbf{K}\mathbf{A} - \mathbf{A}^*\mathbf{K}\mathbf{B}[\mathbf{R} + \mathbf{B}^*\mathbf{K}\mathbf{B}]^{-1}\mathbf{B}^*\mathbf{K}\mathbf{A} + \mathbf{Q},$$

with Potter's method of solution implemented in `dare`.

Optimal least squares control problems have solutions for the coefficient matrices of the optimal system that are expressible in terms of the Riccati equation solution. The formulas are embedded in subroutines that generate the optimal coefficients.

```
# Potter's eigenvector method
# -- references in Laub 1979, Willems' survey.
control[are] := proc(a, b, q, r)
 local hamiltonian, eig_seq, eig_sort, K, n, m, i, j,
     U, V, W, hvectors, h12, R;
  if nargs < 3 then
      ERROR('3 or 4 arguments (a,b,q,r) required')
  fi ;
  n := rowdim(a);
  if n <> coldim(a) then
      ERROR('a must be square')
  fi ;
  if n <> rowdim(b) then
      ERROR('b must match rowdim of a')
  fi ;
  if n <> rowdim(q) or n <> coldim(q) then
      ERROR('q must be the size of a')
  fi ;
  if nargs = 4 then
      h12 := evalm(b &* inverse(r) &* transpose(b))
  else
      h12 := evalm(b &* transpose(b))
  fi ;
  hamiltonian :=
      evalm(stackmartix(concat(a, -h12), concat(-q, -transpose(a))));
  eig_seq := eigenvects(hamiltonian, radical);
  eig_sort := sort([eig_seq], leftrightorder);
```

```
U := matrix(n,n);
V := matrix(n,n);
W := matrix(2*n, n);
m := 1;
j := 1;
while Re(convert(eig_sort[m][1], float)) < 0
 do
  hvectors := convert(eig_sort[m][3], list);
    for i from 1 to eig_sort[m][2]
    do
     W := insertcol(W, j, hvectors[i]);
     j := j+1;
    od;
    m := m+1
 od;
 U := submatrix(W, n+1..2*n, 1..n);
 V := submatrix(W, 1..n, 1..n);
 K := evalm(U &* inverse(V));
 K := convert(eval(K), float);
 RETURN(eval(K))
end;

# basically Potter's method
# as described in Laub's 1979 IEEE-TAC paper
control[dare] := proc(a, b, q, r)
 local hamiltonian, eig_seq, eig_sort, K, n, m, i, j,  U, V, W,
     hvectors, brb, atc;
 if nargs < 3 then
     ERROR('3 or 4 arguments (a,b,q,r) required')
 fi ;
 n := rowdim(a);
 if n <> coldim(a) then
     ERROR('a must be square')
 fi ;
 if n <> rowdim(b) then
     ERROR('b must match rowdim of a')
 fi ;
 if n <> rowdim(q) or n <> coldim(q) then
     ERROR('q must be the size of a')
 fi ;
 if nargs = 4 then
     brb := evalm(b &* inverse(r) &* transpose(b))
 else
```

```
        brb := evalm(b &* transpose(b))
    fi ;
    atc := evalm((inverse(transpose(a))) &* q);
    hamiltonian :=
        evalm(stackmatrix(concat(a + brb &* atc,
                    -brb &* inverse(transpose(a))),
                    concat(-atc, inverse(transpose(a)))));
    eig_seq := eigenvects(hamiltonian, radical);
    eig_sort := sort([eig_seq], axialorder);
    U := matrix(n,n);
    V := matrix(n,n);
    W := matrix(2*n, n);
    m := 1;
    j := 1;
    while abs(convert(eig_sort[m][1], float)) < 1
    do
        hvectors := convert(eig_sort[m][3], list);
        for i from 1 to eig_sort[m][2]
        do          W := insertcol(W, j, hvectors[i]);
            j := j+1;
        od;
        m := m+1
    od;
    U := submatrix(W, n+1..2*n, 1..n);
    V := submatrix(W, 1..n, 1..n);
    K := evalm(U &* inverse(V));
    K := convert(eval(K), float);
    RETURN(eval(K))
end;

# Linear quadratic regulator
# returns [ closed loop matrix, optimal gains]
control[lqr] := proc(a, b, q, r)
 local K, closed_loop, gains, n, R;
  if nargs < 3 then
      ERROR(`3 or 4 arguments (a,b,q,r) required`)
  fi ;
  n := rowdim(a);
  if n <> coldim(a) then
      ERROR(`a must be square`)
  fi ;
  if n <> rowdim(b) then
      ERROR(`b must match rowdim of a`)
```

```
fi ;
if n <> rowdim(q) or n <> coldim(q) then
   ERROR(`q must be the size of a`)
fi ;
if nargs = 4 then
   if rowdim(r) <> coldim(r) or coldim(b) <> rowdim(r) then
    ERROR(`r must be square and conformable with b`);
   fi ;
   R := r;
else
   R := array(identity, 1..n, 1..n);
fi ;
K := are(a, b, q, R);
gains := evalm( inverse(R) &* transpose(b) &* K);
closed_loop := evalm( a - b &* gains);
RETURN(eval([evalm(closed_loop), evalm(gains)]));
end;

# returns discrete time regulator [closed_loop, optimal_gains]
control[dlqr] := proc(a, b, q, r)
 local K, closed_loop, gains, n, R;
  if nargs < 3 then
     ERROR(`3 or 4 arguments (a,b,q,r) required`)
  fi ;
  n := rowdim(a);
  if n <> coldim(a) then
     ERROR(`a must be square`)
  fi ;
  if n <> rowdim(b) then
     ERROR(`b must match rowdim of a`)
  fi ;
  if n <> rowdim(q) or n <> coldim(q) then
     ERROR(`q must be the size of a`)
  fi ;
  if nargs = 4 then
     if rowdim(r) <> coldim(r) or coldim(b) <> rowdim(r) then
      ERROR(`r must be square and conformable with b`);
     fi ;
     R := r;
  else
     R := array(identity, 1..n, 1..n);
  fi ;
```

```
  K := dare(a, b, q, R);
  gains := evalm( inverse(R + transpose(b) &* K &* b ) &* transpose(b) &* K);
  closed_loop := evalm( a - b &* gains);
  RETURN(eval([evalm(closed_loop), evalm(gains)]));
end;

# coefficient matrices for a continuous Kalman Bucy
# filter [closed_loop, gains]
control[lqgf] := proc(a, c, ro, ri)
 local  lqr_list;
  lqr_list := lqr(transpose(a), transpose(c), ri, ro);
  RETURN([eval(transpose(op(1, lqr_list))),
           eval(transpose(op(2, lqr_list)))]);
end;

# coefficient matrices for a discrete Kalman Bucy
# filter [closed_loop, gains]
control[dlqgf] := proc(a, c, ro, ri)
 local  lqr_list;
  lqr_list := dlqr(transpose(a), transpose(c), ri, ro);
  RETURN([eval(transpose(op(1, lqr_list))),
        eval(transpose(op(2, lqr_list)))]);
end;

# coefficient matrices for a Separation Theorem controller
# [filter_coefficients, filter_gains, optimal_control_gains]
control[lqgr] := proc(a, b, c, q, r, ro, ri)
 local lqr_list, lqg_list;
  lqr_list := lqr(a, b, q, r);
  lqg_list := lqgf(a, c, ro, ri);
  RETURN(eval([op(1,lqg_list), op(2, lqg_list), op(2, lqr_list)]));
end;

# coefficient matrices for a Separation Theorem controller
# [filter_coefficients, filter_gains, optimal_control_gains]
control[dlqgr] := proc(a, b, c, q, r, ro, ri)
 local lqr_list, lqg_list;
  lqr_list := dlqr(a, b, q, r);
  lqg_list := dlqgf(a, c, ro, ri);
  RETURN(eval([op(1,lqg_list), op(2, lqg_list), op(2, lqr_list)]));
end;
```

17.3.7 Control Equation Generation

After an optimal control design is produced, it must be simulated and implemented. The routines that follow generate the equations governing the optimal systems in the form of C program code, directly pasteable into the author's control simulation/execution program.

The unfamiliar Maple artifact in this code may be the `lprint` routine. That procedure produces output in typewriter font, rather than pretty-printing it in a fancy graphics mode font.

```
# Separation theorem controller equations output in a format
# past-able into dlxrun/dlxsim
control[lqgr_eqns] := proc(a, b, c, q, r, ro, ri, states, controls, outputs)
 local lqgr_list, states_hat, d_dt_states_hat, eqn_rhs, neqns, i;
  if nargs <> 10 then
   ERROR(`you must specify a, b, c, q, r, ro, ri, states, controls, outputs`);
  fi ;
  lqgr_list := lqgr( a, b, c, q, r, ro, ri);
  states_hat := map(elt->elt.`_hat`, states);
  d_dt_states_hat := map(elt ->`d_dt_`.elt, states_hat);
  eqn_rhs := evalm( lqgr_list[1] &* states_hat
                  + lqgr_list[2] &* outputs)
                  - lqgr_list[3] &* controls;
  neqns := nops(op(3, op(1, states)));
  for i from 1 to neqns do
   lprint(d_dt_states_hat[i], `=`, eqn_rhs[i], `;`);
  od;
  lprint();
  neqns := nops(op(3, op(1, controls)));
  eqn_rhs := evalm( lqgr_list[3] &* states_hat);
  for i from 1 to neqns do
   lprint(controls[i], `=`, eqn_rhs[i], `;`);
  od;
end;

# equations for a discrete time  Separation Theorem design
control[dlqgr_eqns] := proc(a, b, c, q, r, ro, ri, states, controls, outputs)
 local lqgr_list, i, neqns, states_hat, next_states_hat, eqn_rhs;
  if nargs <> 10 then
   ERROR(`you must specify a, b, c, q, r, ro, ri, states, controls, outputs`);
  fi ;
  lqgr_list := dlqgr( a, b, c, q, r, ro, ri);
  states_hat := map(elt->elt.`_hat`, states);
```

```
  next_states_hat := map(elt →'next_'.elt, states_hat);
  eqn_rhs := evalm( lqgr_list[1] &* states_hat
                 + lqgr_list[2] &* outputs)
                 − lqgr_list[3] &* controls;
  neqns := nops(op(3, op(1, states)));
  for i from 1 to neqns do
   lprint( next_states_hat[i], '=', eqn_rhs[i], ';');
  od;
  neqns := nops(op(3, op(1, controls)));
  eqn_rhs := evalm( lqgr_list[3] &* states_hat);
  for i from 1 to neqns do
   lprint(controls[i], '=', eqn_rhs[i], ';');
  od;
end;

# equations of motion for a continuous time linear regulator
control[lqr_eqns] := proc(a, b, q, r, states, controls)
 local lqr_list, i, neqns,  d_dt_states, eqn_rhs;
  if nargs <> 6 then
   ERROR('you must specify a, b,  q, r, states, controls');
  fi ;
  lqr_list := lqr( a, b, q, r);
  d_dt_states := map(elt →'d_dt_'.elt, states);
  eqn_rhs := evalm( a &* states
                 − b&*controls);
  neqns := nops(op(3, op(1, states)));
  for i from 1 to neqns do
   lprint( d_dt_states[i], '=', eqn_rhs[i], ';');
  od;
  neqns := nops(op(3, op(1, controls)));
  eqn_rhs := evalm( lqr_list[2] &* states);
  for i from 1 to neqns do
   lprint(controls[i], '=', eqn_rhs[i], ';');
  od;
end;

# equations of motion for a continuous time Kalman Filter
control[lqg_eqns] := proc(a, c, ro, ri, states, outputs)
 local lqr_list, i, neqns,  d_dt_states, eqn_rhs;
  if nargs <> 6 then
   ERROR('you must specify a, c,  ro, ri, states, outputs');
  fi ;
  lqg_list := lqgf( a, c, ro, ri);
```

```
  d_dt_states := map(elt →'d_dt_'.elt, states);
  eqn_rhs := evalm( lqg_list[1] &* states + lqg_list[2] &* outputs);
  neqns := nops(op(3, op(1, states)));
  for i from 1 to neqns do
   lprint( d_dt_states[i], '=', eqn_rhs[i], ';');
  od;
end;

# equations of motion for a discrete time linear regulator
control[dlqr_eqns] := proc(a, b, q, r, states, controls)
 local lqgr_list, i, neqns,  next_states, eqn_rhs;
  if nargs <> 6 then
   ERROR('you must specify a, b, q, r, states, controls');
  fi ;
  lqgr_list := dlqgr( a, b, q, r);
  next_states := map(elt →'next_'.elt, states);
  eqn_rhs := evalm( a &* states - b &* controls);
  neqns := nops(op(3,op(1, states)));
  for i from 1 to neqns do
   lprint( next_states[i], '=', eqn_rhs[i], ';');
  od;
  neqns := nops(op(3, op(1, controls)));
  eqn_rhs := evalm( lqgr_list[2] &* states);
  for i from 1 to neqns do
   lprint(controls[i], '=', eqn_rhs[i], ';');
  od;
end;

# equations of motion for a discrete time Kalman Filter
control[dlqg_eqns] := proc(a, c, ro, ri, states, outputs)
 local lqg_list, i, neqns,  next_states, eqn_rhs;
  if nargs <> 6 then
   ERROR('you must specify a, c,  ro, ri, states, outputs');
  fi ;
  lqg_list := dlqgf( a, c, ro, ri);
  next_states := map(elt →'next_'.elt, states);
  eqn_rhs := evalm( lqg_list[1] &* states
                + lqg_list[2] &* outputs);
  neqns := nops(op(3, op(1, states)));
  for i from 1 to neqns do
   lprint( next_states[i], '=', eqn_rhs[i], ';');
  od;
end;
```

17.3.8 LTI Realization Calculations

Because of the duality between the controllability and observability conditions, a routine needs only be written to handle one of the cases. Checking observability for the system

$$\frac{d}{dt}\mathbf{x} = \mathbf{A}\,\mathbf{x}, \;\; \mathbf{y} = \mathbf{C}\,\mathbf{x}$$

can the phrased as looking for the common null space of the set of matrices

$$\{\mathbf{C}, \mathbf{CA}, \mathbf{CA}^2, \dots, \mathbf{CA}^{n-1}\}.$$

An equivalent formulation is to look for the common null space of the set of symmetric matrices

$$\{\mathbf{C}'\mathbf{C}, \mathbf{A}'\mathbf{C}'\mathbf{CA}, \dots, \mathbf{A}'^{n-1}\mathbf{C}'\mathbf{CA}^{n-1}\},$$

and this is a numerically much better behaved problem, since the eigenvectors of symmetric matrices are orthogonal to each other.

```
# parameter s is for returning subspace basis
control[cocheck] := proc(a, c, s)
 local i, j, k, n, u, vectors, eigenlist, alpha, m, subspacefound, maxev;
  if nargs < 2 then
      ERROR(`2 or three arguments expected`)
  fi ;
  n := rowdim(a);
  if n <> coldim(a) then
      ERROR(`first argument must be square`)
  fi ;
  if n <> coldim(c) then
      ERROR(`second argument must be conformable with first`)
  fi ;
  u := matrix(n,n);
  for i to n
  do
   for j to n
   do
      if i = j then
        u[i,j] := 1.0
      else
        u[i,j] := 0.0
      fi ;
    od;
  od;
```

```
i := 0;
while i < n
do
 subspacefound := false;
 m := evalm(transpose(u) &* transpose(c) &* c &* u);
 eigenlist := [eigenvects(m, radical)];
 maxev := 0.0;
 for j to nops(eigenlist)
 do
   if convert(eigenlist[j][1], float) > maxev then
     maxev := convert(eigenlist[j][1], float)
   fi ;
 od;
 if maxev < .5*10^(-Digits) then
  u := evalm(a &* u);
  i := i+1;
  next;
 fi ;
 vectors := NULL;
 for j to nops(eigenlist)
 do
   if convert(eigenlist[j][1], float) < (maxev * 10^(-5)) then
     vectors := vectors union convert(eigenlist[j][3], list);
     alpha := evalm(transpose(matrix(vectors)));
     subspacefound := true;
   fi ;
 od;
 if subspacefound = true then
  u := evalm(a &* u &* alpha);
  i := i+1;
 else
   RETURN(true)
 fi ;
od;
vectors := NULL;
for j to nops(eigenlist)
do
  vectors := vectors union convert(eigenlist[j][3], list);
  alpha := evalm(transpose(matrix(vectors)));
od;
u := evalm(u &* alpha);
if nargs > 2 and type(s, name) then
 s := u ;
fi ;
```

```
 RETURN(false)
end;

control[is_controllable] := proc(a,b,c)
 RETURN(cocheck(evalm(transpose(a)),evalm(transpose(b))))
end;

control[is_observable] := proc(a,b,c)
 RETURN(cocheck(a,c))
end;

control[is_minimal] := proc(a,b,c)
 RETURN(is_controllable(a,b,c) and is_observable(a,b,c))
end;


control[init]:=proc()
with(linalg);
end;
```

17.3.9 Initialization

The names of procedures in a package have a restriction: `init` is a reserved name with special meaning. The `init` procedure is executed automatically immediately after the package is loaded by the `with` statement.

The `init` procedure should initialize any variables and data structures needed by the package. In the example below, the code makes sure the `linalg` package is loaded. The control package makes use of many `linalg` routines for doing the state space calculations.

```
control[init]:=proc()
 with(linalg);
end;
```

17.4 Package Creation and Installation

Creating and installing a package is a matter of saving the package table in a binary library archive file (which must be located on the Maple `libpath` for the `with` command to work).

Creation of the package therefore can be done with a simple read of the control package source file after which Maple will quit. The `control.m` will be placed in the first user library location listed in the `libpath` variable sequence. See the discussion of libraries in the following section for information on managing library paths.

```
savelib(control, 'control.m');
quit
```

17.5 How About Help?

Built-in and package-loaded Maple commands differ from one-off user written ones in that help commands (with pasteable examples) are typically provided.

The help information in recent releases of Maple is stored and managed in parallel with procedure libraries and packages.

17.5.1 Maple Libraries

Maple libraries are stored as files[1] under the operating system. There are "system wide" Maple libraries that are set up when Maple is installed. It is not a good idea to change these, and so a user can (and should) set up a personal maple library facility.

This can be done through use of a Maple initialization file.[2] A sample initialization file is the following.

```
# set a personal Maple library
libname := '/home/jon/maple/lib', libname:

plotsetup(x11):
with(plots):
interface(verboseproc=2):

# support saving plots in an eps file, and switching

EPS:= proc(s);
 plotsetup(ps, plotoutput=s, plotoptions='color=cmyk,noborder,portrait');
end:
X11:=proc();
 plotsetup(x11);
end:
```

Here we are concerned with the "libname" assignment which is the first effect of the above file of Maple commands. The "libname" is a global Maple workspace

[1]Perhaps they are referred to as "folders" under systems that are designed to hide the existence of files.

[2]The initialization file is ".mapleinit" in the user home directory on a Unix system, and "maple.init" on other platforms.

variable. When Maple initially starts, it is set to the system wide default value, and the effect of the line is to prepend a personal library directory to the list of library directories.

In order for this to work, in the first place the indicated directory must exist. More than that, the directory must be set up to function as a Maple library directory. The facility that ensures this is the Maple archive manager, called "march". This program must be invoked with the personal archive (and a size parameter) as arguments in order to initialize the archive database and index files. The details of how this is done are platform dependent, but the Maple help system provides guidance in the usual way.

```
> ?march
```

Make sure that a user write-able library directory has been set up with "march", and that the Maple initialization file sets "libname" so that this directory is the default library, that is, the first component of the Maple `libname` variable. If this is not done, attempts to create package library and help file entries will either fail due to permission problems, or modify the default library files on platforms which do not support file permissions.

17.5.2 Creating and Installing Help Files

Providing help information for a package involves two separate processes. In the first place, the files that create the actual help text must be created. After that, the help files must be installed on the system in such a way that they are accessible when Maple is run by the user.

Recent incarnations of Maple present the user with help files that contain "hypertext" links. Earlier versions had help files that were simple text files, augmented with fairly strict formatting conventions about argument lists and example code.

The current help files have an appearance close to the Maple worksheets themselves. The creation tool for the help pages is simply Maple itself. Create a work sheet which largely consists of text areas, but Maple execution groups for the example section. The "hypertext" help links are entered by means of the same input menu that selects other entry modes. This process is illustrated in Figure 17.1.

It is most natural to save the resulting worksheet with the name of the procedure as the filename prefix. Also, when a related group of help files are being created, a "blank" help file containing an empty text group, an empty execution group, and hyperlinks to all of the package procedure names is a useful first step.[3] The help text and examples can then be subsequently pasted into the blank spaces, and edited for appearance.

The result then is a set of Maple worksheet files, one for each package procedure (and perhaps an additional overview.) The problem then is how to link the collected information into the Maple help system.

---

[3]The hyperlink entry menu is tedious, and if a quicker alternative is available, the author was unable to find it.

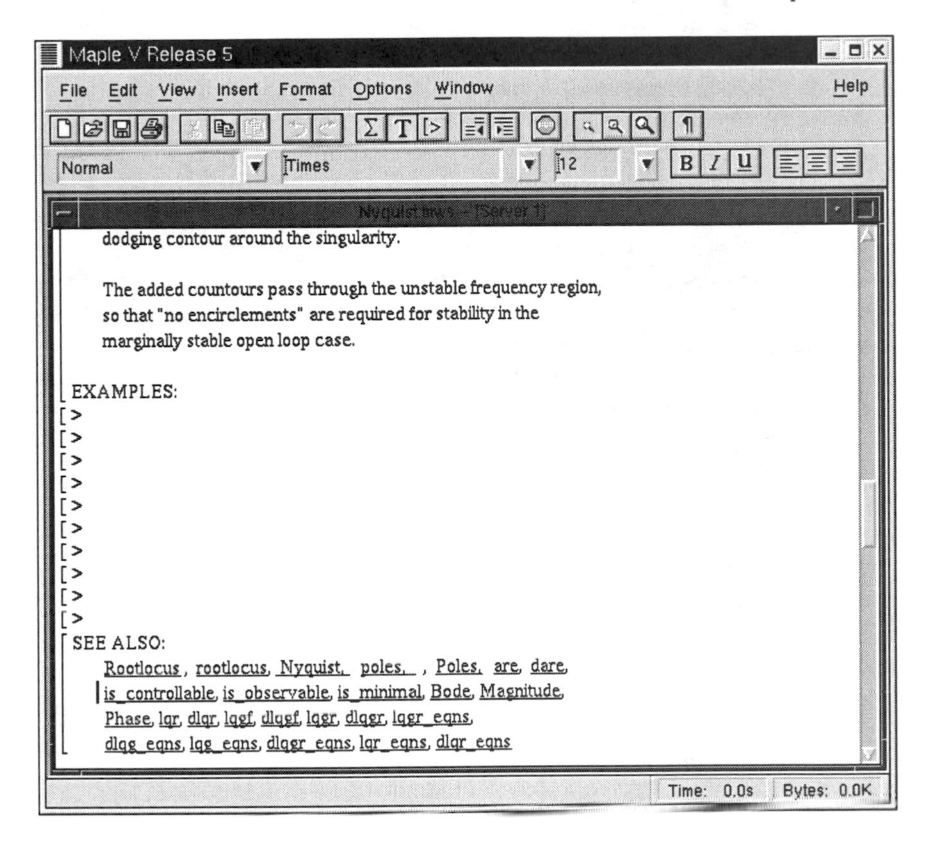

FIGURE 17.1. Maple help creation process

The mechanism for doing this is the library routine `makehelp`.[4]

One invocation of `makehelp` is required for each installed help topic. Given the tedium, it is natural to simply create a Maple command file containing the required statements, and do the installation by means of a Maple `read` command. Such a command file follows:

```
# This assumes that libname[1] is user writable.
# It is written assuming the  help file name is the procedure name.
#

readlib(makehelp):

makehelp( Bode, `Bode.mws`, libname[1]):
makehelp( Magnitude, `Magnitude.mws`, libname[1]):
```

[4]The makehelp help says it uses text files to create the help. It will, but one ends up without sample code, or hyperlinked help. That leaves the conclusion that either only the Maple developers have a help creation tool, or that makehelp might be willing to eat worksheet files.

```
makehelp( Nyquist, 'Nyquist.mws', libname[1]):
makehelp( Phase, 'Phase.mws', libname[1]):
makehelp( Poles, 'Poles.mws', libname[1]):
makehelp( Rootlocus, 'Rootlocus.mws', libname[1]):
makehelp( are, 'are.mws', libname[1]):
makehelp( axialorder, 'axialorder.mws', libname[1]):
makehelp( cocheck, 'cocheck.mws', libname[1]):
makehelp( control, 'control.mws', libname[1]):
makehelp( dare, 'dare.mws', libname[1]):
makehelp( dlqg_eqns, 'dlqg_eqns.mws', libname[1]):
makehelp( dlqgf, 'dlqgf.mws', libname[1]):
makehelp( dlqgr, 'dlqgr.mws', libname[1]):
makehelp( dlqgr_eqns, 'dlqgr_eqns.mws', libname[1]):
makehelp( dlqr, 'dlqr.mws', libname[1]):
makehelp( dlqr_eqns, 'dlqr_eqns.mws', libname[1]):
makehelp( insertcol, 'insertcol.mws', libname[1]):
makehelp( is_controllable, 'is_controllable.mws', libname[1]):
makehelp( is_minimal, 'is_minimal.mws', libname[1]):
makehelp( is_observable, 'is_observable.mws', libname[1]):
makehelp( leftrightorder, 'leftrightorder.mws', libname[1]):
makehelp( lqg_eqns, 'lqg_eqns.mws', libname[1]):
makehelp( lqgf, 'lqgf.mws', libname[1]):
makehelp( lqgr, 'lqgr.mws', libname[1]):
makehelp( lqgr_eqns, 'lqgr_eqns.mws', libname[1]):
makehelp( lqr, 'lqr.mws', libname[1]):
makehelp( lqr_eqns, 'lqr_eqns.mws', libname[1]):
makehelp( plotset, 'plotset.mws', libname[1]):
makehelp( poles, 'poles.mws', libname[1]):
makehelp( rootlocus, 'rootlocus.mws', libname[1]):
```

This code might be included in the source code file for the package itself.[5] The counter suggestion is that so many separate files are involved that it is an awkward prospect. On the other hand, perhaps a series of worksheet windows could be spawned by the package definition code, filled with help text, saved, ... This might actually be possible, but the details are left as an investigation for the interested reader.

17.6 Control Package Demo

Start Maple on your workstation by typing

[5] Operating platform quiz: in how few keystrokes and mouse clicks can the command file be generated?

```
xmaple &
```

or clicking on an appropriate icon. Then load the control package by typing

```
> with(control);
```

```
Warning: new definition for   norm
Warning: new definition for   trace
```

[*Magnitude*, *Nyquist*, *Phase*, *Poles*, *Rootlocus*, *are*, *axialorder*, *cocheck*, *dare*, *dlqg_eqns*, *dlqgf*, *dlqgr*, *dlqgr_eqns*, *dlqr*, *dlqr_eqns*, *help_hook*, *init*, *insertcol*, *is_controllable*, *is_minimal*, *is_observable*, *leftrightorder*, *lqg_eqns*, *lqgf*, *lqgr*, *lqgr_eqns*, *lqr*, *lqr_eqns*, *plotset*, *poles*, *rootlocus*]

This loading message shows that there are both classical (frequency domain) procedures, as well as routines that do state space domain design calculations present in the control package.

Of the routines present in the package, it is probably the frequency response plotting tools that are accessible without special knowledge about the details and concerns of control systems.

Various data processing operations require the calculation of the derivative of a measured signal. This can be accomplished in various ways, including a straightforward finite difference, and by means of filters to smooth the measured signal. The behavior of these operations can be visualized in the frequency domain by use of a Bode plot of the transfer function involved.

The calculation of the finite difference expression in the time domain is the usual difference quotient

$$\frac{f(t) - f(t-T)}{T},$$

Because of the Laplace transform delay law, this operation has a particularly simple transfer function representation.

```
> finitediff := (1 - exp(-s*T))/T;
```

$$finitediff := \frac{1 - exp(-s\,T)}{T}$$

Before any Maple plots can be made, values must be substituted for any parameters in the expressions. The ease with which this can be done under the Maple user interface is one of the motivations for constructing the control package,

```
> Finitediff:= subs(T=.001, finitediff);
```

$$Finitediff := 1000. - 1000.\,exp(-.001\,s)$$

The question is how this compares with a "genuine" differentiation operation. The transfer function of differentiation follows from the differentiation rule for Laplace transforms. That is,

$$\mathcal{L}\{\frac{df}{dt}\} = s\,\mathcal{L}\{f\} - f(0^+).$$

Since we are regarding the function f itself as the input, and the derivative $\frac{df}{dt}$ as the response, the associated transfer function is

$$G(s) = s,$$

and the frequency response is taken as

$$G(i\omega) = i\omega.$$

A Bode plot of the "true" derivative frequency response (try it) gives a straight line as the magnitude response, and a phase (argument of the complex frequency response) that is constant at +90 degrees.

The magnitude Bode plot in Figure 17.2 for the finite difference transfer function is produced by the Maple command

```
> Magnitude(Finitediff, .01 .. 100000);
```

The magnitude response of the finite difference approximation resembles that of the derivative operation over low frequencies. At high frequencies, the finite difference approximation has a strong response that is quite far from (one might say bears no resemblance to) the "true" response.

The corresponding calculations for a "first order filter" approximate derivative follow. The starting place is the differential equation

$$\frac{d}{dt}x = -a\,x + a\,u,$$

which has an associated transfer function

$$G(s) = \frac{a}{s+a}.$$

By combining that transfer function with a multiple of the original input we can obtain a transfer function of the form

$$-a\,\frac{a}{s+a} + a = \frac{a\,s}{s+a}.$$

If a is "large" (candor: relative to the frequency range in use) then the effect of the above is essentially "multiplication by s", or differentiation.

```
> filter := a/(s+a);
```

$$\mathit{filter} := \frac{a}{s+a}$$

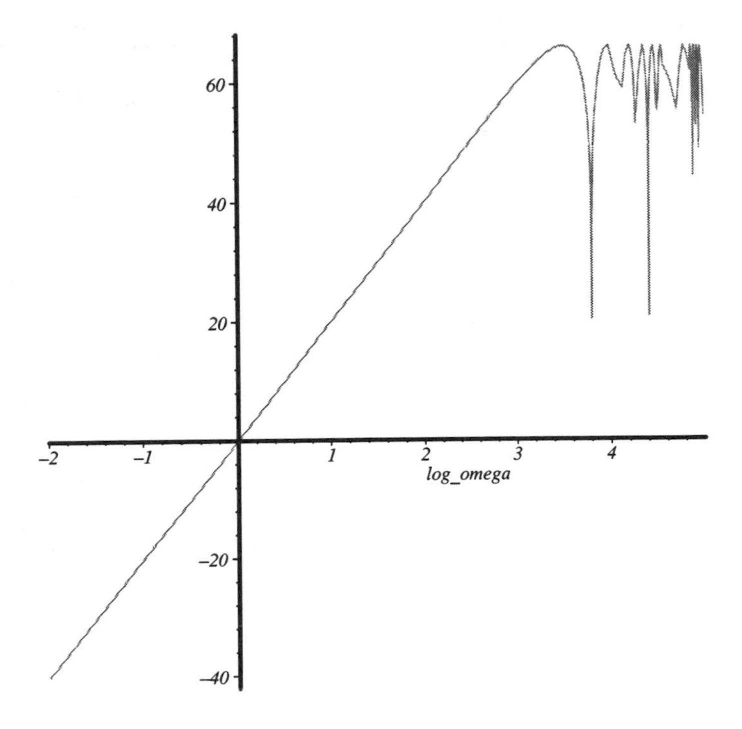

FIGURE 17.2. Finite difference Bode plot

```
> filterder := -a*filter + a;
```

$$filterder := -\frac{a^2}{s+a} + a$$

```
> Filterder := subs(a=10, filterder);
```

$$Filterder := -100\,\frac{1}{s+10} + 10$$

```
> Magnitude(Filterder, .01 .. 10000);
```

The resulting magnitude plot shown in Figure 17.3 approximates the derivative for low values of the frequency variable, and has a uniform response at high frequencies.

For some applications, it is useful to have a processing element that approximates a derivative, but suppresses signals with high frequency content (often high frequency measurement noise is an issue). The constant high frequency gain of

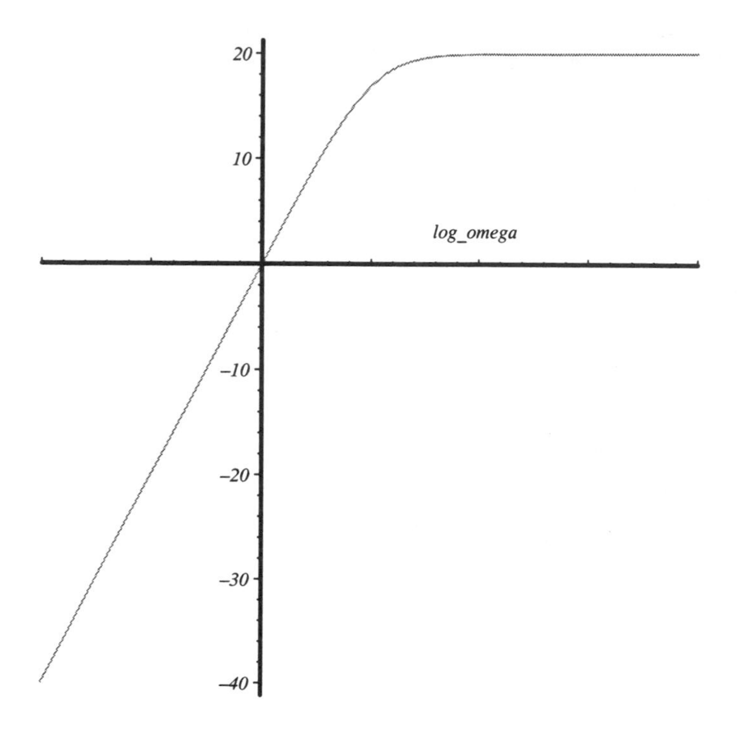

FIGURE 17.3. First order derivative Bode plot

the first order filter approximation can be turned into a roll-off by using a second order filter.

The calculations corresponding to use of a second order filter for the purpose are

```
> filter2 := filter * filter;
```

$$filter2 := \frac{a^2}{(s+a)^2}$$

```
> Filter2 := subs(a=10, filter2);
```

$$Filter2 := 100\,\frac{1}{(s+10)^2}$$

```
> Magnitude(s*Filter2, .01 .. 10000);
```

The corresponding plot is in Figure 17.4.

The Nyquist plot is the (polar complex) locus of the frequency response of the system loop gain transfer function. For proportional PD and PID control of a

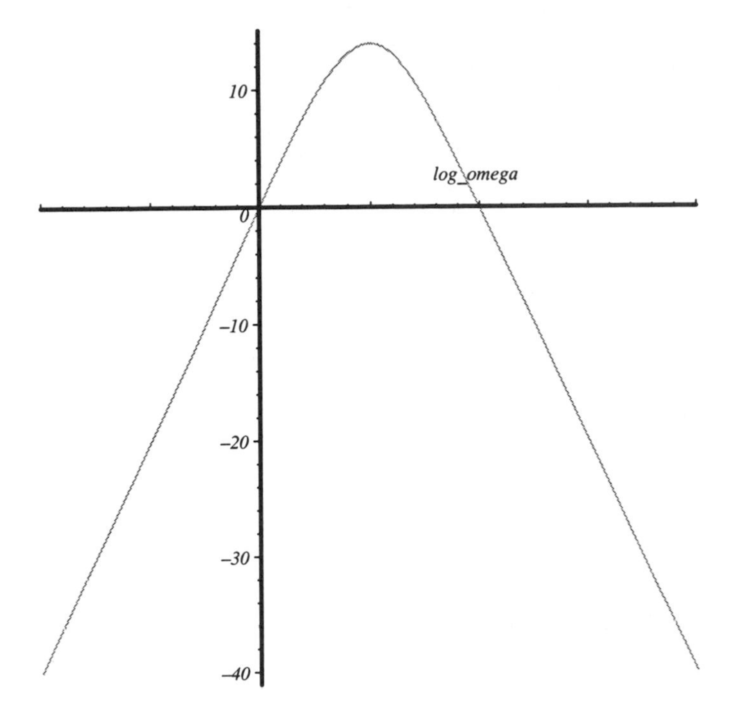

FIGURE 17.4. Second order derivative Bode plot

servo motor, the loop transfer function consists of the product of the motor transfer function

$$G(s) = \frac{b}{s(s+c)}$$

and the transfer function of the "compensator" appearing between the error and the motor terminal input. For the case of proportional control, the compensator is a simple (frequency independent) gain

$$H_P(s) = K,$$

while in the case of PD and PID controllers the ideal forms are

$$H_{PD}(s) = K\,(1 + ds)$$

and

$$H_{PID}(s) = K\,(1 + ds + \frac{e}{s}).$$

The forms of the compensator transform are written normalized relative to the proportional gain K. This is the useful form for the use of the Nyquist criterion,

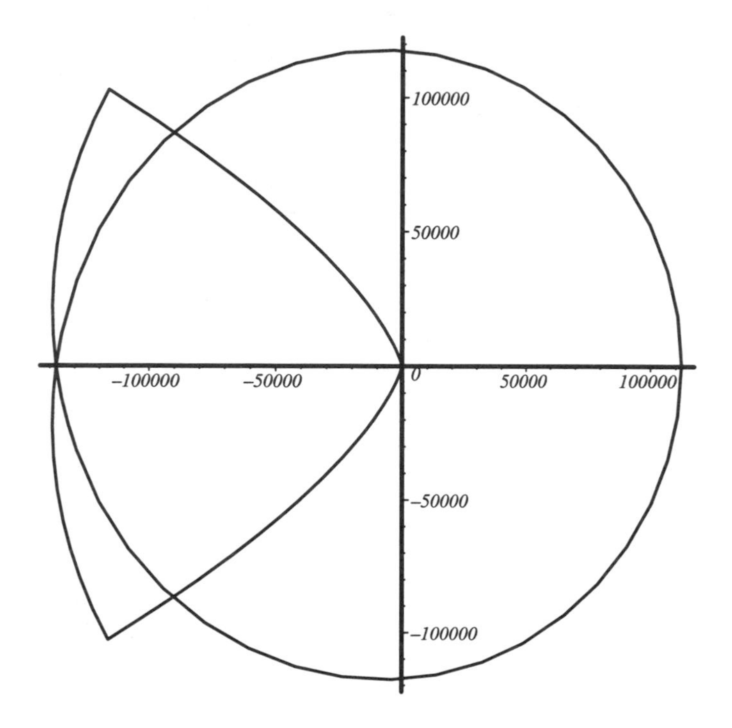

FIGURE 17.5. Nyquist plot for a PID controller

where encirclements of the $-\frac{1}{K}$ are the issue of interest. With those compensators the corresponding loop transforms are (for proportional control)

$$HG(s) = \frac{b}{s(s+c)},$$

(for PD control)

$$HG(s) = \frac{b(1+ds)}{s(s+c)},$$

and (for PID compensation)

$$HG(s) = \frac{b(1+ds+\frac{e}{s})}{s(s+c)}.$$

The K is omitted from the loop gain as the overall gain scale factor whose stability is to be determined.

One way to implement PD control is to use a first order filter to "smooth" the error signal and calculate the error derivative. This replaces the proportional gain in the above expressions with

$$K\frac{a}{s+a}.$$

The loop gain for the filtered proportional controller is

$$HG(s) = \frac{ab}{s(s+c)(s+a)}.$$

Determine for what range of K the filtered proportional controller would be stable using the `Nyquist` routine of the `control` package in Maple. Use the "best guess" parameter set $b = 43.64$, $c = 0.11$, and $a = 15$. The control package will also draw root locus plots. Between the use of root locus and the `poles` routines one can locate the value of K on the margin of stability.

The transfer function associated with the derivative of a first order filter can be readily determined. The governing equation of such a filter is in generic form

$$\frac{d}{dt}x_f = -a\,x_f + a\,x$$

so that the (zero initial condition) Laplace transform version is

$$X_f(s) = \frac{a}{s+a}X(s).$$

Since the filter has zero initial conditions

$$\mathcal{L}\left\{\frac{d}{dt}x_f\right\} = s\,X_f(s) = \frac{a\,s}{s+a}X(s).$$

If the forcing input x also has zero initial conditions this is just

$$\mathcal{L}\left\{\frac{d}{dt}x_f\right\} == \frac{a}{s+a}\mathcal{L}\left\{\frac{d}{dt}x_f\right\},$$

so that the derivative term has the same transfer function as the proportional one. We investigate the stability properties of a filtered PID control scheme by using the control package to produce a Nyquist plot of the loop transfer function. The loop transfer function for a first order filtered PID control scheme is

```
> g1:=(d*s^2+s+e)*b*a/(s*s*(s+c)*(s+a));
```

$$g1 := \frac{(d\,s^2 + s + e)\,b\,a}{s^2\,(s+c)\,(s+a)}$$

```
> g1a:=subs(a=15,d=2,b=46.34,e=5,c=.11,g1);
```

$$g1a := 695.10\,\frac{2\,s^2 + s + 5}{s^2\,(s+.11)\,(s+15)}$$

Because the transfer function has poles at $s = 0$, the range facility of the Nyquist plot routine is used.

```
> Nyquist(g1a,[-10 .. -.1,.1 .. 10]);
```

The corresponding Nyquist plot appears in Figure 17.5.

Part IV

Appendices

Appendix A
Review Problems

In real life, problem solution methods are not generally found three pages back in the text. While the following are not really scrambled, they are at least not labelled with a method. In the absence of other instructions, find the general solution of the following

1. $y\,\frac{dy}{dx} = x$.
2. $2y\,\frac{dy}{dx} = e^x$.
3. $\frac{dy}{dx}\,\sin(y) = x^2$.
4. $x\,\frac{dy}{dx} = (1 - y^2)^{\frac{1}{2}}$.
5. $\sin(y) + (x\,\cos(y) + 3y^2)\frac{dy}{dx} = 0$.
6. $\sin(x)\,\cos(y)\,dx + \cos(x)\sin(y)\,dy = 0$.
7. $3x^2\,ln(|x|)\,dx + \frac{x^3}{y}\,dy = 0$.
8. What a, b make $[y^2\cos(xy^2) + a]\,dx + [bx\,\cos(xy^2) + 3y]\,dy = 0$ exact?
9. Same for $ln(y)\,dx + [a\,x + \frac{x}{y}]\,dy = 0$.
10. When can $\mu(x)$ make $\mu(x)M(x, y)\,dx + \mu(x)N(x, y)\,dy = 0$ exact?
11. Use the above on $\frac{1}{x}\,dx - (1 + xy^2)\,dy = 0$.
12. $\frac{dy}{dt} - y(t) = e^{2t}$.

13. $\frac{dy}{dt} - \frac{a}{t}y(t) = t^m$.

14. $\frac{dy}{dt} + \cos(t)y(t) = \cos(t)$.

15. $\frac{dy}{dt} + e^t y(t) = e^t$.

16. $y\,dx - (3x + y^4)\,dy = 0$.

17. $(x + x^3)\frac{dy}{dx} + 4x^2y = 2$.

18. What is the form of the general solution of

$$\frac{dy}{dx} + 5y = 0?$$

19. What does the general solution of

$$\frac{d^2y}{dx^2} + 25y = 0$$

look like?

20. What is the general solution of

$$\frac{d^2y}{dt^2} + 5\frac{dy}{dt} = 0?$$

21. What are the time constants in the general solution of

$$\frac{d^2y}{dt^2} + 10\,\frac{dy}{dt} + 25\,y = 0?$$

22. What is the form of the general solution of

$$\left(\frac{d}{dt} + 5\right)^3 y = 0?$$

You've never seen that before. What must this mean?

23. What is the (actually, a) particular solution of

$$\left(\frac{d}{dt} + 5\right)^3 y = 0?$$

24. Compute

$$\mathcal{L}\{t\,e^{-2t}\}.$$

25. Evaluate

$$\mathcal{L}\{t^2\, e^{-2t}\}.$$

26. Determine

$$\mathcal{L}\{t\, e^{-2t}\, \sin(3)\}.$$

27. Grind out

$$\mathcal{L}\{t^2\, e^{-2t}\, \cos(5\, t)\}.$$

There are better and worse ways to do this one. Which are which?

28. Inadvertently learn about hyperbolic sines by discovering

$$\mathcal{L}\{\sinh(t)\},$$

where $\sinh(x) = \frac{e^x - e^{-x}}{2}$

29. In order not to slight hyperbolic cosines get a formula for

$$\mathcal{L}\{t\, \cosh(t)\},$$

where $\cosh(x) = \frac{e^x + e^{-x}}{2}$.

30. Discover why Maple is schizophrenic about solution formulas by checking

$$\cosh(i\, x)$$

and

$$\sinh(i\, x),$$

where $i = \sqrt{-1}$.

31. Compute

$$\mathcal{L}\{f(t)\}$$

where

$$f(t) = \begin{cases} 1 & \text{if } 0 < t < T \\ 0 & \text{if } t > T. \end{cases}$$

32. Compute

$$\mathcal{L}\{g(t)\}$$

where

$$g(t) = \begin{cases} 0 & \text{if } 0 < t < T \\ \sin(2\pi\, \frac{t}{T}) & \text{if } T < t < 2T \\ 0 & \text{if } t > 2T. \end{cases}$$

33. Find by undetermined coefficients the particular solution appropriate to

$$\frac{dy}{dt} + 5\,y = e^{-4t}.$$

34. Use undetermined coefficients on

$$\frac{dy}{dt} + 5\,y = e^{-5t}.$$

35. Try your undetermined coefficient patience with the particular solution appropriate to

$$\left(\frac{d}{dt} + 5\right)^2 y = e^{-4t}.$$

36. Find by undetermined coefficients the particular solution for

$$\left(\frac{d}{dt} + 5\right)^2 y = e^{-5t}.$$

37.

$$\frac{dy}{dt} + 5\,y = e^{i\,\omega\,t}.$$

38. Same problem for

$$\frac{d^2y}{dt^2} + 25\,y = e^{-5t}.$$

39. And also for

$$\frac{d^2y}{dt^2} + 25\,y = e^{i\,5\,t}.$$

40.

$$\frac{d^2y}{dt^2} + 25\,y = \sin(5\,t)$$

(You need no additional computations.)

41.

$$\frac{d^2y}{dt^2} + 25\,y = \cos(5t)$$

(Ditto.)

42. What is the general solution of

$$\frac{d^2y}{dt^2} + 25\,y = \sin(5\,t)?$$

43. Find the particular solution of

$$\frac{dy}{dt} + 2\,y = f(t)$$

where $f(t)$ is the function defined in question 31, by means of a convolution calculation.

44. Find the particular solution of

$$\frac{dy}{dt} + 2\,y = g(t)$$

where $g(t)$ is the function defined in question 32, using convolution evaluation to do the computation.

45. Find the particular solution of

$$\frac{dy}{dt} + 2\,y = f(t)$$

where $f(t)$ is the function defined in question 31 by means of Laplace transforms.

46. Find the particular solution of

$$\frac{dy}{dt} + 2\,y = g(t)$$

where $g(t)$ is the function defined in question 32, using Laplace transforms to do the computation.

47. Find the particular solution of

$$\frac{d^2y}{dt^2} + 2\frac{dy}{dt} + y = f(t)$$

where $f(t)$ is the function defined in question 31 by means of a convolution calculation.

48. Find the particular solution of

$$\frac{d^2y}{dt^2} + 2\frac{dy}{dt} + y = g(t)$$

where $g(t)$ is the function defined in question 32, using convolution evaluation to do the computation.

49. Find the particular solution of

$$\frac{d^2y}{dt^2} + 2\frac{dy}{dt} + y = f(t)$$

where $f(t)$ is the function defined in question 31 by means of Laplace transforms.

50. Find the particular solution of

$$\frac{d^2y}{dt^2} + 2\frac{dy}{dt} + y = g(t)$$

where $g(t)$ is the function defined in question 32, using Laplace transforms to do the computation.

51. Calculate the convolution

$$(f * g)(t)$$

where $f(t)$ and $g(t)$ are defined in questions 31 and 32 above, by explicitly evaluating the convolution integral.

52. Calculate the convolution

$$(f * g)(t)$$

where $f(t)$ and $g(t)$ are defined in questions 31 and 32 above, by use of the Laplace transform convolution theorem. You better get the same answer you ended up with in question 51.

53. Expand in partial fractions

$$\frac{1}{(s+\pi)(s+e)(s+1)}.$$

54. Expand in partial fractions

$$\frac{1}{(s^2+\pi^2)(s+e)(s+1)}.$$

55. Expand in partial fractions

$$\frac{1}{s^2(s+1)}.$$

56. Expand in partial fractions

$$\frac{s+2}{s^2(s+1)^2}.$$

57. Expand in partial fractions

$$\frac{1}{s^2\,(s^2+2\,s+4)}.$$

58. The companion matrix associated with the polynomial

$$p(s) = s^n + p_{n-1}\,s^{n-1} + p_{n-2}\,s^{n-2} + \ldots p_0$$

is

$$\begin{bmatrix} 0 & 1 & 0 & \cdots \\ 0 & 0 & 1 & \cdots \\ \vdots & \vdots & \ddots & \vdots \\ -p_0 & -p_1 & \cdots & -p_{n-1} \end{bmatrix}.$$

Expand by Cramer's rule to show that the characteristic polynomial of the companion matrix is just the polynomial it was built from.

59. A system of two coupled linear homogeneous equations of the form

$$\frac{d}{dt}\mathbf{x} = \mathbf{A}\,\mathbf{x}$$

has the following properties. The characteristic polynomial of **A** is

$$(s+1)(s+14),$$

while the set of eigenvectors of **A** is

$$\left\{\begin{bmatrix} 2 \\ 1 \end{bmatrix}, \begin{bmatrix} 1 \\ 2 \end{bmatrix}\right\}.$$

What is the form of the general solution of the system of equations ?

60. Calculate

$$e^{\begin{bmatrix} 1 & 2 \\ -2 & 1 \end{bmatrix} t}.$$

61. Is the system of equations

$$\frac{d}{dt}\mathbf{x} = \begin{bmatrix} 1 & 2 \\ -2 & 1 \end{bmatrix}\mathbf{x}$$

a stable one?

62. What is required of the system of equations

$$\frac{d}{dt}\mathbf{x} = \begin{bmatrix} a & b \\ c & d \end{bmatrix}\mathbf{x}$$

in order that it be an asymptotically stable one? Marginally stable?

63. Is the set of equations in 59 above stable, asymptotically stable, or unstable?

64. The characteristic polynomial for the coefficient matrix of a four dimensional set of coupled linear, homogeneous differential equations is

$$\left(s^2 + 2s + 4\right)^2.$$

Is the set of equations stable, asymptotically stable, or unstable?

65. A set of six coupled linear homogeneous differential equations has the characteristic polynomial

$$\left(s^2 + 4\right)^2 (s + 2)^2$$

associated with its coefficient matrix. A determined hunt for eigenvectors turns up a mere three of them, in spite of repeated checking of the computations. Show that the system of equations has to be unstable.

66. A set of six coupled linear homogeneous differential equations has the characteristic polynomial

$$\left(s^2 + 4\right)^2 (s + 2)^2$$

associated with its coefficient matrix. A quick consultation with Maple in pursuit of eigenvectors turns up five of them. Prove that if the coefficient matrix has all real coefficients in it, then the system of equations is stable.

Hint: If a matrix is real valued, then complex conjugate eigenvalues have complex conjugate eigenvectors. (Prove this?) This problem separates sheep from other fauna.

67. Find a Lyapunov function for

$$\frac{dx}{dt} = -x(t).$$

68. When will

$$V\left(y(t), \frac{dy}{dt}\right) = a\,(y(t))^2 + b\left(\frac{dy}{dt}\right)^2$$

be a Lyapunov function for the equation

$$\frac{d^2y}{dt^2} + 2\frac{dy}{dt} + 4y = 0?$$

You can do this "on a physical basis".

69. A model for the "central reaction" of a certain chemical kinetics problem is the following system of differential equations:

$$\begin{aligned} \frac{dx}{dt} &= a - (b+1)\,x + x^2\,y \\ \frac{dy}{dt} &= b\,x - x^2\,y. \end{aligned}$$

   (a) Find the equilibrium point of this system of equations.

   (b) For what values of the parameters a, b is the equilibrium point stable, and for what values is it unstable?

   When the equilibrium is unstable, this system has a periodic, self-sustaining oscillation. Solutions starting near the equilibrium move "outward" to the oscillatory cycle. Numerical approaches are in order to see the oscillations.

70. A RC filter subjected to a periodic square wave input is modeled by the equation

$$RC\,\frac{dv}{dt} + v = sw(t),$$

   where the $2T$ periodic forcing function is described by

$$sw(t) = \begin{cases} 1 & \text{if } 0 < t \leq T \\ 0 & \text{if } T < t \leq 2T \end{cases},$$

   with $sw(t + n\,2T) = sw(t)$ otherwise. Find the explicit formula for the periodic solution of the above system as a function of time.

Appendix B

Laplace Transform Table

$f(t)$	$\mathcal{L}\{f(t)\} = F(s)$
$f(t)$	$\int_0^\infty e^{-st} f(t)dt$
$e^{at} f(t)$	F(s-a)
$U(t-T)f(t-T)$	$e^{-sT} f(s)$
$\frac{df}{dt}$	$sF(s) - f(0^+)$
$tf(t)$	$-\frac{d}{ds}F(s)$
$(g * f)(t) = \int_0^t g(t-\tau)f(\tau)d\tau$	$F(s)G(s)$
e^{at}	$\frac{1}{s-a}$
$U(t)$	$\frac{1}{s}$
$\delta(t)$	1
t	$\frac{1}{s^2}$
$\frac{t^2}{2}$	$\frac{1}{s^3}$
$\frac{t^{n-1}}{(n-1)!}$	$\frac{1}{s^n}$
$\sin(\Omega t)$	$\frac{\Omega}{s^2+\Omega^2}$
$\cos(\Omega t)$	$\frac{s}{s^2+\Omega^2}$
$\sinh(at)$	$\frac{a}{s^2-a^2}$
$\cosh(at)$	$\frac{s}{s^2-a^2}$

$U(\cdot)$= Heaviside unit step function

$$U(t) = \begin{cases} 1 & t \geq 0 \\ 0 & t < 0 \end{cases}$$

References

[1] E. Hairer, S.P. Norsett, and G. Wanner, *Solving Ordinary Differential Equations I*, Springer-Verlag, 1990.

[2] W. H. Pess, S. S. Teuklosky, W. T. Vetterling, *Numerical Recipes in C: The Art of Scientific Computing*, Cambridge University Press, Cambridge, 1993.

[3] C. Runge, "Uber die numerische auflösung von differential gleichungen", *Mathematische Annalen*, 46, pp. 167–178, 1895.

[4] W. Kutta, "Beitrag zur nächerungsweisen integration totaler differential gleichungen", *Zeitschrift fur Math. und Phys.*, 46, pp. 435–453, 1901.

[5] I. R. Epstein, J. A. Pojman, *An Introduction to Nonlinear Chemical Dynamics: Oscillations, Waves, Patterns, and Chaos*, Oxford University Press, Oxford, 1998.

[6] P. M. DeRusso, R. J. Roy, C. M. Close, *State Variables for Engineers*, John Wiley and Sons, New York, 1965.

[7] E. P. Popov, *Introduction to the Mechanics of Solids*, Prentice-Hall, Inc., Englewood Cliffs, N.J., 1968.

[8] R. Bellman, R. Kabala, eds., *Mathematical Trends in Control Theory*, Dover, New York, 1964.

[9] W. Hahn, *Theory and Application of Lyapunov's Direct Method*, Prentice-Hall, Englewood Cliffs, N.J., 1963.

[10] J. P. Lasalle, "An Invariance Principle in the Theory of Stability", in *Differential Equations and Dynamical Systems*, J.K. Hale and J. P. Lasalle, eds, Academic Press, New York, 1967.

[11] Laplace, P. M, *Traité de Méchanique Celeste*, J. B. M. Duprat, Paris, 1801.

[12] B. Kernighan and D. Ritchie, *The C Programming Languange*, Prentice-Hall, Englewood Cliffs, N.J., 1988.

[13] W. B. Monagan, K.O.Geddes, K. M. Heal, G. Labahn, S. Vorkoeretter, J. S. Devitt, M. L. Hansen, D. Redfern, K. M Rickard, *Maple V Programming Guide*, Springer-Verlag, New York, 1995.

[14] B. W. Char, K. O. Geddes, G. H Gonnet, B. L. Leong, M. B. Monagan, S. M. Watt, *Maple V Language Reference Manual*, Springer-Verlag, New York, 1991.

Index

are you kidding, 338
arguments, array, 24
arguments, table, 24
arrays as arguments, 24
ASIL, 204
asymptotic stability, 204
asymptotic stability in the large, 204

basic Jordan block, 181
basis of solution space, 118
basis, solution set, 118
beam sign conventions, 36
birth rate, 57
Bode plots, 363
built-in end, 42

cantilever beam, 43
capacitance, 33
capacitor, 173
capacitors, 32
causality, 164
chain rule, 325, 331
change of variable, 55
characteristic equation, 126
circuits, 141
classic blunder, 25
closed form solutions, 73
closed loops, 58
coefficient, heat transfer, 48
commuting factors (not), 326
companion matrix, 113, 192
compiling, 355
complex conjugates, 150
complex variables, 143
conservation law, 30
constant coefficients, 112, 118, 119, 281
constructive existence, 75, 119
contour plots, 59
contour plotting, 7
convolution theorem, 166
convolutions, 163
coupled equations, 44
coupled masses, 32
creating help files, 377
critical damping, 129

d, 326
damping coefficient, 315
damping ratio, 129

data structures, 284
death rate, 57
debug, 26
decay rate, 66
definition
  convolution, 165
  fundamental set, 118
  Heaviside function, 147
  integrating factor, 67
  Laplace transform, 141
  linear independence, 116
  matrix exponential, 189
  solution, 74
  solution set, 118
  unit step function, 147
  vector Laplace transform, 190
  weighting pattern, 69
  Wronskian determinant, 117
delay law, 148
delayed unit step, 147
derivative, transform of, 153
DEtools, 51
dh, 331
diagonalization, 177
diffusion problem, 259
displacement field, 39
distinct roots, 150
dummy procedures, 334

ecological dynamics, 57
ecology, 57
eigenvalues, 178
eigenvector components, 179
eigenvectors, 178
electric circuits, 32
empty Iris plot, 316
energy, 34
enzyme reaction, 34
equations of order n, 111
equations, systems of, 282
equilibrium point, 58, 206
Euler's equations, 127
Euler's formula, 144
Euler's method, 81, 82
Euler–Bernoulli beam, 301
Euler–Bernoulli beams, 35
existence and uniqueness of solutions, 76
existence of solutions, 73
existence theorems, 73
explicit method, 329
exponential bound, 142
exponential shift, 145
expression tree, 9, 10

feedback, 361
Fick's law, 276
first method, Lyapunov, 210
first order system, 45, 123
first order systems, 45, 112
Floquet representation, 240, 242
Floquet theory, 239
flow field, 44
flow problems, 30
food web, 58
forced response, 69
forcing function, 65, 111
Fourier transforms, 143
frequency domain, 253
frequency response, 224
fundamental frequency, 129
fundamental set, 117
fundamental solution set, 118
fundamental solutions, 158

gate capacitance, 48
general solution, 124
Grammian matrix, 116
growth rate, 66

harmonic oscillator, 31, 45, 157
heat equation, 259
heat flux, 260
heat transfer, 48, 49
Heaviside cover up, 150
Heaviside function, 147
hello world, 356
help files, 377
homogeneous, 66
homogeneous equation, 112

homogeneous form, 112
horse choker, 338
hyperlink, 379

identity alias, 315
identity matrix, 310
impedance, 141
implicit equation, 59
implicit methods, 329
inductance, 33
inductors, 32, 173
inhomogeneous equation, 111
initial condition response, 69
initial value problem, 120
integrating factor, 67
interacting species, 57
interconnect-ability, 245
interface, 6, 315
interpreter, 328
inttrans, 311

Jordan canonical form, 181
Jordan form, 181
Jordan form theorem, 183, 310

kinetics, 34
Kirchoff's laws, 248
Kirchoff's voltage law, 248
KVL, 248

Lagrangian mechanics, 216
laplace, 311
Laplace domain, 253
Laplace transform existence, 142
Laplace transform routines, 311
Laplace transforms, 281
Laplace's equation, 262
Laurent series, 148
Leibniz, 29
level curves, 7, 59
level sets, 59
libpath, 376
library archive, 376
linear algebra, 45, 115, 118
linear equations, 45, 65
linear independence, 119
linear models, 65
linearized equation, 236
Lipschitz condition, 75
listing procedures, 5
loop of death, 25
lprint, 371
Lyapunov equation, 213
Lyapunov theory, 210

makehelp, 379
Maple, 51
Maple matrices, 307
mass action, 34
mass balance, 34
mathematical model, 73
matrix exponential, 189
mechanics, 30, 111
mixing problems, 30
MOS transistor, 48
MULT, 328
multiple plot windows, 363

Newton, 29
Newton's law of cooling, 260
nodes, 9
non-linear equations, 45
non-linear problems, 51
non-unique solutions, 74
nops, 10
normal stress, 36
normalization, 333
numerical solution, 34
numerical solutions, 51
Nyquist plot, 361

Ohm's law, 32, 247
op, 10
operand count, 10
operands, 10
operational methods, 141
order, 44
order conditions, 323
ordinary differential equations, 46
overdamping, 200

partial differential equation, 259
partial differential equations, 46
partial fractions, 148
particular solution, 124
periodic solutions, 58
phase plane, 198, 316
phasor, 224, 245
phasor response, 224
physical processes, 73
Picard iteration, 76, 123
piecewise function, 147
pin joint, 42
plot structures, 284
predator-prey, 57
procedure listing, 5
procedures, tracing, 26
proper rational function, 148
pulse function, 147
pyramid scheme, 65

radioactive decay, 30
ramp function, 144
RC circuit, 147, 154
RC impedance, 252
reaction balance, 34
reaction waves, 34
recursion, 331
remember tables, 334
resistors, 32
resolvent, 192
resolvent of a matrix, 192
resonance, 134, 229, 361
RK condition code, 338
RL impedance, 253
RLC circuit, 34
RLC series circuit, 47
Runge–Kutta, 81, 281
Runge–Kutta tableau, 92, 282

saddle, 203
salt tanks, 49
scalar-vector equivalence, 113
second law, 29
second method, Lyanupov, 210
second order tensor, 326
separable equations, 52
separated solution, 269
separation constant, 269
separation of variables, 269
shear failure, 49
shear stress, 36
sign conventions, beam, 36
similarity transformation, 177
solid mechanics, 35
solution (definition), 74
solution basis, 159
solution set basis, 118
solution, general, 124
solution, particular, 124
species, 57
spring mass, 31
spring mass system, 126
spring-mass, 111
stability, 204, 361
standard representation, 113
state vector, 113
stationary point, 206
stiff problems, 34, 329
stiffness matrix, 173
strain, 39
stress, 134
summation convention, 325, 333
switching speed, 48
symbolic subscripts, 302
systems of equations, 282

t-multiplication, 146
tableau, Runge–Kutta, 92
tables as argument, 24
Taylor series, 148, 282
tensor, 324
tensor analysis, 324
theoretical/explicit existence, 118
time constant, 126, 161
time invariant systems, 169
trace, 26
tracing procedures, 26
transfer coefficient, heat, 49
trial solution, 131
truncation order, 281

ugly code, 302
undetermined coefficients, 130, 160
unit step, 147

Vandermonde matrix, 128
variation of parameters, 67
vector equation, 46
vector equations, 112
vector space, 118
verboseproc, 5
vibrating string, 261
Volterra, 57

wave equation, 261
weighting function, 164
weighting pattern, 69, 70
with command, 376
Wronskian, 116
Wronskian determinant, 117

Young's modulus, 40

Other Birkhäuser Titles with MAPLE

Lynch. *Dynamical Systems with Applications using Maple*, 0-8176-4150-0, $49.95, 2000

Parlar. *Interactive Operations Research with Maple: Methods and Models*, 0-8176-4165-3, $59.95, 2000

Enns/McGuire. *Nonlinear Physics with Maple for Scientists and Engineers*, 2nd ed., 0-8176-4119-X, $69.95, 2000

Rovenski. *Geometry of Curves and Surfaces with Maple*, 0-8176-4074-6, $49.95, 2000

Baylis. *Electrodynamics: A Modern Geometric Approach*, 0-8176-4025-8, $49.50, 1999

Stroeker/Kaashoek. *Discovering Mathematics with Maple*, 3-7643-6091-7, $44.50, 1999

Yeargers/Shonkwiler/Herod. *An Introduction to the Mathematics of Biology: With Computer Algebra Models*, 0-8176-3809-1, $64.50, 1996

Baylis. *Theoretical Methods in the Physical Sciences: An Introduction to Problem Solving Using Maple V*, 0-8176-3715-X , $51.95, 1994

Lopez. *Maple via Calculus: A Tutorial Approach*, 0-8176-3771-0, $21.95, 1994

Lopez. *Maple V: Mathematics and Its Applications*, 0-8176-3791-5, $51.95, 1994

Lee. *Mathematical Computation with Maple V: Ideas and Applications*, 0-8176-3724-9, $47.95, 1993